DIALYSIS ACCESS

CURRENT PRACTICE

DIALYSIS ACCESS

CURRENT PRACTICE

EDITORS

J. A. AKOH
DERRIFORD HOSPITAL, PLYMOUTH, UK

N. S. HAKIM
IMPERIAL COLLEGE SCHOOL OF MEDICINE AT
ST. MARY'S HOSPITAL, UK

ICP

Imperial College Press

Published by

Imperial College Press
57 Shelton Street
Covent Garden
London WC2H 9HE

Distributed by

World Scientific Publishing Co. Pte. Ltd.
P O Box 128, Farrer Road, Singapore 912805
USA office: Suite 1B, 1060 Main Street, River Edge, NJ 07661
UK office: 57 Shelton Street, Covent Garden, London WC2H 9HE

British Library Cataloguing-in-Publication Data
A catalogue record for this book is available from the British Library.

DIALYSIS ACCESS
Current Practice

ISBN 1-86094-169-9

Printed in Singapore.

FOREWORD

As we enter the 21st century, it is striking to remember that clinical dialysis has been possible for only the last four decades. As with transplantation, there has been tremendous evolution. Patients with end-stage renal disease (ESRD) now have a number of alternatives for therapy — haemodialysis, peritoneal dialysis and transplantation. And care of the patient with ESRD requires an integrated program where patients can transfer from one modality to another.

Central to the care of the ESRD patient is a comprehensive dialysis program. For an individual patient, dialysis may be the primary therapy, may be done in preparation for transplantation, or may be initiated after failure of a kidney transplant. Critical to a successful dialysis program is a thorough understanding and appreciation of vascular access.

It is likely that most patients with ESRD, including those waiting for a cadaver kidney transplant, will spend years receiving dialysis. Thus, long-term planning is essential. Unfortunately, all too often, dialysis access surgery is not given appropriate attention. It must be appreciated that as a means for long-term dialysis, such surgery is life-saving. Importantly, there are a limited number of surgical options for any one patient. And unless the access surgery is done properly, the options may be rapidly used up.

This volume provides the global perspective necessary for planning, initiation, and long-term care of dialysis access. Detailed are the algorithms for the initial choice of access and how to care for access-related morbidity. Important consideration is given to reoperative surgery and to access surgery for patients with technically challenging issues. The indispensable roles of both diagnostic and interventional radiology and of nursing care of dialysis access sites are delineated.

Written by individuals experienced in the field, this volume provides insight into this difficult area. The required team approach — nephrology, surgery, nursing and radiology — is readily apparent. This volume, edited by two experts in the field, summarises current practices and provides a stepping stone for the future.

Arthur Matas MD
University of Minnesota Hospital
Minneapolis USA

LIST OF CONTRIBUTORS

Jacob A Akoh FMCS, FWACS, FRCSEd, FRCSEd(Gen), FICS
Consultant General Surgeon
Plymouth Hospitals NHS Trust
Level 03, Derriford Hospital
Derriford Road
Plymouth PL6 8DH, UK

Murat A Akyol MD, FRCS
Consultant Surgeon
Scottish Liver Transplant Unit
Royal Infirmary of Edinburgh
1 Lauriston Place
Edinburgh EH3 9YW, UK

Aghiad Al-Kutoubi MD, FRCR, DMRD
Professor and Chairman
Department of Diagnostic Radiology
American University of Beirut Medical Centre
Beirut, Lebanon

Kenneth L Brayman MD, PhD
Associate Professor of Surgery
Director of Renal Transplantation
University of Pennsylvania Medical School
Children's Hospital
Philadelphia, PA 19104
USA

Paul W Chamney BEng(Hons), PhD, AMIEE
Department of Electrical and Electronic Engineering
University of Hertfordshire
Hatfield
Herts AL10 9AB, UK

Adil Eltayar DIC, MSc, FRCS
Clinical Fellow
Transplant Unit
St Mary's Hospital
London W2 1NY, UK

Joanne Emery RGN
Nursing Sister
Haemodialysis Unit
Lister Hospital
Stevenage SG1 4AB, UK

Ken Farrington MD, FRCP
Consultant Nephrologist
Department of Renal Medicine
Lister Hospital
Stevenage
Herts SG1 4AB, UK

Oswald N Fernando FRCS, FRCSEd
Consultant Transplant Surgeon
Transplantation Unit
Royal Free Hospital
Pond Street
London NW3 2QG, UK

Albert G Hakaim MD, MSc, FACS
Associate Professor of Surgery
Section of Vascular Surgery
Mayo Clinic
4500 San Pablo Road
Jacksonville, FL 32224
USA

Nadey S Hakim MD, PhD, FRCS, FACS, FICS
Consultant Surgeon
Surgical Director Transplant Unit
St Mary's Hospital
London W2 1NY, UK

Philip Korsah MB, ChB, DA, FFARCSI
Research Fellow
Clinical Shock Study Group
Western Infirmary, Dumbarton Road
Glasgow G11 6NT, UK

Paul A Lear FRCS
Consultant Vascular and Transplant Surgeon
Renal Transplant Unit
Southmead Hopspital
Westbury-on-Trym
Bristol BS10 5NB, UK

Derek Manas BSc, MBBCh, FCS(SA)
Consultant Surgeon
Liver Unit
Level 5 Ward 12
The Freeman Hospital
Newcastle Upon Tyne NE7 7DN, UK

Andrew Nicolaides MS, FRCS
Professor of Surgery
Academic Surgical Department
Vascular Unit
St Mary's Hospital
London W2 1NY, UK

W Andrew Oldenburg MD
Chief, Section of Vascular Surgery
Mayo Clinic
4500 San Pablo Road
Jacksonville, FL 32224
USA

Nick Pace MB, ChB, FRCA, MRCP, MPhil
Consultant Anaesthetist
Dept of Anaesthesia
Western Infirmary
Dumbarton Road
Glasgow G11 6NT, UK

Keith M Rigg MD, FRCS
Consultant Surgeon
Nottingham City Hospital
Hucknall Road
Nottingham NG5 1PB, UK

Gabriel Szendro MD, FRCS
Consultant Vascular Surgeon/Senior Lecturer
Soroka Medical Centre and Faculty of Health Sciences
Ben-Gurion University
Beer-Sheva, Israel

P. Wallace MSN, RN
University of Pennsylvania Medical School
Children's Hospital
Philadelphia, PA 19104
USA

CONTENTS

DIALYSIS ACCESS

CURRENT PRACTICE

CHAPTER 1

DIALYSIS ACCESS: PAST, PRESENT AND FUTURE

Jacob A Akoh FRCSEd (Gen)
Plymouth Hospitals NHS Trust
Derriford Hospital
Plymouth PL6 8DH, UK

1.1. Introduction

Prior to the establishment of renal replacement as a viable therapeutic option, end-stage renal disease (ESRD) was a critical terminal illness. Access to the circulation made haemodialysis possible. Improved survival of haemodialysis patients coupled with the inability to provide enough renal transplants for the growing ESRD population has resulted in an increase in the average length patients spend on dialysis. This in turn means that vascular or peritoneal accesses are required to function for longer periods of time. By 1993, approximately 160 000 patients were maintained on long-term haemodialysis in the United States with an estimated 8–10% rise in prevalence rate per year (1). Vascular access complications are the largest single cause of morbidity in the chronic haemodialysis population (2). The interval between access placement and the need for a procedure to restore access patency is decreasing (3,4). Revision of failing access is expensive (4) and often of poor outcome (5). A reliable and durable vascular access, an absolute necessity for efficient haemodialysis, continues to be a challenge for surgeons and other members of the health care team.

1.2. Historical Milestones

1.2.1. *Extracorporeal dialysis*

Haemodialysis had its humble beginings in the early 1920s when temporary access to the circulation was achieved by the insertion of arterial and venous cannulae. The first dialysis in a human lasted 15 minutes. Vascular cannulas were inserted into the left radial artery and antecubital vein under local anaesthesia (6). This and subsequent dialysis conducted by Georg Haas did not have any therapeutic benefit due to the short duration, small dialysate volumes and low blood flows involved. Nevertheless, it pointed the way for further technological advancements culminating in the development of the rotating artificial kidney by Willem (Pim) Johan Kolff and Hendrik Berk (7). By connecting metallic or glass cannulae to an artificial kidney, Kolff and Berk showed that ESRD patients could be sustained for prolonged periods by intermittent filtering of the blood.

1.2.2. *External arteriovenous shunt*

Chronic dialysis was not possible until the introduction of the external arteriovenous (AV) shunt in 1960 by Quinton, Dillard and Scribner (8). Through a cutdown Teflon or Silastic cannulae were inserted into either the radial artery and cephalic vein in the wrist or the posterior tibial artery and the great saphenous vein in the ankle. This provided a direct access to the circulation and made outpatient haemodialysis feasible. This access modality was soon dogged by multiple complications which shortened its lifespan. The major complications encountered were clotting, infection, dislodgement of cannula with arterial bleeding, pressure necrosis of skin overlying the subcutaneous Teflon, septicaemia and pulmonary embolism. The cost of methodical daily access care was enormous. This regular, almost ritualistic care of the external shunt often produced psychologic fixations in patients. There had to be a better method of access.

1.2.3. *Native arteriovenous fistula*

Helped by their previous experience with venipuncture techniques for haemodialysis, but disappointed by the long-term results of external shunts, Brescia and co-workers explored the possibility of creating an AV fistula to assure a flow rate of 250–300 ml/min. In 1966, they published their initial experience in the *New England Journal of Medicine* (9). Brescia *et al.* used their surgically created fistula the day after surgery and observed, “As time has passed, we have noted that the vessels have become even more prominent and thick-walled, making venipuncture even easier; this ‘arterialisation’ is caused by prolonged exposure to arterial pressure and flow.” This breakthrough ensured a repeated and routine access to the circulation.

The native AV fistula overcame many of the problems associated with the Scribner shunt and became the primary mode of vascular access (1,10,11). Although troubled by a high primary failure rate (12,13), once matured, they have excellent long-term function with low rate of complications (13–15). Fistulas that never developed adequately for dialysis, or thrombosed before use, were regarded as primary failures (4,14). Antecubital fossa fistulas were introduced to deal with situations where forearm vessels were found to be unsuitable, intrinsically inadequate or following failure of wrist AV fistulas.

Although AV fistulas were initially needled soon after they were created, it became clear that premature needling resulted in excessive access loss from either haematoma formation or excessive external pressure to stop haemorrhage after dialysis. It is now good clinical practice to leave the AV fistula undisturbed for a period of 4 to 8 weeks to allow for arterialisation of the venous limb before using it for dialysis (1,16,17).

As a direct consequence of this development, there was a rapid growth and expansion of ESRD program in the Western world. Advances in technology and changes in the demographics of ESRD patients led to the adoption of alternatives to native fistula such as biological or synthetic grafts and permanent indwelling central venous catheters.

1.2.4. *Grafts*

Biological grafts

The rising number of ESRD patients who were not suitable for AV fistulas brought about the search for alternative access modalities. Commonly cited reasons for this include unsuitable vessels particularly in elderly and diabetic patients; destruction of cephalic veins due to indiscriminate use for phlebotomy; late referral for vascular access (11). Autologous vein (18,19), allogeneic vein (20), umbilical cord vein (21) and bovine carotid artery (22,23) were all used for haemodialysis with mixed results. These materials are of limited availability, expensive and of variable size and quality. Surgical procedures for placement of biological accesses are also more demanding. The experience of many centres with autogenous vein is variable with patency rates ranging from 35 to 66% at two years (24,25). While autogenous vein grafts are still favoured in many centres in the UK (19), they are hardly used these days in the US where expanded polytetrafluoroethylene (PTFE) is widely used (1).

Synthetic grafts

Due to a high rate of thrombosis and pseudoaneurysm formation with biological grafts, attention turned to synthetic grafts. Dacron® (E.I. du Pont de Nemours and Co, W) was introduced because of its promise in vascular surgery, but the theoretical advantage of a transmural fibrillar structure, which would encourage tissue growth and provide greater durability for recurrent cannulation was not bourne out in practice. Expanded polytetrafluoroethylene (PTFE, Gore-Tex; W.L. Gore & Assoc, Inc, Flagstaff, AZ Impra; Impra Inc, Tempe, AZ), a fluorocarbon polymer with similar qualities to Dacron®, performed much better, and became the prosthetic graft of choice (1,10,17). This access modality became so popular that in some centres in the US, more than 80% of established haemodialysis patients were using expanded PTFE grafts (26). Unfortunately, these synthetic graft materials were attended with great morbidity, infection and thrombosis being the chief culprits. With complication-free access survival of 9 to 16 months (4), most patients experienced frequent hospital admissions for salvage procedures with consequently high utilisation of resources. The search for an ideal vascular access continued.

1.2.5. *Central venous catheters*

The idea to use central venous catheters (CVC) for dialysis came from the successful use of silastic catheters for chemotherapy and parenteral nutrition. Subclavian vein (SCV) cannulation for haemodialysis was first described by Erben *et al.* in 1969 (27). They used two single lumen catheters inserted into both SCVs, the same SCV or one inserted into a SCV and the other into a femoral vein. In the early 1980s, double lumen catheters were introduced. Initially, this was in the form of an "arterial cannula" and an inner coaxial "venous cannula", which was replaced for each dialysis session (28). Catheter technology evolved from the use of stiff tetrafluoroethylene, polyurethane materials to soft silicone rubber. Similarly, catheter design changed from coaxial lumina to two lumina separated by a midline septum and then to two distinct lumina lying side by side. The main disadvantages of the double lumen catheters are large calibre, stiffness and liability to poor flow. To overcome these problems, the Tesio Twin-Cath (Medcomp, Harleysville, PA), which consists of two independent single lumen silicone catheters with holes around the distal 4 cm, were introduced in the mid 1980s (29).

CVC were initially introduced for acute haemodialysis but soon became an established method of access for long-term haemodialysis. As renal services expanded, more elderly people with peripheral vascular disease were recruited on to dialysis and the proportion of patients with exhausted vascular accesses increased. This created an increased role for CVC. These catheters provide blood flow rates of 200 to 300 ml/min (30–32). The advantages offered by CVC are many. Ease of insertion, replacement or removal, immediacy of use and absence of haemodynamic stress to the cardiovascular system. Avoidance of venipuncture carries great appeal for many patients with needle phobia. However use has important morbidity. Insertion complications occur in about 7% of procedures (33) and include arterial puncture, haemothorax, pneumothorax, anatomical damage to trachea, brachial plexus, superior vena cava and myocardium. Thrombosis, malposition, central vein stenosis and infection are some of the late complications. Inspite of this, central venous catheters are gaining popularity in more and more units. In a recent survey of renal units in the UK, 60 out of 66 used permanent central venous catheters for long-term haemodialysis (34). Because of the attendant increase in access complications, the National Kidney Foundation expert panel recommended that less than 10% of patients should be maintained on chronic haemodialysis by means of catheters (35).

Although the subclavian route was initially popular (36,37), the high incidence of stenosis (38–40) led to the conclusion that it is sub-optimal for long term access. The right IJV is now the preferred site for CVC insertion.

1.2.6. *Peritoneal access*

Peritoneal dialysis which started as a holding procedure for haemodialysis, has now become an established form of renal replacement therapy especially in children. Human peritoneal dialysis started in 1938 when Rhoads used intermittent peritoneal dialysis to treat two patients with nephrotic syndrome (41). The introduction of the concept of equilibrium long dwell peritoneal dialysis in 1976, later known as chronic ambulatory peritoneal dialysis (42), was a landmark development. As reported in the 1997 USRDS Annual Data Report, there was an overall increase in the use of peritoneal dialysis from 1984 to 1994 (43). Now about 11% of all dialysis patients are maintained on peritoneal dialysis (44).

Insertion of a catheter is needed to perform peritoneal dialysis. Catheters made of latex rubber were initially used (45), followed by the development of nylon (46) and finally silicone catheters (47). In 1968, Tenckhoff and Schecter (48) invented a silicone catheter with two Dacron cuffs. Cuffs served to create a bacterial barrier in the subcutaneous tunnel. Thus the Tenckhoff catheter overcame the main obstacle to chronic peritoneal dialysis. Various catheter designs have been employed (49–52) without significant influence on technical failures.

The cumulative survival of Tenckhoff catheters is impressive in the first 2–3 years (53). Long term studies in patients on CAPD for more than five years show that only a small percentage of the starting population remain on it (54,55) due to peritonitis, catheter related factors and inadequacy of dialysis and death.

1.3. Access Morbidity

Complications from vascular access accounted for 15% of hospital admissions among US haemodialysis patients with an estimated cost of $150 million in 1990 (2). In an earlier review of Medicare ESRD patient data obtained from 1984 to 1986, Feldman *et al.* (3) found that 15–16% of hospital stays among prevalent ESRD patients were associated with vascular-access-related morbidity. In some units, admission for vascular access related morbidity is as high as 30% (5). Vascular access morbidity continues to be the largest single cause of morbidity among ESRD patients. Black race, old age, female sex and diabetic mellitus as cause of ESRD were found to be independent risk factors for access related morbidity (3). The prevalence of dialysis treatments with delivered Kt/V (K = dialyser urea clearance, t = duration of dialysis, V = volume of distribution or total body water) less than 1.2, which signifies inadequate dialysis, is about 28% (56,57). Overcoming barriers to dialysis delivery not only improves the adequacy of dialysis but also improves patient survival and reduces hospitalisation.

Thrombosis and infection are the two leading causes of access failure (58). Thrombosis could result from vascular obstruction or be due to stenosis in the venous circuit. With respect to prosthetic arteriovenous fistula, stenosis results from neointimal hyperplasia. The aetiology of intimal hyperplasia has not been fully worked out but it is thought to be due to increased intraluminal pressure, turbulent flow and calcification in the vessel. Antecedent catheter insertion into the subclavian vein is an important cause of outflow obstruction of an access in the ipsilateral arm. Subclavian venous stenosis which complicates CVC use in 19–53% (38–40) not only prevents successful replacement of venous catheters but precludes successful placement of AV fistula in the ipsilateral arm.

Fistula infection accounts for about 20% of fistula complications and is the second leading cause of fistula failure (59). Considering how frequently

AV fistula or grafts are cannulated, it is somewhat surprising that this complication is not more common. Fistula infection has a devastating effect on vascular access. It is either due to a breakdown in aseptic techniques and can often be traced to dialysis staff (11), poor patient hygiene or dermatitis involving the skin covering the access. Bacteraemia should be treated as a quality assurance issue. Any individual staff associated with a high incidence should be identified and retrained.

Strategies to reduce vascular access morbidity include identification of patients in whom native fistula creation is a viable option and early detection of access dysfunction combined with prompt surgical or radiological intervention.

1.4. 2000 and Beyond

As at 31 December 1996, 283 932 ESRD patients were receiving renal replacement therapy in the US (44). Of these, 72% were being treated with either haemodialysis or peritoneal dialysis. There is no doubt that kidney transplantation is the treatment of choice for patients with ESRD. However, the increase in waiting time for cadaveric organs coupled with the low transplantation rates will likely ensure that dialysis remains the primary method of renal replacement for the near future. Dialysis access planning may therefore need to look 20 years into the future for the patient, who barring a transplant, may remain on dialysis for a long time.

1.4.1. *Requirements of modern dialysis*

The most important determinant for effective haemodialysis is a reliable means of repetitive access to large vessels capable of providing rapid extracorporeal flow. Whatever type of access is chosen must meet certain criteria for adequate vascular access. These criteria (58) include:

- A reliable access to the circulation. Thrombosis and infection are not acceptable as a frequent occurrence.
- Sufficient extracorporeal flow to not impair the efficiency of dialysis.

- Trouble free access, not requiring frequent or costly intervention to keep it functioning.

To ensure selection of the appropriate operation, preoperative assessment including arterial evaluation, vein mapping by duplex examination (where necessary) and venous outflow testing should be rigorously adhered to (60).

In centres where access service is provided by general and/or vascular surgeons, whose priorities lie elsewhere, frequent conflicts between nephrologists and surgeons regarding the timing for formation or dealing with access complication are common. In the larger nephrology and transplant units, there is increasing need for the appointment of surgeons with a special interest in vascular access. Such surgeons are more likely to treat vascular access issues with the interest, attention and promptness required. The need for experienced surgical skill in establishing satisfactory AV fistula and grafts cannot be overemphasised. Access surgery is a challenging but frustrating type of surgery. The surgeon knows from the start that the access procedure will eventually fail and that alternatives will become necessary. Some surgeons feel unduly responsible for such failures. Two important virtues will stand the access surgeon in good stead — a good eyesight and perseverance.

Nursing skill in the care and use of access is another important requirement. It is they who are in contact with the patients on a regular basis and who will call the attention of medical staff to access problems.

1.4.2. *Choice of access modality*

It is considered by many that a native AV fistula is the optimum access modality for most ESRD patients. Wrist (radio-cephalic) and elbow (brachio-cephalic) primary fistula have lower morbidity associated with their creation, excellent patency once established, improved performance over time with low complication rates compared to other access types (11,12,17,26,61–63). But not all patients can have primary elbow or wrist fistula. Each patient has to be assessed individually and the decision on which access procedure to employ based on the interplay of relevant factors. Hirth and co-workers (2) published a paper based on a large number of patients drawn from random, national samples of ESRD patients starting haemodialysis in 1986–1987 ($n = 2741$) and 1990 ($n = 1409$). They found that 56% of patients had synthetic grafts 30 days after starting haemodialysis in 1986–1987 compared to 65% in 1990. This practice had a strong regional variation, which persisted after correction for clinical determinants, meaning that non-clinical factors were also at play. Non-clinical determinants harbour potential deviations from optimal care with implications for increased patient morbidity. This changing pattern of access modality is not uniform throughout the world. Synthetic grafts are less popular outside the USA.

The choice of vascular access may reflect the attitude of surgeons and nephrologists in charge of patient care. It is easy to speculate that regional and international variations may be due to the distribution of surgeons proficient in performing different access procedures; differences and changes in the subspecialties of the surgeons performing vascular access; the type of training received and where applicable the financial remuneration involved. The importance of surgical skill and workload is brought out by a recent study from Austria which found the strongest predictor of fistula failure to be which surgeon performed the procedure (64).

Efficient dialysis requires among other things, a high flow rate of undialysed blood. This may be directly related to the type of vascular access (arteriovenous fistula, graft or catheter). In a large prospective study from Cleveland, Ohio, multivariate techniques were used to examine the relative importance of a large number of potential barriers (including access type) to delivery of dialysis (56). Haemodialysis by use of CVC instead of AV fistula or graft was associated with a decrease in delivered *Kt/V* by approximately 0.2. Access type is therefore an independent variables affecting delivery of dialysis.

If a wrist or elbow fistula cannot be created, a synthetic AV graft or transposition of the basilic vein is the next choice. Only when these secondary access modalities are unavailable is there justification for long term use of CVC. To increase the proportion of patients having AV fistula, it is necessary to educate physicians and patients about the importance of protecting potential access sites and early referral to the nephrologist before dialysis becomes imminent. Currently only 23–58% of patients would have seen a nephrologist before initiation of dialysis. Early referral will allow deliberate psychosocial preparation and choice of dialysis modality (65).

1.4.3. *Dialysis access adequacy*

Preserving access function and long-term patency are essential for efficient dialysis delivery. Vascular access thrombosis and/or stenosis are the most common cause of haemodialysis access impairment or loss (11,13,26,59,62). The reduction of blood flow through an access, which inevitably accompanies a developing occlusion, leads to recirculation and reduction of effective solute clearance. Both factors decrease the adequacy of treatment. Uncorrected stenosis is associated with eventual AV graft thrombosis (66).

Action is required to extend access use life by establishing protocols to monitor accesses at risk and audit programmes to ensure standards of care are being met. Methods for the detection of incipient access failure (Table 1.1.) combined with percutaneous or surgical interventions have been shown to prolong access life (67). Furthermore, AV grafts revised electively have a more prolonged survival than grafts revised at the time of thrombosis (68). Any surveillance protocol employed should deploy simple and inexpensive but sensitive and specific methodology that can be applied easily by dialysis staff. Data on clinical parameters, dialysis adequacy and results of monitoring tests must be accurately documented and reviewed regularly for them to be useful.

Table 1.1. Methods of monitoring vascular access patency.

Methods
Clinical assessment
Venous dialysis pressures
Recirculation studies
Dialysis adequacy (*Kt*/*V*, SRI)
Duplex Doppler ultrasound
Intravascular ultrasound
Fistulography

Physical examination is a useful method of screening for a failing access. Prolonged bleeding following needle withdrawal, swelling in the fistula arm and pain with dialysis point to access dysfunction. Alteration in the character of the palpable thrill or conversion of a thrill to a pulse indicates stenosis. Due to increased velocity over a stenotic area, a localised increase in the pitch of the bruit suggests stenosis. The dialysis nurse can easily assess these physical parameters on a weekly basis before starting dialysis sessions.

Access patency can be assessed by measurement of the pressure in the venous return line of the dialyser circuit. When monitored regularly using standardised protocol, dynamic venous pressure (DVP) provides a sensitive and cost effective method of detecting access stenosis (69). If the DVP exceeds 150 mm Hg on three consecutive dialysis sessions, fistulography is recommended. Although static dialysis pressure (venous dialysis pressure at zero blood flow) is thought to be more strongly predictive of outflow stenosis than dynamic pressure measurements, extra equipment is required for its measurement. Measuring DVP does not incur additional cost and is not time consuming.

Narrowing anywhere in the access circuit can reduce flow and thereby limit blood flow to the dialyser. The dialyser may however continue to pump at a rate that exceeds the fistula inflow rate. To achieve this, recirculation is required. Access recirculation is the immediate return of venous (dialysed) blood to the dialyser, effectively short-circuiting the patient. Several methods exist for measuring recirculation: urea dilution method, urea modelling, thermal dilution, ultrasound dilution, optical density and bedside occlusion/pressure measurements (67,70). Using the two-needle slow flow or stop flow technique, recirculation [*R*] is calculated from the formula:

$$R = C_s - C_a/C_s - C_v,$$

where C_s is systemic blood urea nitrogen (BUN), C_a is arterial BUN and C_v is venous BUN (67). Experience with monthly recirculation studies coupled with fistulography in patients with consistently high levels (> 15%) has shown that elevated recirculation ratios correctly identify patients with significant venous stenosis (71). Urea recirculation ratios, however, depend on factors such as needle placement, extracorporeal blood flow, hypotension, cardiac output, intravascular volume depletion, venous stenosis and arterial stenosis making it non-specific for detecting access dysfunction (11).

The most commonly used methods of calculating dialysis delivery are based on urea kinetic measurements. *Kt*/*V* provides a mathematical quantitation of dialysis (72). However, rebound of urea after dialysis, changes in body fluid volume and recirculation during dialysis all combine to increase the margin of error of the calculated from the real *Kt*/*V*. Moreover, *K* is calculated by the manufacturer under *in vitro* conditions and may be less than the real *K*. A more accurate method for determining the dose of dialysis, solute removal index (SRI urea), is calculated using the formula:

$$\text{SRI} = V_d \times C_d \times 100/V_o \times C_o,$$

where V_d = volume dialysate effluent; C_d = urea level in dialysate effluent; V_o = total body water before dialysis and C_o = blood urea before dialysis. Currently, dialysis is planned to achieve a minimum *Kt*/*V* of 1.2. The optimum *Kt*/*V* is 1.4 (SRI 80%) for non-diabetics and 1.6 (SRI 85%) for diabetics (73). Irrespective of the method used in calculating the delivered dose of dialysis, unexplained decrease in their values must raise suspicion about access dysfunction.

The use of colour-flow Doppler ultrasonography in predicting haemodialysis access flow is rapidly gaining ground (74,75). This technique is non-invasive, painless, portable and, in expert hands, reproducible. While its ability to detect or predict prosthetic grafts at risk of stenosis is proven, its role with native AV fistulas is being investigated (74). Gadallah *et al.* (76) in a comparative study, showed close correlation between Doppler ultrasound and fistulography in diagnosing anatomic stenosis, thus enhancing its role as a screening test. The use of ultrasonography as a routine screening test must however await a prospective randomised trial. This will answer questions about its efficacy and cost-effectiveness in a non-selected group as opposed to current reports about its use in patients suspected to have or be developing stenosis.

Intravascular ultrasound gives precise information about the size of a stenotic lesion. It is particularly useful in assessing angioplasty results and deciding who requires intervention with atheterectomy catheters.

Routine angiography screening of all accesses is uneconomical and not practicable. Using clinical and quality assurance parameters as indications for performing fistulography, Schwab and co-workers demonstrated improved longevity of access sites and a three fold decrease in thrombosis (69). Using a slightly different surveillance protocol, Cayco *et al.* (77) reported a decreased thrombosis rate in patients with AV grafts. There is no doubt that the way forward is regular monitoring of accesses with the aim of early detection of dysfunction and prompt intervention to preserve or salvage the access. Adoption of such a protocol will result in fewer access related hospitalisation, lower access replacement rates and improved access survival (69,71). To this end, all significant stenosis (> 50%) detected should be corrected by angioplasty or surgical revision before actual fistula failure.

Patient-related factors

It is common knowledge that while some patients are prone to developing thrombosis, stenosis or other access complications, others seem resistant to them. Evidence accrued from various sources suggests that multiple factors are involved in access complications. Analysis of these factors may allow the selection of the most appropriate access procedure for individual patients (26). Such factors as small vessels (14), diabetes mellitus, black race, age > 64 (3), patients with hypercoagulable states, erythropoietin (78,79) have been found to increase the risk of vascular access thrombosis. Patients with frequent episodes of access failure need to be investigated for any behavioural characteristics that may be responsible. Patients and their caregivers should be educated about simple emergency procedures and basic care of the access. This must include immediate reporting of any symptoms and sign of infection or absence of bruit to dialysis personnel.

1.4.4. *Recent developments*

Technical innovations aimed at: preventing stenosis of graft-vein anastomosis; dealing with aneurysmatic AV fistula; diagnosis and treatment of fistula, graft or catheter dysfunction by interventional radiology; and prevention of catheter related sepsis by use of implanted ports or antimicrobial coated catheters are recent developments which are covered in various sections of this book.

The use of Doppler ultrasonography in the Access Clinic not only improves patient selection but the selection of the most appropriate procedure for vascular access.

The National Kidney Foundation Dialysis Outcome Quality Initiative guidelines (35) and the increasing importance being given to "evidence-based medicine" are likely to have significant impacts on the provision of dialysis access in the future.

1.5. Conclusion

With improvements in obtaining access to the circulation, dialysis moved from a halfway technology to a full contributor to the therapeutic armamentarium of the nephrologist. Modern technology is being applied all the time to medical practice. New synthetic grafts, new central venous catheters like the Dialock implantable device and kink-free catheters are being evaluated for use. The future of renal replacement therapy is bright and will only be limited by cost pressure in a changing health service environment. For the full potential of tomorrows science to reach the bedside, collaboration between industry, the nephrologist researcher and health care financers is required (80).

In the future, the volume of patients requiring vascular access for dialysis will undoubtedly increase. The proportion of elderly patients with peripheral vascular disease and other co-morbid factors will also increase. So will the number of long-term dialysis patients who have exhausted all sites for both primary and secondary vascular access procedures. This will happen in the face of dwindling organ donor resources. At present, the weakest link in the chain of management of patients on chronic haemodialysis is the provision of a reliable and durable vascular access. The lessons of the past must challenge us to formulate carefully considered clinical practice guidelines for dialysis access provision in the future. To combat this problem as we enter the new millennium, action is required on several fronts: patient evaluation prior to access placement, monitoring and maintenance, prevention of infection, when to intervene for access complications, optimal approaches for treating complications and potential quality of care standards (35). Peritoneal and haemodialysis strategies are interdependent and should be considered in concert.

These recently published National Kidney Foundation Dialysis Outcome Quality Initiative (NKF-DOQI) guidelines made preservation of veins in prospective dialysis patients and early native AV fistula creation the main focus of their recommendations. Those involved in dialysis provision or the management of patients with end-stage renal failure must familiarise themselves with it. Construction of a primary AV fistula was the best choice 30 years ago; it is still the best choice today and no doubt will be the best option for the foreseeable future. The changing demographics of dialysis patients will ensure that alternate means of access will gain greater prominence and continue to drive technology towards the development of the ideal central venous catheter and prosthetic graft.

References

1. Shaffer, D. (1995). Lessons from vascular access procedures for haemodialysis. *Surgical Oncology Clin N Am*, **4**, 537–548.
2. Hirth, R.A. *et al.* (1996). Predictors of type of vascular access in haemodialysis patients. *JAMA*, **276**, 1303–1308.
3. Feldman, H.I. *et al.* (1993). Haemodialysis vascular access morbidity in the United States. *Kidney Int*, **43**, 1091–1096.
4. Feldman, H.I., Kobrin, S. and Wasserstein, A. (1996). Haemodialysis vascular access morbidity (editorial). *J Am Soc Nephrol*, **7**, 523–535.
5. Chazan, J.A., London, M.R. and Pono, L.M. (1995). Long-term survival of vascular access in a large chronic haemodialysis population. *Nephron*, **69**, 228–233.
6. Gottschalk, C.W.and Fellner, S.K. (1997). History of the science of dialysis. *Am J Nephrol*, **17**, 289–298.
7. Kolff, W.J. and Berk, H.T.J. (1944). The artificial kidney: A dialyser with a great area. *Acta Med Scand*, **117**, 121–131.
8. Quinton, W.E., Dillard, D. and Scribner, B.H. (1960). Cannulation of blood vessels for prolonged haemodialysis. *Trans Am Soc Artif Intern Organs*, **6**, 104–113.

9. Brescia, M.J. *et al.* (1966). Chronic haemodialysis using venipuncture and a surgically created arteriovenous fistula. *N Engl J Med*, **275**, 1089–1092.
10. Kapoian, T. and Sherman, R.A. (1997). A brief history of vascular access for haemodialysis: An unfinished story. *Seminars in Nephrology*, **17**, 239–245.
11. Fan, P-Y. and Schwab, S.J. (1992). Vascular access: Concepts for the 1990s. *J Am Soc Nephrol*, **3**, 1–11.
12. Kinnaert, P. *et al.* (1977). Nine years' experience with internal arteriovenous fistulas for haemodialysis: A study of some factors influencing the results. *Br J Surg*, **64**, 242–246.
13. Winsett, O.E. and Wolma, F.J. (1985). Complications of vascular access for haemodialysis. *South Med J*, **78**, 513–517.
14. Reilly, D.T., Wood, R.F.M. and Bell, P.R.F. (1982). Prospective study of dialysis fistulas: problem patients and their treatment. *Br J Surg*, **69**, 549–553.
15. Bonalumi, U. *et al.* (1982). Nine years' experience with end-to-end arteriovenous fistula at the anatomical snuffbox for maintenance haemodialysis. *Br J Surg*, **69**, 486–488.
16. Koo Seen Lin, C. and Taube, D. (1994). Elbow arteriovenous fistulas for chronic haemodialysis (letter). *Br J Surg*, **81**, 1696.
17. Harland, R.C. (1994). Placement of permanent vascular access devices: Surgical considerations. *Adv Ren Replace Ther*, **1**, 99–106.
18. May, J. *et al.* (1969). Saphenous vein arteriovenous fistula in regular dialysis treatment. *N Engl J Med*, **280**, 770.
19. Jenkins, A. McL., Buist, T.A.S. and Glover, S.D. (1980). Medium-term follow-up of forty autogenous vein and forty polytetrafluoroethylene (Gore-Tex) grafts for vascular access. *Surgery*, **88**, 667–672.
20. Kestlerova, M., Kocandrle, V. and Vrubel, J. (1978). The use of autologous and allogeneic grafts for arteriovenous fistulas in chronic haemodialysis. *Czech Med*, **1**, 169–179.
21. Kester, R.C. (1978). Early results with human umbilical cord vein allografts for haemodialysis. *Br J Surg*, **65**, 609–610.
22. Chinitz, J.L. *et al.* (1972). Self-sealing prosthesis for arteriovenous fistula in man. *Trans Am Soc Artif Intern Organs*, **18**, 452–455.
23. Rosenthal, J.J. *et al.* (1975). Problems with bovine heterografts for haemodialysis. *Am J Surg*, **130**, 182–188.
24. May, J., Harris, J. and Fletcher, J. (1980). Long-term results of saphenous vein graft arteriovenous fistula. *Am J Surg*, **140**, 387–390.
25. Morgan, A. and Lazarus, M. (1975). Vascular access for dialysis. *Am J Surg*, **129**, 432–439.

26. Windus, D.W. (1993). Permanent vascular access: A nephrologist's view. *Am J Kidney Dis*, **21**, 457–471.
27. Erben, J. *et al.* (1969). Experience with routine use of subclavian vein cannulation in haemodialysis. *Proc Eur Dial Transplant Assoc*, **6**, 59–64.
28. Twardowski, Z.J. (1995). Advantages and limits of the jugular catheter approach. *Nephrol Dial Transplant*, **10**, 2178–2182.
29. Canaud, B. *et al.* (1986). Internal jugular vein cannulation using 2 silastic catheters: A new, simple and safe long-term vascular access for extracorporeal treatment. *Nephron*, **43**, 133–138.
30. Shusterman, N.H., Kloss, K. and Mullen, J.L. (1988). Successful use of double-lumen, silicone rubber catheters for permanent haemodialysis access. *Kidney Int*, **35**, 887–890.
31. Moss, A.H. *et al.* (1990). Use of silicone dual lumen catheter with a Dacron cuff as a long term vascular access for haemodialysis patients. *Am J Kidney Dis,* **16**, 211–215.
32. Uldall, R. *et al.* (1993). A new vascular access catheter for haemodialysis. *Am J Kidney Dis*, **21**, 270–277.
33. Surratt, R.S. *et al.* (1991). The importance of preoperative evaluation of the subclavian vein in dialysis access planning. *Am J Roentgenol*, **156**, 623–625.
34. Kumwenda, M.J., Wright, F.K. and Haybittle, D.J. (1996). Survey of permanent central venous catheters for haemodialysis in the UK. *Nephrol Dial Transplant*, **11**, 830–832.
35. NKF-DOQI (1997). Clinical practice guidelines for vascular access. *Am J Kidney Dis*, **30**, S150–S191.
36. Jones, C.E. and Walters, G.K. (1992). Efficacy of the supraclavicular route for temporary haemodialysis access. *South Med J*, **85**, 725–728.
37. Uldall, P.R. *et al.* (1979). A subclavian cannula for temporary vascular access for haemodialysis or plasmapheresis. *Dial Transplant*, **8**, 963–968.
38. De Moor, B., Vanholder, R. and Ringoir, S. (1994). Subclavian vein haemodialysis catheters: Advantages and disadvantages. *Artif Organs,* **18**, 293–297.
39. Schillinger, F. *et al.* (1991). Post catheterisation vein stenosis in haemodialysis: Comparative angiographic study of 50 subclavian and 50 internal jugular accesses. *Nephrol Dial Transplant*, **6**, 722–724.
40. Cimochowski, G.E. *et al.* (1990). Superiority of the internal jugular over the subclavian access for temporary dialysis. *Nephron*, **54**, 154–161.
41. Freischlag, J.A. (1996). Peritoneal dialysis. In *Vascular Access Principles and Practice*, (ed. S.E. Wilson), pp. 262–270 Mosby: St Louis.

42. Popovich, R.P. *et al.* (1978). Chronic ambulatory peritoneal dialysis. *Ann Int Med*, **88**, 449–456.
43. US Renal Data System: USRDS 1997 Annual Data Report. The National Institute of Health, National Institute of Diabetes and Digestive and Kidney Diseases, Bethesda, MD 1997.
44. US Renal Data System: USRDS 1998 Annual Data Report. The National Institute of Health, National Institute of Diabetes and Digestive and Kidney Diseases, Bethesda, MD 1998.
45. Boen, S.T. *et al.* (1962). Periodic peritoneal dialysis in the treatment of chronic uraemia. *Trans Am Soc Artif Intern Organs,* **8**, 256–265.
46. Weston, R.E. and Roberts, M. (1965). Clinical use of stylet-catheter for peritoneal dialysis. *Arch Intern Med*, **115**, 659–662.
47. Gutch, C.F. and Stevens, S.C. (1966). Solstice catheter for peritoneal dialysis. *Trans Am Soc Artif Intern Organs*, **12**, 106–107.
48. Tenckhoff, H. and Schecter, H. (1968). A bacteriologically safe peritoneal access device. *Trans Am Soc Artif Intern Organs*, **14**, 181–187.
49. Di Paolo, N. *et al.* (1996). A new self-locating peritoneal catheter. *Perit Dial Int*, **16**, 623–627.
50. Shah, G.M. *et al.* (1990). Peritoneal catheters: A comparative study of column disc and Tenckhoff catheters. *Int J Artif Organs*, **13**, 267–272.
51. Lye, W.C. *et al.* (1996). A prospective randomized comparison of the Swan neck, coiled, and straight Tenckhoff catheters in patients on CAPD. *Perit Dial Int,* **16**(Suppl. 1), S333–S335.
52. Eklund, B. *et al.* (1997). Peritoneal dialysis access: Prospective randomised comparison of single-cuff and double-cuff straight Tenckhoff catheters. *Nephrol Dial Transplant*, **12**, 2664–2666.
53. Weber, J. *et al.* (1993). Survival of 138 surgically placed straight double-cuff Tenckhoff catheters in patients on continuous ambulatory peritoneal dialysis. *Perit Dial Int*, **13**, 224–227.
54. Faller, B. and Lamier, N. (1994). Evolution of clinical parameters and peritoneal function in a cohort of CAPD patients followed over 7 years. *Nephrol Dial Transplant,* **9**, 280–286.
55. Issad, B. *et al.* (1996). 213 elderly uraemic patients over 75 years of age treated with long-term peritoneal dialysis: A French multicentre study. *Perit Dial Int,* **16**(Suppl. 1), S414–S418.
56. Seghal, A.R. *et al.* (1998). Barriers to adequate delivery of haemodialysis. *Am J Kidney Dis*, **31**, 593–601.

57. Held, P.J. *et al.* (1994). Haemodialysis therapy in the United States: What is the dose and does it matter? *Am J Kidney Dis*, **24**, 974–980.
58. Schwab, S.J. (1994). Assessing the adequacy of vascular access and its relationship to patient outcome. *Am J Kidney Dis*, **24**, 316–320.
59. Butterly, D.W. (1994). A quality improvement program for haemodialysis vascular access. *Adv Ren Replace Ther*, **1**, 163–166.
60. Rutherford, R.B. (1997). The value of noninvasive testing before and after haemodialysis access in the prevention and management of complications. *Seminars Vasc Surg,* **10**, 157–161.
61. Burdick, J.F. and Maley, W.R. (1997). Update on vascular access for haemodialysis. *Advances in Surgery*, **30**, 223–232.
62. Palder, S.B. *et al.* (1985). Vascular access for haemodialysis. Patency rates and results of revision. *Ann Surg*, **202**, 235–239.
63. Ryan, J.J. and Dennis, M.J.S. (1990). Radiocephalic fistula in vascular access. *Br J Surg,* **77**, 1321–1322.
64. Prisch, F.C. *et al.* (1995). Parameters of prognostic relevance to the patency of vascular access in haemodialysis patients. *J Am Soc Nephrol*, **6**, 1613–1618.
65. Schmidt, R.J. *et al.* (1998). Early referral and its impact on emergent first dialyses, health care costs, and outcome. *Am J Kidney Dis,* **32**, 278–283.
66. Windus, D.W. *et al.* Optimization of high-efficiency haemodialysis by detection and correction of fistula dysfunction. *Kidney Int*, **38**, 337–341.
67. Depner, T.A. (1994). Techniques for prospective detection of venous stenosis. *Adv Ren Replace Ther*, **1**, 119–130.
68. Sands, J.J. and Miranda, C.L. (1995). Prolongation of haemodialysis access survival with elective revision. *Clin Neph*, **44**, 329–333.
69. Schwab, S.J. *et al.* (1989). Prevention of haemodialysis fistula thrombosis. Early detection of venous stenosis. *Kidney Int*, **36**, 707–711.
70. Sherman, R.A. and Kapoian, T. Recirculation, urea disequilibrium, and dialysis efficiency: Peripheral arteriovenous versus central venovenous vascular access. *Am J Kidney Dis*, **29**, 479–489.
71. Collins, D.M. *et al.* (1992). Fistula dysfunction: Effect on rapid haemodialysis. *Kidney Int*, **41**, 1292–1296.
72. Gotch, F.A. and Sargent, J.A. (1985). A mechanistic analysis of the National Cooperative Dialysis Study (NCDS). *Kidney Int,* **28**, 526–534.
73. Shohat, J. and Boner, G. (1997). Adequacy of haemodialysis 1996. *Nephron*, **76**, 1–6.

74. Bay, W.H. *et al.* (1998). Predicting haemodialysis access failure with colour flow Doppler ultrasound. *Am J Nephrol*, **18**, 296–304.
75. Köksoy, C. *et al.* (1995). Predictive value of colour Doppler ultrasonography in detecting failure of vascular access grafts. *Br J Surg*, **82**, 50–52.
76. Gadallah, M.F. *et al.* (1998). Accuracy of Doppler ultrasound in diagnosing stenosis of haemodialysis arteriovenous access as compared with fistulography. *Am J Kidney Dis*, **32**, 273–277.
77. Cayco, A.V. *et al.* (1998). Reduction in arteriovenous graft impairment: Results of a vascular access surveillance protocol. *Am J Kidney Dis*, **32**, 302–306.
78. Muirhead, N., Lanpacis, A. and Wong, C. (1992). Erythropoietin for anaemia in haemodialysis patients: Results of a maintenance study (The Canadian Erythropoietin Study Group). *Nephrol Dial Transplant,* **7**, 811–816.
79. Churchill, D.N. *et al.* (1994). Probability of thrombosis of vascular access among haemodialysis patients treated with recombinant human erythropoietin. *J Am Soc Nephrol*, **4**, 1809–1813.
80. Henderson, L.W. (1996). Dialysis in the 21st century. *Am J Kidney Dis*, **28**, 951–957.

CHAPTER 2

MODALITY SELECTION AND PATIENT OUTCOME

Ken Farrington MD, FRCP
Department of Renal Medicine
Lister Hospital
Stevenage
Herts SG1 4AB, UK

2.1. Introduction

About 40 years ago, the diagnosis of end-stage chronic renal failure was tantamount to a death sentence. Since then, there has been an explosion of therapeutic possibilities which has transformed the outlook for patients with this condition. This chapter briefly examines the various treatment modalities now available and emphasises the potential contribution of appropriate patient selection and adequate pre-dialysis preparation, to outcome.

2.2. Treatment Modalities in End-Stage Renal Failure

The therapeutic options include haemodialysis, peritoneal dialysis in various guises and renal transplantation.

2.2.1. *Renal transplantation*

A well-functioning renal transplant is widely regarded as the optimal therapy of end-stage renal failure. Improved immunosuppression has led to an increasing success rate, but transplantation is far from a panacea. Diminishing supplies of cadaveric organs have caused an increase in waiting lists in spite of more rigorous patient selection to qualify for a place on that list. The changing demography of the dialysis population with increasing numbers of elderly patients and patients with non-renal co-morbidity has added to the "immunological" wreckage of previous transplant eras to condition the rapid growth of an "untransplantable"group of patients. In many units, this group constitutes well over 50% of the dialysis population. It is unlikely that even the longed-for success of xeno-transplantation will do much to change the number. The sole future of many patients with end-stage renal failure is that of a dialysis patient.

2.2.2. *Haemodialysis*

The emergence of haemodialysis in the 1960s as a viable treatment for end-stage chronic renal failure was made possible by innovations in vascular access technology (1). These and subsequent innovations fuelled the huge growth in the use of regular haemodialysis as a treatment for end-stage chronic renal failure through the developed world and beyond. This growth is still continuing. In parallel with this growth, there has been considerable diversification both in the available technologies and the methods for provision. Haemodialysis can be "centre-based" in units located in hospitals or in free-standing facilities in cities, towns or rural areas. Self-supervised haemodialysis performed in the patient's home was pioneered in the late 1960s (2), largely to cope with increasing numbers. For a number of decades, home haemodialysis became the preferred option for many patients particularly in the UK. In recent years, chronic ambulatory peritoneal dialysis (CAPD), has almost displaced haemodialysis from the home and back to the centre or satellite unit.

Short hours haemodialysis, evolved as another coping mechanism, inadvertently spawned the concept of dialysis adequacy by dint of its association with excess mortality especially in the USA (3). Now in many units, the dose of dialysis is prescribed and monitored utilising methods such as urea kinetic monitoring (4). In addition to considerations of solute, questions have been raised about the capacity of short hours treatments to adequately remove excess fluid and thereby facilitate control of blood pressure. Cardiovascular morbidity and mortality are high in the dialysis population and hypertension is an important determinant.

Dialysis-associated amyloidosis was described in the 1980s as a sometimes crippling complication of long-term haemodialysis treatment (5). It was shown to be associated with progressive accumulation of beta-2-microglobulin (a large "middle molecule" of 11 800 daltons). This compound was not removed by standard "low flux" cuprophane membranes. Newer synthetic high-flux membranes such as polysulphone and polyacrylonitrile offer the prospect of reduced accumulation of such molecules especially when employed in techniques utilising convective membrane (haemofiltration) or combined diffusive and convective transport (haemodiafiltration). Cuprophane membranes have been shown to be bio-incompatible in other ways such as activation of complement components and initiating cytokine release. Synthetic membranes are more biocompatible but more expensive making reuse protocols necessary to control expense. Data on the potential long-term benefits is still accruing. Other improvements in the biocompatibility of the haemodialysis process such as the use of bicarbonate rather than acetate buffer and the use of ultrapure water have already achieved wider acceptance.

The modality of haemodialysis currently thus encompasses a wide church of different techniques and methods of provision. The only real certainty is that its evolution will continue.

2.2.3. *Peritoneal dialysis*

In peritoneal dialysis, there is exchange of solute and water between the peritoneal capillary blood and dialysate which is instilled into the peritoneal dialysis cavity. In the past 40 years, the technique has evolved from an intermittent treatment requiring percutaneous insertion of a disposable catheter prior to each treatment session to a variety of more or less continuous treatments, the most popular of which is the technique of CAPD. The major factor in this evolution was the introduction of the permanent indwelling silicone rubber catheter (Tenckhoff catheter) (6).

2.2.4. *(i) CAPD*

In CAPD (7), a prescribed volume of dialysis solution is instilled into the peritoneal cavity by gravity from a commercially prepared plastic bag via a transfer system through the indwelling Tenckhoff catheter. After a period of equilibration, the fluid is drained (thus completing one exchange) and the process is repeated. A total of 3 to 5 exchanges per day are required which usually includes an overnight exchange with a longer dwell time. The technique is relatively simple, requiring no complex or expensive equipment. These features fuelled its rapid growth in popularity especially in the UK where it became the dominant mode of dialysis therapy in the 1980s. The introduction of disconnect systems (8) incorporating the "flush before fill" principle reduced the incidence of its major complication, peritoneal infection, introduced by touch contamination occurring when connections are broken and remade. Nevertheless, peritonitis is still a significant problem and a major cause of technical failure. Others include abdominal wall hernia, fluid leaks, catheter migration and peritoneal membrane damage causing hyperpermeable and hypopermeable states and peritoneal membrane sclerosis.

Another significant problem came to prominence during the early 1990s when it became apparent by application of urea kinetic modelling techniques that CAPD adequacy was critically dependent on residual renal function. As residual renal function is lost, an increase in the volume or frequency of exchanges is required to maintain adequacy (9), thus increasing the complexity and expense of the treatments. For many patients, especially those of above average body size, the technique becomes non-viable when residual renal function has been lost. For all these reasons, technique survival of CAPD is poor. Only 1 to 4% of patients commencing CAPD continue on that regime for eight years (10). A number of techniques have evolved to prolong the technique survival of peritoneal dialysis. In particular, automated peritoneal dialysis (APD) can enhance peritoneal clearances and perhaps reduce peritonitis rates.

2.2.4. *(ii) APD*

This comprises a number of techniques (11) which utilise a machine to control delivery and drainage of dialysate allowing use of larger volumes with shorter dwell times to enhance peritoneal clearance. Dialysis is performed at home during the night (nightly intermittent pd, tidal pd). In some cases, daytime CAPD exchanges are also required to maintain adequacy thus increasing further the complexity and expense (continuous cyclic pd, high dose continuous cyclic pd). In spite of these disadvantages, APD programmes currently are expanding rapidly.

2.3. Pattern of Dialysis Provision

2.3.1. *Demography*

There are significant geographical variations in the pattern of dialysis employed in the treatment of end-stage chronic renal failure. Demographic and geographical factors clearly play a role but economic and political factors are also important determinants of the pattern as well as the availability of treatment. In the United Kingdom, the pattern of provision has changed dramatically over the past 20 years (Fig. 2.1). The advent of CAPD saw a dramatic decline in the home haemodialysis programme with hospital haemodialysis retained as a rescue mode for patients failing other modalities. In the 1990s, home haemodialysis continued to decline, the EDTA Registry reported only six patients commencing on this modality during 1990–92 in the UK out of a reported total of 5508 new starts (12). CAPD has also declined during this period as a result of adequacy considerations and industry driven investment in centre-based haemodialysis facilities. The APD programme though still numerically small is growing rapidly. In the United States, by contrast, the pattern of provision has remained fairly constant (Fig. 2.2), certainly over the most recent ten-year period reported in the United States Renal Data Systems (US RDS), though the figure does not do justice to the pronounced shift from CAPD to APD which has taken place during that time (13). Centre-based haemodialysis is by far the commonest treatment modality in the USA. In Europe, there are variations in the

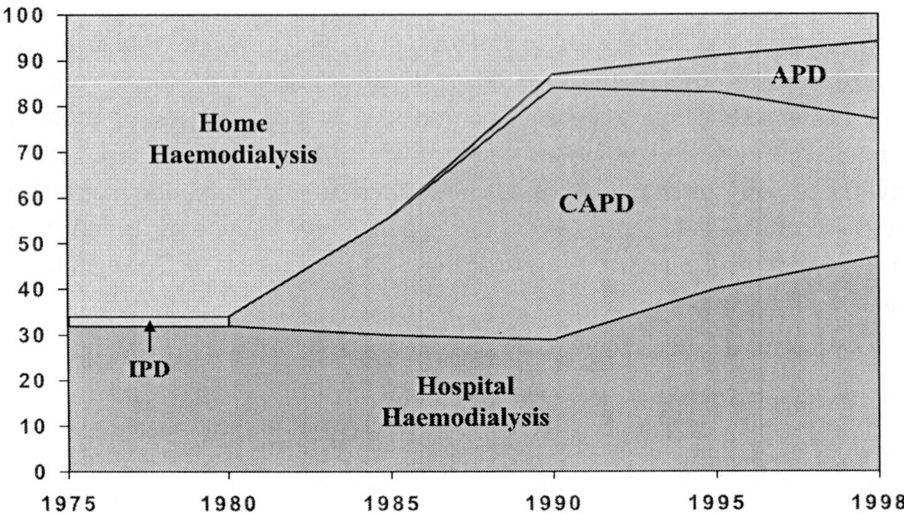

Fig. 2.1. Dialysis treatment modality in the UK. Schematic representation of changes in treatment modes.

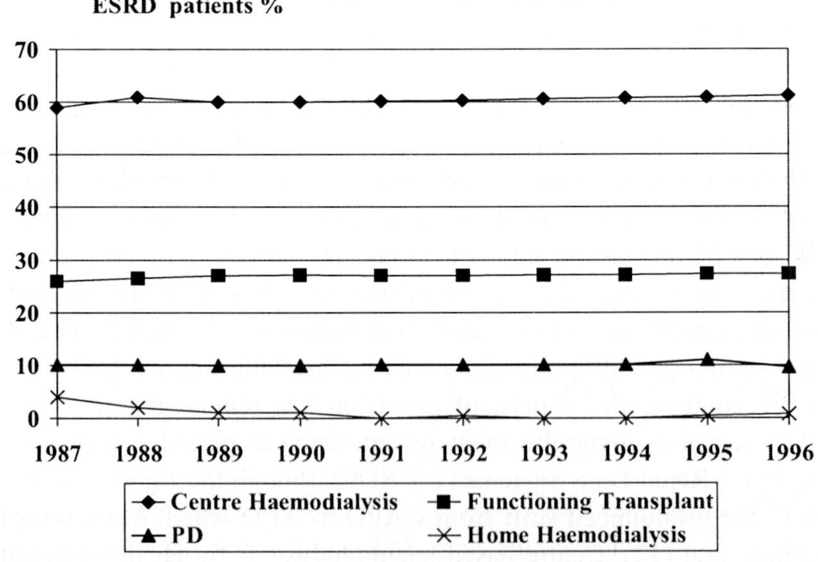

Fig. 2.2. Percent distribution of ESRD patients by treatment modality and year 1987–1996 in the USA. (ESRD = End-stage renal disease). Data from USRDS.

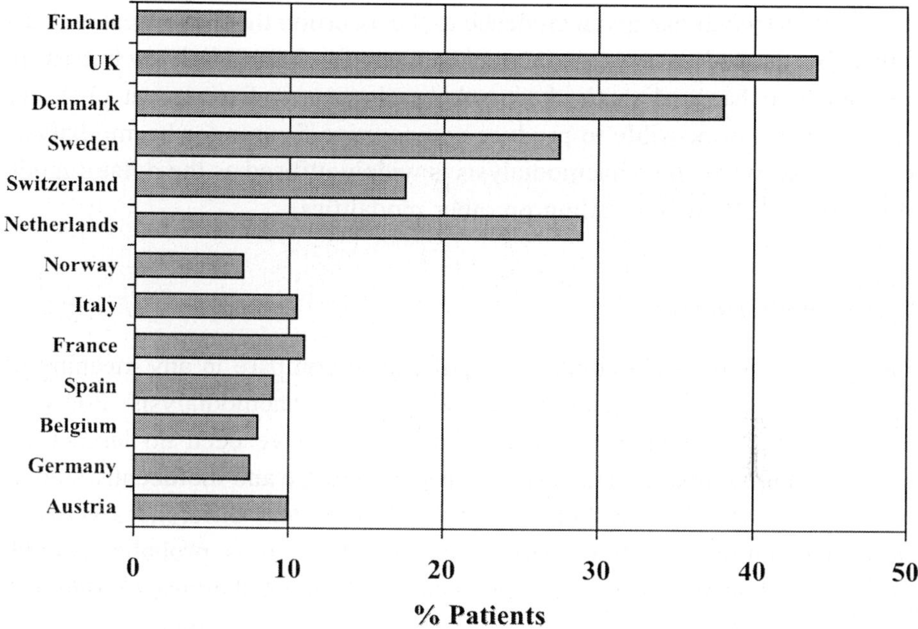

Fig. 2.3. Percentage ESRD patients treated by PD in Europe 1995.

proportion of patients with end-stage renal failure treated by peritoneal dialysis (Fig. 2.3). There are many reasons for this, including differences in clinicians' expertise and in their attitudes towards the indications for peritoneal dialysis, economic factors including reimbursement practices, and differences in patient attitude towards their illness and their involvement in their own treatment (14).

2.4. Outcomes of Dialysis

2.4.1. *Technique survival*

The poor technique survival of CAPD has already been alluded to. It is not yet possible to say whether the increased use of APD regimes to enhance adequacy will result in improved peritoneal dialysis technique survival. Poor techniques survival has given credence to the assertion that peritoneal dialysis cannot be viewed as a potential replacement for haemodialysis but as an alternative to haemodialysis during the early years of treatment (14). By contrast, it is not possible to produce comparative figures for haemodialysis technique survival since haemodialysis is widely utilised as the default mode of treatment for patients failing on other modalities.

2.4.2. *Patient survival*

For the reasons outlined above, it is difficult to compare in any meaningful way patients' survival on peritoneal dialysis and haemodialysis. For very many practical as well as ethical reasons, there have been no large-scale prospective randomised trials. Data from single-centre and multicentre studies and analyses of Registry Data do not show consistent differences between the two techniques with respect to survival (15). It is probably safe to conclude that in the early years of renal replacement therapy, mortality is similar in patients treated by peritoneal dialysis and by haemodialysis when treatment modes have been selected on practical grounds.

2.4.3. *Patient morbidity*

Both types of treatment have an associated spectrum of complications, but hospital admission rates are probably higher in patients receiving peritoneal dialysis than those receiving haemodialysis (16), largely due to the incidence of peritonitis. Quality of life assessments are similar in both groups but both are inferior to those obtained in patients successfully transplanted (17).

2.5. Planning for End Stage

2.5.1. *Referral of patients for dialysis*

Late referral for dialysis occurs in about 45% of patients taken onto the chronic programme (18,19), and it is an important cause of early morbidity and mortality (Fig. 2.4). It has been estimated that 33–60% of late referrals

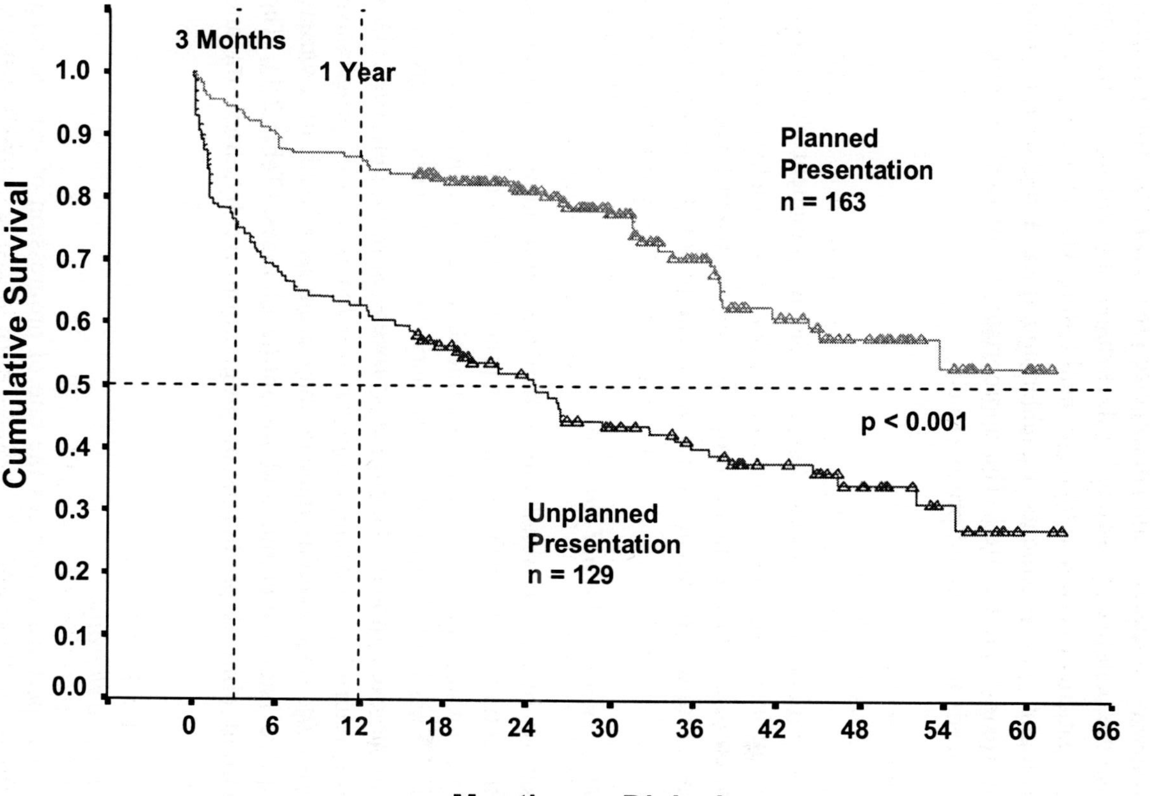

Fig. 2.4. Kaplan–Meier survival curve based on method of referral. Each step represents one death and each triangle denotes a survivor at the latest follow up. $p < 0.001$ (log Rank test).

are unavoidable due to asymptomatic slowly progressive disease, fulminant progressive disease or patient non-compliance (18). There is clearly a need to evangelise the benefits of early referral. Perhaps, as worrying is the high proportion of patients with severe chronic renal failure who are never referred to a nephrologist. These patients are older and have a higher non-renal co-morbidity than those referred. The appropriateness of this rationing by non-referral needs to be studied prospectively.

2.5.2. *Pre-dialysis treatment*

There are a number of important facets in the management of the pre-dialysis patient including preservation of residual renal function, treatment and prevention of complications of chronic renal failure, and preparation for end-stage renal failure treatment. These often require concurrent attention.

2.5.2. *(i) Preservation of renal function*

The progression of chronic renal failure depends on a number of factors, including the activity of the primary disease process. Many of these primary processes are not amenable to specific treatment. When a critical amount of renal substance has been destroyed by a primary disease process, progressive loss may occur even though the primary process may be quiescent. This "point of no return" may occur quite early in the course of progressive renal failure (20). This ongoing damage may occur as a result of a variety of maladaptive haemodynamic and metabolic changes (Table 2.1). The most important clinical predictors of progression are hypertension and proteinuria.

Anti-hypertensive therapy

Treatment of hypertension reduces the rate of progression of chronic renal failure in both diabetic and non-diabetic renal diseases. Tight control is particularly beneficial (21). Angiotensin-converting enzyme inhibitors

Table 2.1. Secondary factors associated with progression of chronic renal failure.

Systemic hypertension
Intra-glomerular hypertension
Glomerular hypertrophy
Proteinuria
Phosphate retention
Hyperlipidaemia
Secondary tubulo interstitial damage
Metabolic acidosis
Increased ammonia production
Retained uraemic toxins

(ACEIs) seem to confer additional benefit probably mediated by reduction of intraglomerular pressures due to lowered efferent arteriolar tone. This effect of ACEIs is best established in the diabetic nephropathy of Type I disease (22), and is also apparent in non-diabetic renal disease (23). The evidence is less tight in Type II diabetic nephropaths.

Anti-proteinuric therapy

The anti-proteinuric effect of ACEIs may have an important additional inhibitory effect on progression. Other anti-hypertensives are less efficacious as anti-proteinuric agents with the possible exceptions of diltiazem and verapamil (24), and beta blocker and diuretic combinations (25). Angiotensin II receptor antagonists and ACEIs are probably equally effective.

Hazards of ACEIs

The potential benefits of ACEIs are great but there are potential hazards. Patients with critical stenosis of both renal arteries or the renal artery supplying a single functioning kidney may experience an abrupt reduction in renal function with ACEIs. Renal function must be monitored closely when ACEIs are prescribed to patients with co-existent macrovascular disease.

Protein restriction

The use of dietary protein restriction to retard the progression of chronic renal failure remains controversial. Experimentally, protein restriction protects against glomerular hypertrophy and scarring by reducing hormone-induced efferent arteriolar resistance and hence intraglomerular pressures and by diminishing cytokine-induced matrix accumulation. To date though, the largest clinical studies of moderate protein restriction have not documented significant benefits (21,26), raising questions about whether protein restriction produces any additional benefit in patients on ACEIs and with well-controlled blood pressure. The risks of inducing malnutrition have been highlighted by the demonstration that protein intake is reduced spontaneously as renal failure progresses (27). Though these risks may be offset by keto-acid and amino-acid supplementation of low-protein diet, there is probably insufficient evidence at present to justify routine prescription of low-protein diets to pre-dialysis patients.

Other factors

There is experimental evidence to suggest that metabolic acidosis, phosphate retention and hyperlipidaemia contribute to the progression of chronic renal failure. There are no substantive clinical studies on the effect of controlling these factors on the progression of chronic renal failure. However, there is a wealth of clinical evidence to suggest that control of these factors may be important in retarding the development of some of the major complications of chronic renal failure (see below).

2.5.2. *(ii) Treatment and prevention of complications of chronic renal failure*

Anaemia

The pathogenesis of the anaemia/chronic renal failure is multifactorial though the major factor is undoubtedly impaired renal production of erythropoietin. The commercial availability of recombinant erythropoietin has transformed the management of anaemia of chronic renal disease and in the process has redefined the uraemic syndrome. Initially, its use was confined to the dialysis population but it is increasingly used successfully in the pre-dialysis phase (28). Who should be treated, at what stage, and with what target haemoglobin, have yet to be defined. Correction of iron deficiency is important to maximise the benefits of erythropoietin therapy. Exacerbation of hypertension is a potential problem but early fears about accelerated progression of renal failure appear to have been unfounded. There may be a slightly increased risk of vascular access thrombosis.

Bone disease

A number of diverse disorders including hyperparathyroidism, osteomalacia and adynamic bone disease constitute renal osteodystrophy. Secondary hyperparathyroidism is the commonest and is an early feature of progressive renal failure. Phosphate retention, hypocalcaemia and impaired renal production of calcitriol are the major pathogenic factors, and therapy is aimed at controlling each of these (29). Control of plasma phosphate requires the use of oral phosphate binders (usually calcium carbonate or calcium acetate), compliance with dietary phosphate restriction is difficult to achieve. The same agents may be used as calcium supplements. Calcitriol may be used to correct hypocalcaemia or to suppress PTH directly. The dose for the latter purpose should really exceed 0.25 μg daily. There is little experience with intermittent dosing schemes in the pre-dialysis period. Hypercalcaemia is a real risk especially when calcitriol and calcium compounds are used together. It is nephrotoxic and should be avoided by rigorous monitoring. The optimal target PTH is still debated but is probably twice the upper limit of normal in advanced CRF. Over-suppression risks adynamic bone disease. The importance of regular monitoring of plasma calcium, phosphate and PTH levels cannot be overstated in patients taking these agents.

Metabolic acidosis

Metabolic acidosis stimulates metabolism of the bone and muscle and when severe promotes cardiovascular instability (30). Treatment is with sodium bicarbonate to maintain plasma bicarbonate levels in the range of 20 to 24 mmols/l or the initiation of dialysis if acidosis coexists with severe fluid overload or other indicators of end stage. Sodium overload leading to extracellular fluid volume expansion is a potential risk which is often avoidable by simultaneous prescription of loop diuretics. Correction of acidosis may also be helpful in the treatment of mild hyperkalaemia.

Accelerated cardiovascular disease

Patients with CRF are at high risk for the development of cardiovascular disease. It is a logical but as yet unproven strategy to begin to tackle known risk factors, (extrapolated from general population studies), early in the course of progressive CRF (Table 2.2). In patients for whom transplantation is contemplated, cardiovascular evaluation including echocardiography and coronary arteriography may be indicated. This is particularly relevant in older patients (over the age of 50) and in all diabetics.

Table 2.2. Cardiovascular risk factors prevalent in CRF.

Condition	Treatment
Hypertension	Anti-hypertensive therapy
LVH	Anti-hypertensive therapy ACEIs Correction of anaemia
Hyperlipidaemia	Statins
Smoking	Counselling Nicotine replacement
Menopause	Oestrogen replacement
Homocystinaemia	Folic acid B vitamins
Thrombogenic factors	Anti-platelet therapy

2.5.3. *Selection for dialysis*

2.5.3. *(i) Patient selection*

The prevalence of end stage renal failure increases dramatically with increasing age, a pattern common to many other degenerative disease processes. The elderly constitute the fastest growing population group in most Western societies. These factors translate into a massive potential demand for renal replacement therapy in an increasingly elderly and infirm population, generating a huge and escalating economic burden. It is not surprising that the unrestricted availability of renal replacement therapy is being questioned.

It is generally accepted that implicit rationing is widely practised. Formal criteria for acceptance on to renal replacement programmes based on functional capacity and coexistent extra-renal co-morbidity had been proposed but have proven difficult to justify when applied retrospectively (19). There is an urgent need for prospective work. Pending this, patients will continue to be assessed individually for their likelihood of benefit. Using this type of approach, Hirsch *et al.* found that 25% of all patients referred for dialysis were not accepted for treatment (31).

The decision to offer dialysis treatment or not is clearly of momentous importance to the individual patient, his/her family and carers, as well as to the renal unit concerned. It demands a comprehensive assessment and multidisciplinary involvement. We have found a team approach which involves the patient his/her family and carers, and includes assessment of the patient in his/her own home, to be of great benefit. It is important to stress that dialysis only replaces aspects of renal function and is not a cure for other ills which may relate to disease in other organs.

If a decision not to proceed to dialysis is made with the patient, and his/her family and/or carers, it is crucially important to offer continued support including effective symptomatic treatment and timely terminal care.

In most cases, the decision will be made to proceed to dialysis when it becomes necessary. This signals the requirement for an information and support system with different objectives. Plans must be made to choose the appropriate modality of treatment and make the necessary preparations including the establishment of the appropriate form of access at the appropriate time. Patients and their families require information and education to allow their full participation in this process. A multidisciplinary approach is vital in this context too.

Late referral often with the patient requiring urgent initiation of dialysis is highly disadvantageous to the process described above. It removes from all parties the opportunity for reflection on the appropriateness of treatment, and decisions made in this context can be regretted later.

2.5.3. *(ii) Choice of modality* (*Table 2.3*)

Patient preference

Patient preference plays a major role in modality selection but may be overridden by other practical imperatives. Centre-based haemodialysis and peritoneal dialysis (CAPD and APD) are currently the commonest available modalities. In many respects, they lie at the opposite end of the treatment spectrum. Patients on CAPD must take full responsibility for the performance of their own treatment. The pay-off is a degree of flexibility with regard to the precise timing of the exchanges and the facility to dialyse virtually anywhere. In addition, fluid allowances may be more liberal, at least early in the course of CAPD. In contrast, patients on centre-based haemodialysis are relieved of most of the responsibility for the performance of their treatment but the price is a requirement to travel sometimes considerable distances, rain, hail or shine to keep thrice-weekly appointments. Taking a holiday is much more difficult. Patients may perceive unacceptable distortions of body image by the physical nature of the access required particularly in CAPD. Patients with different temperaments will choose different modalities.

Age and social factors

Age by itself probably has little bearing on the choice of modality. However, older patients with little or no close family support may struggle to sustain

Table 2.3. Selection of treatment modality.

1. Patient preference
2. Age
3. Social factors
   — mental state
   — employment status
   — family and social support
   — distance of home from centre
   — transport
4. Medical factors
   — relative contraindications to peritoneal dialysis
     previous abdominal surgery
     multiple hernia
     gross obesity
     visual impairment
   — relative contraindications to haemodialysis
     potential difficulties with vascular access
     severe cardiac disease
     autonomic neuropathy
     uncontrolled diabetic retinopathy
5. Mode of presentation
     planned or unplanned

self-treatment and be better treated by centre-based haemodialysis. CAPD, and increasingly APD, tend to be favoured by patients in regular employment. The precise demands of the job, the attitudes and the relationships with the employer are all significant factors in deciding the optimal initial modality in the individual patient. Situations also have a habit of changing and it is important to maintain a degree of flexibility with regard to modality selection as the patient's dialysis career progresses.

Geography undoubtedly plays a role in the choice of modality. Patients living considerable distances from the centre may be more likely to choose (or be advised to choose!) peritoneal dialysis as the initial modality, especially if they do not have their own transport.

Medical factors

There are few absolute contraindications to particular modalities. Previous major abdominal surgery, multiple abdominal hernia and gross obesity mitigate against the choice of CAPD as the initial mode of therapy. Potential vascular access problems, such as severe arterial disease and obliteration of forearm veins, might tend to mitigate against initial choice of haemodialysis if there are no contraindications to peritoneal dialysis. Such problems exercise the invention of the access surgeon and creative solutions are frequently found.

Patients with severe cardiac disease may not tolerate the rapid fluid shifts which can occur on haemodialysis. Fluid removal by CAPD and APD is gentler and these modalities may be more suitable for such patients. Most patients with diabetes would do equally well on haemodialysis or peritoneal dialysis. However, patients with advanced autonomic neuropathy tolerate rapid fluid shifts poorly and peritoneal dialysis may be preferable. In addition, patients with uncontrolled proliferative retinopathy should not be subjected to the regular heparinisation which haemodialysis entails. Severe visual impairment does not necessarily preclude self-treatment but it does make it less likely to be successful especially if the patient has no family or social support.

Mode of presentation

Patients presenting for dialysis in an unplanned fashioned bypass the planning phase. The initial mode of therapy in these patients is usually haemodialysis through temporary vascular access. Such patients still account for 40% of patients commencing dialysis. Subsequently, there is a natural reluctance to change modality and the initial modality will often become the long-term preference.

2.5.4. *Creation of access*

2.5.4. *(i) Haemodialysis*

The creation of adequate vascular access is the most crucial aspect of a patient's preparation for haemodialysis. Extensive guidelines have been prepared covering aspects of placement, monitoring and management of complications. The guidelines are based on evidence as interpreted by the National Kidney Foundation Dialysis Outcomes Quality Initiative (DOQI) (32) as well as on the work group's opinions. The following account is based on DOQI recommendations predominantly focusing on the fashioning of primary access.

Preferred access

Where possible, a wrist radiocephalic fistula is widely regarded as the optimal access. These fistulas are associated with much less vascular steal than more proximal fistulae and, once established, can provide good blood flow which can increase over time. A major drawback is the vulnerability of forearm veins to cannulation and venepunction during previous hospitalisations. This might obliterate the veins, rendering them unusable for access procedures or may cause poor development of the fistula. Forearm vasculature needs to be protected in patients who are destined for dialysis. A further drawback is that such fistulae often require a considerable time, (frequently many months) to mature enough to allow repeated cannulation. They should be fashioned three to six months before the projected time for initiation of dialysis. For many of these reasons, it may be necessary to resort to other types of primary accesses, including in order of preference, an elbow brachio-cephalic fistula, an arterio-venous graft composed of synthetic materials (eg PTFE), and cuffed tunnelled central venous catheters.

The latter, especially when placed in subclavian veins, are associated with a high rate of central vein thrombosis and stenosis which might prejudice later attempts to create access. Their primary use should be discouraged. Unfortunately, many patients still present late for dialysis. Many such patients will require central venous access by default. The internal jugular route is then preferred since stenotic complications may be less than with subclavian catheters. Where only short-term central venous access is required (less than three weeks) non-cuffed, percutaneously inserted catheters are often preferable. Right-sided catheters are preferred to left-sided ones since the right-sided great veins offer a more direct route to the caval atrial junction. Tunnelled catheters are also required by patients in whom all other access options have been exhausted.

Timing of access placement

Referral for access placement should allow sufficient time for assessment, surgery and access maturation. Radiocephalic fistulas should be constructed at least three to four months prior to the anticipated need and longer if there are doubts about the potential success of the primary attempt. This allows time for a second attempt before dialysis initiation. AV grafts should be fashioned three to six weeks before the anticipated need. Central venous catheters should not be inserted until the need for dialysis is immediate.

Assessment for access placement

This is best carried out in close collaboration between physicians and surgeons preferably in a dedicated access clinic. An adequate history is vital and should include identification of dominant arm, details of previous vascular access procedures, details of previous central venous catheterisation and pacemaker insertion, details of previous surgery or trauma to arms, neck and chest. History of co-morbid conditions including diabetes mellitus, cardiac problems and malignancy is also important. All these factors may have a major influence on the optimal type and location of access placement. Physical examination of the cardiovascular system should be carried out with particular reference to peripheral arterial supply and venous drainage. Assessment of the character of peripheral pulses, upper limb blood pressure measurement, and an Allen test of the integrity of the arterial supply to the hand, may need to be supplemented by hand-held Doppler studies. Tourniquet venous mapping allows selection of the appropriate vein. Localised oedema and the presence of collateral veins may indicate venous obstruction. Such abnormalities may require additional investigation by venography or formal Doppler studies. These may be particularly useful in cases in which there have been previous procedures involving the central veins of the neck and thorax.

Monitoring and maintenance of vascular access

The early detection of significant access stenosis, when combined with an effective programme for correction of the problems identified, can improve patency and prevent access loss due to thrombosis. Regular physical examination can be a useful screen looking for altered characteristics of pulse, thrill or bruit related to the access. Prolonged bleeding after needle withdrawal in the absence of over-anticoagulation, or persistent swelling of the arm may indicate venous obstruction. Upward trends in serial measurement of static or dynamic venous pressures during dialysis should also raise suspicion of venous stenoses.

Unexplained decreases in the measured dose of dialysis delivered, obtained by regular monitoring of Kt/V or urea reduction ratio, demand further investigation to exclude access recirculation resulting from access stenosis.

Significant access recirculation (>10%), detected by the stop-flow method or saline or thermal dilution methods (33) after confirmation of correct needle placement, requires further investigation by fistulography to anatomically define stenosis prior to angioplastic or surgical correction. Regular screening of intra-access flow can be carried out by ultrasonic dilution, doppler flows or magnetic resonance angiography, though not all these hold the promise of widespread clinical utility.

There are thus a wide variety of monitoring methods of varying complexity. Those chosen will depend on local availability and expertise.

2.5.4. *(ii) Peritoneal dialysis*

Timely pre-assessment is useful particularly in patients who have had previous abdominal surgery or who have known or suspected abdominal hernia. Hernia repair before catheter insertion can prevent more serious problems later which might require temporary abandonment of the mode. There is little hard data on the optimal time for catheter placement prior to commencement of dialysis. There is wide variation in catheter types and placement techniques which sometimes include primary omentectomy. Dialysate leaks seem to be less likely if dialysis initiation is postponed for two weeks after catheter implantation.

2.6. Initiation of Dialysis

The time to start renal replacement therapy in a patient with end-stage chronic renal failure depends on a variety of interacting factors. The chief considerations are clinical, mainly relating to the presence of symptoms of uraemia or fluid overload, and biochemical, traditionally blood urea and serum creatinine measurements. Over the last decade, the practice of reliance on these parameters to assess dialysis adequacy has been undermined. Low blood urea levels may indicate poor nutrition rather than adequate dialysis. Low serum creatinine concentrations may indicate low muscle mass as a result of malnutrition rather than adequate dialysis. The inverse correlation between serum creatinine and mortality suggests that wasting may be the more common cause (34). To overcome these problems, kinetic methods such as urea kinetic modelling have been established as more reliable tools to assess dialysis adequacy. In recent years, these methods have also been applied to pre-dialysis patients to help define the optimal time to initiate dialysis.

In this context, normalised urea clearance (Kt/V) at dialysis initiation has been shown to be at a level that would indicate severe under-dialysis in regularly dialysed patients, suggesting a systematic delay initiation of dialysis based on standard criteria. This might contribute to early morbidity and mortality on dialysis (35). Indeed, NKF-DOQI guidelines now suggest that dialysis should be initiated when weekly Kt/V urea falls below 2.0, unless clinical features of uraemia or fluid overload are completely absent, body weight is stable, and nPNA (normalised protein equivalent of total nitrogen appearance) is greater than or equal to 0.8. The latter is an indication of nutritional status based on urinary urea excretion (36). These guidelines have the advantage of aligning thinking about the adequacy of residual renal function in patients approaching end stage and dialysis adequacy, but as yet lack a firm evidence base. The assumption that residual renal urea clearance and dialysis urea clearance are equivalent markers of adequacy is difficult to justify. In addition, since many patients still present for dialysis late and in extremis, they may not be tackling the root cause of early dialysis mortality.

References

1. Brescia, M.J., Cimino, J.E., Appel, K. and Hurwich, B.J. (1966). Chronic haemodialysis using venepuncture and a surgically created arteriovenous fistula. *N Engl J Med*, **275**, 1089.
2. Baillod, R.A., Comty, C., Ilahi, M., Konotey-Ahulu, F.I.D., Sevitt, L. and Shaldon, S. (1965). Overnight haemodialysis in the home. *Proc Eur Dial Transplant Assoc*, **2**, 99.
3. Kjellstrand, C.M. (1985). Short dialysis increases morbidity and mortality. *Contrib Nephrol*, **44**, 65–77.
4. Gotch, F.A. and Sargeant, J.A. (1985). A mechanistic analysis of the National Cooperative Dialysis Study (NCDS). *Kidney Int*, **28**, 526–34.
5. Charra, B., Calemard, E., Vizam, M. *et al.* (1984). Carpal tunnel syndrome, shoulder pain and amyloid deposits in long-term haemodialysis patients. *Proc Eur Dial Transplant Assoc*, 21, 291–295.
6. Tenckhoff, H. and Schecter, H. (1968) A bacteriologically safe peritoneal access device. *Trans Am Soc Artif Intern Organs*, **14**, 181.
7. Popovich, R.P., Moncrief, J.W., Nolph, K.D, Ghods, A.J., Twardowski, Z.J. and Pyle, W.K. (1978) Continous ambulatory peritoneal dialysis. *Ann Int Med*, **88**, 449.
8. Maiorca, R., Cantaluppi, A., Cancarini, G.C. *et al.* (1983). Prospective controlled trial of Y connector and disinfectant in CAPD. *Lancet (ii)* 642–645.
9. Tattersall, J.E., Doyle, S., Greenwood, R.N. and Farrington, K. (1994). Maintaining adequacy in CAPD by individualising the dialysis prescription. *Nephrol Dial Transplant*, **9**, 729–752.
10. Gokal, R. and Oreopoulos, D. (1996). Is long term CAPD possible? *Perit Dial Int*, **16**, 553–555.
11. Gokal, R. and Mallick, N.P. (1999). Peritoneal dialysis. *Lancet*, **353**, 823–828.

12. Annual Report on Management of Renal Failure in Europe XXV111. EDTA Registry (1997).
13. USRDS 1998 Annual Data Report.
14. Mignon, F., Michel, C. and Viron, B. (1998). Why so much disparity of PD in Europe? *Nephrol Dial Transplant*, **13**, 1114–1116.
15. Vonesh, E.F. and Moran, J. (1999). Mortality in end-stage renal disease. A reassessment of differences between patients treated with haemodialysis and peritoneal dialysis. *J Am Soc Nephrol*, **10**, 354–365.
16. Habach, G., Bloembergen, W., Mauger, E., Wolfe, R. and Port, F. (1985). Hospitalisation among United States dialysis patients: Haemodialysis versus peritoneal dialysis. *J Am Soc Nephrol*, **5**, 1940–1948.
17. Gokal, R. (1993). Quality of life in patients undergoing renal replacement therapy. *Kidney Int*, **40** (Suppl. 38), S23–S27.
18. Obrador, G.T. and Pereira, B.J.G. (1998). Early referral to the nephrologist and timely initiation of renal replacement therapy: A paradigm shift in the management of patients with chronic renal failure. *Am J Kidney Dis*, **31**, 398–417.
19. Chandna, S.M., Schultz, J., Lawrence, C., Greenwood, R.N. and Farrington, K. (1999). Is there a rationale for rationing chronic dialysis? A hospital-based cohort study of factors affecting survival and morbidity. *Br Med J*, **318**, 217–223.
20. Maschio, G., Oldrizzi, L. and Rugiu, C. (1991). Is there a point of no return in progressive renal disease? *J Am Soc Nephrol*, **2**, 832.
21. Klahr, S., Leven, A.S., Bech, G.J. *et al.* (1994). The effects of dietary protein restriction and blood pressure control on the progression of chronic renal disease. *N Engl J Med*, **330**, 877.
22. Lewis, E.J., Hunsicker, L.G., Bain, R.P. and Rohde, R.D. (1993). The effect of angiotensin converting enzyme inhibition on diabetic nephropathy. *N Engl J Med*, **329**, 1456.
23. Maschio, G., Alberti, D., Janin, G. *et al.* (1996). Effect of the angiotensin converting enzyme inhibitors benazepril on the progression of chronic renal insufficiency. *N Engl J Med*, **334**, 939.
24. Gansevoort, R.T., Sluiter, W.J., Hemmelder, M.H., de Zeeuw, D. and de Jong, P.E. (1995). Anti-proteinuric effects of blood pressure lowering agents. A meta-analysis of comparative trials. *Nephrol Dial Transplant*, **10**, 1963–1974.
25. Kloke, H.J., Wetzels, J.F., van Hamersfeld, H.W. *et al.* (1993). Angiotensin converting enzyme inhibition and the combination of a beta blocker and a

diuretic are equally effective in lowering proteinuria in patients with glomerulonephritis. *Nephrol Dial Transplant*, **8**, 808.

26. Locatelli, F., Alberti, D., Graziani, S. *et al.* (1991). Prospective, randomised, multicentre trial of effect of protein restriction progression of chronic renal insufficiency. *Lancet*, **337**, 1299.
27. Ikizler, T.A., Greene, J.H., Wingard, R.L. *et al.* (1995). Spontaneous dietary protein intake during progression of chronic renal failure. *J Am Soc Nephrol*, **313**, 771.
28. Teehan, B.P., Krantz, S., Stone, W.A. *et al.* (1991). Double-blind placebo controlled study of the therapeutic use of recombinant erythropoietin for anaemia associated with chronic renal failure in pre-dialysis patients. *Am J Kidney Dis*, **18**, 50–59.
29. Coburn, J.W. and Elangovan, L. (1998). Prevention of metabolic bone disease in the pre-end-stage renal disease setting. *J Am Soc Nephrol*, **9**, S71–S77.
30. Franch, H.A. and Mitch, W.E. (1998). Catabolism in uraemia: The impact of metabolic acidosis. *J Am Soc Nephrol*, **9**, S78–S81.
31. Hirsch, D.J., West, M.L., Cohen, A.D. and Jindal, K.K. (1994). Experience with not offering dialysis to patients with a poor prognosis. *Am J Kidney Dis*, **23**, 463–466.
32. NKF-DOQI Clinical Practice Guidelines for Vascular Access. (1997). *Am J Kidney Dis (Suppl.)*, **30**(4) (Suppl. 3), S150–S191.
33. Tattersall, J., Farrington, K. and Greenwood, R.N. (1998). Adequacy of dialysis in Oxford Textbook of Clinical Nephrology. *Oxford Medical Publications*, 2075–2087.
34. Lowrie, E.G. and Lew, N.L. (1990). Death risk in haemodialysis patients. The predictive value of commonly measured variables and an evaluation of death rate differences between facilities. *Am J Kidney Dis*, **15**, 458–482.
35. Tattersall, J.E, Greenwood, R.N. and Farrington, K. (1995). Urea kinetics and when to commence dialysis. *Am J Nephrol*, **15**, 283–289.
36. NKF-DOQI. Clinical Practice Guidelines for Peritoneal Adequacy. Guideline 1. Dialysis Initiation. National Kidney Foundation, New York.

CHAPTER 3

CONTINUOUS QUALITY IMPROVEMENT THROUGH CLINICAL PATHWAYS

K. L. Brayman MD, PhD and P. Wallace MSN, RN
University of Pennsylvania Medical School
Children's Hospital
Philadelphia, PA 19104, USA

3.1. Introduction

Over the past decade, health care providers have witnessed an increase in the number of different strategies used to achieve their quality objectives. Total quality management, case management, clinical pathways, outcomes management, and patient-centred care units are a sampling of the many practices being adopted by quality-conscious health care organisations. The focus on assuring and improving the quality of medical care and service has increased further during the late 1990s, as witnessed by intensified legislative and health policy activity, consumer advocacy, and industry debate on the subject of health care quality (1). Quality improvement represents a systematic approach of continuously improving processes of critical importance to customers and is typically described in terms such as total quality management or continuous quality improvement. Although not as highly evolved as in other industries, quality improvement in health care has become an important strategy for achieving and maintaining market share.

Continuous quality improvement (CQI) emerged from the industrial sector as an effective package of theory and practical tools to reduce errors in the production process (1). Based on the teachings of pioneers such as Deming,

Crosby and Juran, CQI has been applied in health care through the leadership of many health care organisations and individuals.

As the health care industry evolves and multiple stakeholders demand greater accountability for high quality, providers are approaching quality improvement initiatives with a greater degree of commitment. However, the gaps between known best practice and current practice remain strikingly large and widespread. For example, the National Committee on Quality Assurance found that on a nationwide basis, the managed care industry is performing below 80% on all but one of its clinical performance measures (2). These include the use of beta blockers after acute myocardial infarction, cervical and breast cancer screening, diabetic retinal examination, and childhood immunisation. Effective interventions that can save lives, prevent disabilities, and improve quality of life are not consistently implemented in health care.

Books and articles that address health care quality abound, but providers remain uncertain about how to best measure and improve clinical quality as health care is still experimenting with various clinical quality improvement techniques. Due to the increasing burden of tightening reimbursement controls, health care organisations no longer can afford to embrace every quality initiative that comes their way but must select those methods *most likely to succeed.*

3.2. Continuous Quality Improvement

CQI uses data-driven interventions to test and improve processes by which health care and services are delivered as a means to meet or exceed customer needs. CQI is used throughout health care as a tool to achieve desired results, such as reducing system failures, minimising unnecessary variations in patient care practices, enhancing collaboration among health care team members, achieving better patient satisfaction, improving clinical outcomes, and reducing costs. CQI involves answering the following questions:

- What do health care customers need and value?
- Are these needs being met by current health care delivery processes?

- How can current processes be improved to more effectively and/or efficiently reach desired outcomes?

CQI has been described and applied in a variety of ways although six key success factors form the foundation for all of its activities:

- Leadership commitment to quality
- Customer-focused
- Multidisciplinary team approach
- Data-driven
- Balanced performance measurement system
- Integration and alignment of management systems

3.2.1. *Continuous quality improvement and clinical pathways*

With the aforementioned goals in mind, clinical path methods are a natural fit with any health care organisation's quality improvement efforts. Broadly defined, clinical paths are "collaboratively developed hypotheses that describe what the health care team believes is the best way to manage patients" (3). This involves mapping out the activities of care for a defined patient population over a selected period of time (e.g. what needs to be done during a hospital stay for renal transplant patients). The overall goals are to improve outcomes by assuring that no aspect of care is inadvertently overlooked and to improve efficiency by identifying an optimal timing and coordination of activities (4). The quality management tenets of meeting customer needs, focusing on process, reducing variation, and involving those who work in the process are all reflected in clinical path methodology. Clinical paths address the needs of the external customers (patients and families) through their focus on the outcomes of the care process. Further, these methods also address the needs of internal customers (health care professionals) because they help members of a care team see how their activities fit together with those of their colleagues. Finally, clinical path development methodologies call for the use of multidisciplinary teams to make the process of care explicit and to engage in lively debates aimed at reducing inappropriate variation.

Despite the logic of the clinical path approach in the care management process, the results from the application of these methods in health care have been mixed. Qualitatively, a recent article by Lord (5) describes the various positive benefits for patients, physicians, and nurses. But there remains a general problem of acceptance. Some clinicians view these attempts to explicitly map out the process of care as a restriction on their judgement. Advocates of these methods must constantly rebut the charge that clinical pathways represent "cookbook medicine".

There are also mixed reports regarding the quantitative benefits of clinical path methods. There have been numerous case studies citing the length of stay cuts of 5 to 40%, cost reductions of as much as one-third, significant improvements in readmission rates, elimination of unnecessary testing, to name a few (6–11). Specifically in vascular surgery, pathways have significantly reduced the use of arteriography, general anaesthesia, intensive care unit (ICU) use, average length of stay (ALOS), cost/case without accompanying increases in complication rates or mortality (12–14). But such glowing reports also have their counterbalance in the literature. For example, Falconer *et al.* (15) reported the results of a controlled study involving 53 stroke patients whose care was managed via a clinical path. Their investigation failed to show a significant decrease in costs or improvement in outcomes compared with those who were not managed via clinical paths. However, a critique by Schriefer (16) notes that a pathway requires as much diligence in implementation strategies as those involved with the design of best practice care. Well designed but poorly implemented clinical pathways results in the failure of health care to utilise all of the power of continuous quality improvement methods (17,18).

3.3. Application of the Plan–Do–Check–Act (PDCA) Model of CQI with Clinical Pathway Development

The decision to adopt a path-based clinical management strategy for the organisation requires a long-term commitment to seek out ways of improving clinical processes and requires a fundamental shift in the way patient care is delivered. This requires committed leadership with involvement at all levels from the CEO to direct care providers. CQI programmes are most successful when leadership demonstrates a personal involvement in promoting CQI values. Strong collaboration between medical staff and administrative leadership is essential for a successful path-based management strategy.

One of the most widely used models for CQI is Deming's PDCA cycle of continuous improvement — Plan, Do, Check, Act — which is particularly well suited for path-based management.

3.3.1. *Plan*

> *CQI initiatives begin with a planning stage in which data are analysed to measure the effectiveness of existing processes by identifying discrepancies between current practice and known best practice. Appropriate use of data can focus attention on the outcomes that are in need of improvement. For example, data analysis can help a CQI team determine that patients are dissatisfied with the amount of time it takes to be seen in the physician's office or that the infection rates for renal transplant cases significantly exceeds established benchmarks.*

The first level of analysis involves an evaluation of an organisation's improvement opportunities prior to the selection of specific pathway teams. Pre-determined selection criteria will help justify spending the time and resources needed to develop and implement a pathway. Examples include but are not limited to:

- Quality issues from analysis of mortality, morbidity, complications, readmission rates and patient satisfaction
- High volume conditions
- High clinical or financial risk
- High interest — significant interest in developing a pathway from physicians, payers, or specific care areas
- Length of stay and cost are higher than benchmark data of peer organisations
- Predictability of the care process
- Care processes associated with wide variations in clinical practice

Critical for an effective path-based management strategy is the use of a multidisciplinary team to design the appropriate clinical content. This team must include representatives from all disciplines involved in the relevant activities and sites where care is delivered. If the pathway is to encompass the pre-operative office visit or home care post-discharge, it is essential that staff from these areas be involved in the design of the pathway. The use of multidisciplinary teams is necessary for a variety of reasons, including:

- Teams share knowledge of clinical processes that cross formal departmental boundaries
- Teams bring together people from different staff areas and even different systems. This fosters collaboration and commitment among people who might otherwise never meet each other
- Teams facilitate discussion
- No single person can view the entire progression of clinical care and it is precisely the whole process that needs to be examined in the design of a clinical pathway
- Team members promote implementation of clinical pathways (19)

Key members for the success of the pathway are the physician "champion" and the nurse co-leader. The physician champion is an opinion leader whose role is, in part, to influence physician practice and stimulate change in physician behaviour. The physician champion must be willing to be an active leader in developing, implementing and maintaining the pathway.

Pathways are not meant to codify current practice but rather to identify a strategy for *improving* the care process. During the initial pathway team meeting, the team members establish goals for the pathway. These goals may aim to:

(i) reduce resource utilisation for specific groups of patients,
(ii) reduce the incidence of variation in clinical care, and/or
(iii) improve selected clinical outcomes.

The objective of the next level of analysis to gain a better understanding of the specific care process that has been selected. Examination of internal practice patterns including an analysis of patient flow can begin with a chart audit supplemented with the use of process analysis and data management tools. Focus on the areas with the most significant variation. Examples include but are not limited to:

- Average length of stay
- Average costs
- Costs by department (lab, radiology, pharmacy, etc.)
- Delayed or missed services due to hospital system problems
- Operating room time
- Timing of key events during the process of care (such as catheter removal or initiation of physical therapy)
- Intensive care unit length of stay

The use of available data sources to identify the key areas where variation in practice occurs provides a springboard for a healthy debate on the evidence-basis for this variation. Incorporation of the most recent evidence-based approaches to care (clinical practice guidelines, research studies, "best practice" white papers) are essential ingredients for promoting optimal patient care management. Several guideline developers have created systems for grading the recommendations based on the level of evidence that supports them. Factors which drive these rating systems include the quality of evidence combined with other factors such as the burden of suffering, risk versus benefit of treatment, clinical effectiveness studies, cost, risk of side effects, magnitude of the condition in the population and practical application of a recommendation.

Regardless of whether a rating system is used, the following questions must be answered when evaluating evidence for incorporation into a clinical pathway:

> Is the evidence valid and reliable?
> Are the patients in the studies similar to those seen in the clinical practice that is adopting the pathway?
> Was all relevant evidence considered?
> Were important studies omitted?

3.3.2. *Do*

Once an opportunity for improvement has been identified, the CQI team develops and tests new care or delivery processes to replace the existing practice. These interventions may be as simple as patient or clinician education or as complex as developing and implementing practice guidelines or pathways.

After a synthesis of the available literature and analysis of key data elements, it is time to develop a draft of the pathway by determining key process and outcome measures, and reaching agreement about the treatment steps and sequencing. Key elements for incorporation into the design of a clinical pathway include the clinical outcomes that can be realistically achieved within the predetermined time frame. These outcomes should be achievable by 80% or more of the patients to be placed on the pathway. Daily or intermediate outcomes will help provide the framework for sequencing critical interventions and determining the timeline.

Traditionally, pathways were developed in a matrix format and outlined key clinical care activities (e.g. assessments, consults, testing, medications, activity, etc.) on a day-based approach. However, the goals of each pathway team will determine the most appropriate format and clinically relevant progression throughout the episode of care. For example, an ambulatory surgery pathway may use hours as an appropriate time line. Alternatively, for some medical conditions, a phase or outcome-based approach may be more meaningful. An algorithm-type format may be more meaningful if multiple decision points are placed in the design of the pathway. An algorithm-type style mirrors more closely the way physicians are trained to think about patients and their treatments and may be more readily embraced by physicians. The scope of the pathway may be determined by the level of system integration and may be based solely in a hospital setting or may span the entire continuum of care. With a broadened scope, it becomes important to design a pathway with the focus of cost-efficiency across the multiple settings. Designing a care process that provides the best outcomes over time will be in the best interest of the patient, and often be the least expensive strategy

(20). One last component of the planning phase includes establishing a mechanism for regular review of the pathway to maintain clinical relevancy and consistency with new medical knowledge.

Well-designed pathways address common clinical situations and are practical and simple to follow. The more complex a pathway or guideline is, the less likely it is to be evidence-based, applicable to large numbers of patients, and amenable to implementation and clinical review (21). The implementation phase of the pathway process requires as much attention, if not more, than the design phase in order for pathways to achieve their desired outcomes. A pathway must be effectively implemented for a true change in clinical practice to occur. Simply disseminating the pathway to clinicians is not sufficient to achieve compliance with the pathway. To effectively change clinician behaviour, guidelines must be integrated into daily routines. Key steps requiring attention in the implementation phase include:

Assure stakeholder involvement

Prior to implementation, have the draft of the clinical pathway reviewed by key stakeholders who were not on the design team and are involved in the delivery of that specific clinical process of care. This review may include all MDs with patients in that particular case type within the past year, key department managers/experts/directors whose resources are impacted by the design of the pathway, e.g. nursing, social services, physical therapists, pharmacists, etc. and appropriate administrators. When feedback is provided to the design team, it is critical that the communication loop between the pathway team and the reviewer be closed. The reviewers' comments can be incorporated into the pathway if new and pertinent information is brought to the attention of the team but if the comments are *not* to be used, the reviewer is provided an explanation of the rationale and evidence used by the team in their decision. The value of enlisting the support of department directors and key clinicians cannot be underestimated in assuring successful implementation of the pathway.

Improved work processes

The implementation of a revised clinical care process also provides an opportunity to revise and streamline other work processes which can facilitate pathway implementation. Tailored to the needs and existing systems of each institution, examples of these work processes include but are not limited to:

— computer-generated prompts and reminders make it easy for clinicians to "do the right thing at the right time"
— computerised order entry
— changes in supporting documentation systems
— automating the variance data collection process
— standardised order sets
— developing patient versions of the pathways to facilitate education and participation

If the evidence supporting the pathway recommendations is compelling but practical obstacles to implementation exist, the stated goals will not be met. Anticipating and responding to those barriers is an essential component of the pathway implementation strategy.

Education

One of the essential ingredients to successful pathway implementation is acceptance of the academic and clinical legitimacy of the pathways. Key users of the newly designed care process need to understand the multiple sources which informed the pathway design, such as literature searches for best practice, national guidelines and expert opinion. It is the responsibility of all team members to actively participate in the education of their colleagues. Academic "detailing" includes visits to medical offices, patient care units, or facilities by a respected colleague who introduces and promotes the pathways to clinical staff through educational seminars.

Another key element of the educational process is the responsibility to inform the clinical staff of the purpose of the pathway. Pathways are designed to be best practice protocols for common clinical conditions. The diagnosis and treatment plan for a specific patient is based on the clinical information obtained in the history and physical examination. The decision to place a patient on the pathway rests with the physician. Pathways are tools to use after the physician has placed the patient on the pathway. In addition, pathways are not intended to be inflexible cookbook strategies applied in the absence of consideration of a patient's individual clinical needs. To that end, the pathways need to be designed and implemented in a manner which supports modification as indicated by patient condition.

Alignment of management systems

For CQI to be optimally effective, goals must be aligned strategically, operationally, clinically and financially to facilitate achievement of quality outcomes. Linking financial and non-financial incentives to overall organisational goals provides a stimulus for quality improvement and organisational success and minimises conflicts in priorities.

If clinical pathways are an organisational goal for quality improvement, all individuals accountable for implementation of pathways must work according to the same agenda and be rewarded for achieving its goals. Alignment of systems requires building accountability for implementation across planning and operational lines. In addition to financial incentives, it may be necessary to revise roles and responsibilities of involved staff members.

3.3.3. *Check*

The success of CQI programmes relies on developing interventions effectively and efficiently through a multidisciplinary team effort in which success or failure of the interventions is evaluated through statistical analysis of multiple data points. CQI efforts are grounded in data collection and analysis. Using data effectively is critical to designing and evaluating best practices, and in CQI, this is best achieved through a system that can measure multiple outcomes, such as patient satisfaction, clinical results, resource use and cost-effectiveness.

Providing feedback may stimulate discussions about the pathway, thereby educating those caregivers who are intended to use it and possibly improving compliance with the recommendations of the pathway.

One of the mechanisms for providing feedback related to pathway performance is through monitoring the variances of key interventions or outcomes. A *variance* is defined as a deviation from the expected plan/goals/outcomes/interventions as outlined on the pathway. A variance may be positive if anticipated outcomes are reached faster than planned, or negative, if there are interruptions in reaching anticipated outcome which increase the planned length of stay. The purposes of variance monitoring are to identify causes of variances and any trends and patterns in variances and to promote resolution of major variances that are inhibiting high-quality and cost-effective care. Key issues for each pathway team to discuss will include:

- What variances will we want to analyse?
- Who will be responsible for documenting variance to the pathway?
- How will accuracy of the variance be monitored?
- Who will monitor and report variances and their trends?
- Who will evaluate variance trends and initiate changes to the pathway, if necessary?

For pathways, this involves an analysis of pathway variance data and administrative/patient outcome data in order to answer the questions:

- Did patients who stayed on the path have better outcomes than patients who varied from the path?
- Did system problems cause unnecessary path variances?

One strategy for tracking the pathway impact is through the use of new management tools such as "dashboards" or "instrument panels" or "balanced scorecards" which offer an organisation-wide strategy for monitoring performance measurement and management at multiple levels. These tools can provide a graphic or tabular snapshot of the performance within a particular service or area across multiple dimensions. Using these measurement tools can be compared to a pilot using the data on an airplane console to assess the performance of an aircraft. A balanced measurement system can:

- Integrate multiple data elements into meaningful information
- Provide a monitoring system that identifies improvement opportunities, recognises accomplishments, identifies resource requirements and sets priorities
- Facilitate evaluation of multiple dimensions of performance, including clinical quality, cost and customer value
- Provide feedback to individual care providers regarding their performance

3.3.4. *Act*

Finally, the CQI team acts on the results of its initial effectiveness studies. If an intervention does not improve targeted outcomes, other interventions are developed and tested and the process loop begins again. If an intervention does improve outcomes, the team determines how to implement the pathway in other sites and practices. To ensure continued success, follow-up monitoring and data analysis are necessary.

3.4. Clinical Pathways for Vascular Access Devices

Vascular access surgery is a high frequency procedure for chronic haemodialysis patients. Substantial funds are spent annually to treat access-related complications and maintain haemodialysis vascular accesses. The data presented by Becker *et al.* (22) suggests that a vascular access care pathway can reduce hospital days and costs while achieving acceptable outcomes for access surgery. Tests and treatments based on shared national data can improve the lives of dialysis patients and identify the most efficient and cost-effective treatments available. Clinical pathways, designed and tested by surgeons using statistical process control, can set the standard of care in vascular access surgery. As detailed above, such a pathway involves a standardised plan of pre-operative, peri-operative and post-operative care with clear guidelines for treatment and detailed sequence of actions for monitoring outcome and variance from outcome.

As the medical community wrestles with the challenges of managed care, purchaser power, the "practitioners of surgery and the managers of care of dialysis patients" must take the initiative. The key link person in the pathway should be a clinical nurse specialist who not only educates the patients, but monitors patient progress along the pathway and variance from the pathway. All healthcare providers involved in the care of dialysis patients need to be involved, to maintain high standards and to protect patient interests as these pathways evolve.

The issues discussed earlier in this chapter about critical pathway development and the recently publicised National Kidney Foundation Dialysis Outcome Quality Initiative guidelines form very important ingredients for the development of clinical pathways for vascular access. The establishment of the National Institute for Clinical Excellence in the United Kingdom will give added momentum to the development of clinical pathways.

References

1. Chassin, M.R. and Galvin, R.W. (1998). The urgent need to improve health care quality. Institute of Medicine National Roundtable on Health Care Quality. *JAMA,* **280**, 1000–1005.
2. The State of Managed Care Quality, 1998. National Committee on Quality assurance. Available at http://www.ncqa.org/news/report98.htm.
3. Spath, P. (1993). *Succeeding with Critical Paths*. Brown-Spath and Associates, Forest Grove, OR: p. 3.
4. Laffel, G. (1995). From the editor. *Quality Management in Health Care*, **3**, v–vi.
5. Lord, J. (1993). Practical strategies for implementing continuous quality improvement. *Managed Care Quarterly*, **1**, 43–52.
6. Archer, S.B. *et al.* (1997). Implementation of a clinical pathways decreases length of stay and hospital charges for patients undergoing total colectomy and ileal pouch/anal anastomosis. *Surgery*, **122**, 699–703.
7. Bing, M.L. *et al.* (1977). Implementing a clinical pathway for the treatment of Medicare patients with cardiac chest pain. *Best Pract Benchmarking Health*, **2**, 118–122.
8. Bradshaw, B.G., Liu, S.S. and Thirlby, R.C. (1998). Standardized perioperative care protocols and reduced length of stay after colon surgery. *J Am Coll Surg*, **186**, 501–506.

9. Gottlieb, L.D. *et al.* (1996). Clinical pathway for pneumonia: Development, implementation, and initial experience. *Best Pract Benchmarking Health*, **1**, 262–265.
10. Holtzman, J., Bjerke, T. and Kane, R. (1998). The effects of clinical pathways for renal transplant on patient outcomes and length of stay. *Med Care*, **36**, 826–834.
11. Leibman, B.D. *et al.* (1998). Impact of a clinical pathway for radical retropubic prostatectomy. *Urology*, **52**, 94–99.
12. Collier, P. (1997). Do clinical pathways for major vascular surgery improve outcomes and reduce cost. *J Vasc Surg*, **26**, 179–185.
13. Back, M.R. *et al.* (1997). Improving the cost-effectiveness of carotid endarteretomy. *J Vasc Surg*, **26**, 456–462.
14. Calligaro, K.D. *et al.* (1995). Impact of clinical pathways on hospital costs and early outcome after major vascular surgery. *J Vasc Surg*, **22**, 649–660.
15. Falconer, J.A. *et al.* (1993). The critical path method in stroke rehabilitation: Lessons from an experiment in cost containment and outcome improvement. *Quality Rev Bull*, **19**, 8–16.
16. Schriefer, J. (1995). Managing critical pathway variances. *Quality Management in Health Care*, **3**, 30–42.
17. Luttman, R.J., Laffel, G.L. and Pearson, S.D. (1995). Using PERT/TM to design and improve clinical processes. *Quality Management in Health Care*, **3**, 1–13.
18. Coffey, R.J., Othman, J.E. and Walters, J.I. (1995). Extending the application of critical path methods. *Quality Management in Health Care*, **3**, 14–29.
19. University Hospital Consortium (1993). *Clinical Process Analysis and Pathway Development: A Resource Manual.* Oak Brook, Illinois.
20. King, J.O. (1993). The relevance of practical experience to American hospitals (commentary). *Frontiers of Health Services Management*, **10**, 48–50.
21. Jones, K. (1999). How clinical practice guidelines can improve medical practice. *Seminars in Medical Practice*, **2**, 3–10.
22. Becker, B.N. *et al.* (1997). Care pathway reduces hospitalisation and cost for haemodialysis vascular access surgery. *Am J Kidney Dis*, **30**, 525–531.

Further Reading

American Medical Association (1999). *Clinical Practice Guidelines Directory.* The Association; Chicago.

Boyle, P.J. and Callahan, D. (1998). Physicians' use of outcomes data: Moral conflicts and potential solutions. In: *Getting Doctors to Listen: Ethics and Outcomes Data in Context* (ed. P.J. Boyle, pp. 3–20) Georgetown University Press: Washington DC.

Green, M.S., Rubenstein, E. and Amit, P. (1982). Estimating the effects of nosocomial infections on the length of hospitalization. *J Infect Dis,* **145**, 667–672.

Grimshaw, J.M. and Russell, I.T. (1993). Effect of clinical guidelines on medical practice: A systematic review of rigorous evaluations. *Lancet*, **342**, 1317–1322.

Hyams, A.L. *et al.* (1995). Practice guidelines and malpractice legislation: A two-way street. *Ann Intern Med*, **122**, 450–455.

Institute of Medicine (1992). Guidelines for clinical practice: From development to use (eds. M.J. Field and K.N. Lohr). National Academy Press; Washington DC.

Jarvis, W.R. (1996). Selected aspects of the socioeconomic impact of nosocomial infections: Morbidity, mortality, cost and prevention. *Infect Control Hosp Epidemiol*, **17**, 552–557.

McDonald, C.J. *et al.* (1984). Reminders to physicians from an introspective computer medical record: A two-year randomized trial. *Ann Intern Med*, **100**, 130–138.

Orentlicher, D. (1998). Practice guidelines: A limited role in resolving rationing decisions. *J Am Geriatr Soc*, **46**, 369–372.

Overhage, J.M. *et al.* (1997). A randomised trial of "corollary orders" to prevent errors of omission. *J Am Med Inform Assoc*, **4**, 364–375.

Reiser, S.J. (1994). The ethical life of health care organizations. *Hastings Cent Rep*, **24**, 28–35.

Roddy, S.P. *et al.* (1998). Reduction of hospital resources utilisation in vascular surgery: A four year experience. *J Vasc Surg*, **27**, 1066–1075.

Salzman, M.B. and Rubin, L.G. (1995). Intravenous catheter-related infections. *Adv Pediatr Infect Dis*, **10**, 337–368.

Saver, B.G. (1996). Whose guideline is it, anyway? *Arch Fam Med*, **5**, 32–34.

Tacconelli, E. *et al.* (1997). Central venous catheter-related sepsis in a cohort of 366 hospitalized patients. *Eur J Clin Microbiol Infect Dis*, **16**, 203–209.

The challenge and potential for assuring quality health care for the 21st century. Washington (DC): US Department of Health and Human Services for the

Domestic Policy Council. Publication No. IM 98-0009. June 17, 1998. Available at http://www.ahcpr.gov/qual/21stcena.htm.

Weingarten, S. *et al*. (1998). Can practice guidelines safely reduce hospital length of stay: Results from a multicenter interventional study. *Am J Med*, **105**, 33–40.

Wilson, M.C. *et al*. (1995). Users' guide to the medical literature. VIII. How to use clinical practice guidelines. B. What are the recommendations and will they help you in caring for your patients? The Evidence-Based Medicine Working Group. *JAMA*, **274**, 1630–1632.

Zehr, K.J. *et al*. (1998). Standardized clinical care pathways for major thoracic cases reduce hospital costs. *Ann Thorac Surg*, **669**, 914–919.

CHAPTER 4

THE ACCESS CLINIC

Derek M Manas BSc, MBBCh, FCS(SA)
The Freeman Hospital
Newcastle upon Tyne NE7 7DN, UK

4.1. Introduction

Although there are many crucial issues facing both surgeons and physicians caring for patients with end-stage renal failure (ESRF), none are as important as the planning and timing of appropriate angioaccess. Recent advances in peri-operative care, improvement in haemodialysis (HD), widespread acceptance of synthetic polytetrafluoroethylene (PTFE) grafts in HD and the introduction of "user-friendly" percutaneous central vein catheter kits, have allowed many high-risk, elderly and diabetic patients to now benefit from HD. As a result, the task of ensuring long-term patency and function of any vascular access procedure has become a significant challenge. By and large, patients requiring access surgery are elderly and have complex co-morbid pathology. The access clinic is the ideal setting to assess, counsel and plan their management. Gone are the days of unplanned "quick fix" operations performed by uninitiated junior surgeons. In the modern era of angioaccess, all patients need a pre-operative assessment, and some will require pre-operative imaging in order to individualise and select an appropriate access procedure that will ensure adequate function with a low complication rate. Follow-up and access maintenance are becoming increasingly important and there can also be part of the role of a well-organised access clinic. This chapter covers some important aspects of patient assessment, pre-operative imaging, the individualised selection of an appropriate access procedure and follow-up techniques in the clinic setting.

From the outset, it is important to distinguish between patients who require acute HD and those in whom the need for HD can be predicted. Acute HD access is often needed in patients who are critically ill with haemodynamic instability and acute sepsis necessitating hospital admission. These patients are in need of emergent treatment and are often poor surgical candidates. Percutaneous central venous catheterisation will often suffice. Patients whose reciprocal creatinine plot suggests that dialysis is imminent should be referred to the access clinic where assessment and planning for vascular access can be commenced.

4.2. Role of the Access Clinic

- To establish an easy route for referring nephrologists to have their patients seen by the surgical team. This allows for timely referrals
- To allow for a formal topographical assessment of venous and arterial anatomy to be undertaken and to assess the suitability of superficial veins for endogenous arteriovenous fistula (AVF)
- A setting for pre-dialysis patients to be counselled about their disease, to get to know the surgical team, and to come to terms with the fact that dialysis is imminent
- To discuss with patients, all the surgical options available and the importance of vein preservation (see below)
- The opportunity to select patients who require further pre-operative imaging, such as venography, angiography, and venous and/or arterial Doppler assessment
- Follow-up assessments, early in the post-operative period as well as later on when complications occur
- The access clinic can also provide the opportunity for follow-up of patients with complicated vascular access

There are no hard and fast rules with respect to the timing of access placement. In general, the aim of any access procedure is to provide a reliable and repetitive entry into large blood vessels that facilitate efficient, effective dialysis. This usually means that 250 to 500 ml/min of blood flow needs to be achievable with minimal recirculation of undialysed blood (< 4%) and a low rate of late thrombosis. To achieve these ideals, it is important that nephrologists caring for ESRF patients refer them for surgery when their creatinine clearance is < 25 ml/min, or their serum creatinine is > 4 mg/dl or within one year of the anticipated need for HD (1). A new primary fistula should be allowed to mature for at least one month, and ideally three to four months, prior to cannulation (1). Both the size and anatomical qualities of venous and arterial components of primary AV fistulae will influence the maturation time. If one embarks on an aggressive policy of attempting primary fistulas on all new referrals, one should allow sufficient time to perform a second procedure should the first attempt fail, thus avoiding the need for temporary access.

4.3. Pre-operative Clinical Evaluation

4.3.1. *Examination prior to permanent access selection*

A history must be taken and a physical examination of the patient's venous, arterial and cardiopulmonary systems must be performed. Assessment of these various systems will help with the decision as to type of access to be performed, the most appropriate location, and what diagnostic tests are necessary prior to surgery. Table 4.1 outlines the aspects of a good history and physical exam.

Evaluation of topographical venous anatomy is done by physical examination and/or colour Doppler studies. With a venous tourniquet in place, the entire course of the cephalic vein is examined for. Ideally, this vein is seen coursing from the wrist to well beyond the cubital fossa and occasionally up to the shoulder. Besides being present, the vein should be superficial enough to facilitate easy needling, wind around the outer aspect of the forearm to allow the inserted needle to "sit" comfortably, and be of a calibre that gentle percussion of the distended vein at the wrist should transmit a wave that can be felt or heard using a hand-held Doppler, higher up in the

Table 4.1. Physical examination prior to dialysis access formation.

History	Relevance
Previous CVC/pacemaker	Associated with CV stenosis
CCF	Angioaccess may alter CO and other haemodynamics
Previous venous/arterial line	May have damaged target vessels
Diabetes mellitus	Associated atherosclerosis
Coagulation disorder/anticoagulant therapy	Hypercoagulability or problems with Haemostasis
Previous failed access	Limit sites, influence plan if cause of failure still present
Heart valve prosthesis	Infection rate
Comorbid disease — PVD, CAD, malignancy, life expectancy	Certain access may not be justified
Dominant arm	Non-dominant arm — will reduce negative impact on life
Examination — arterial system	
Character	Poor quality pulse — influence site
Doppler evaluation	Poor quality trace — influence site
Allen test	May C/I the creation of radial AVF
Bilateral arm BPs	Will help to locate the best side
Examination — venous	
Unilateral oedema	Venous outflow obstruction
Collateral veins	Venous outflow obstruction
Tourniquet test with venous mapping	Allows selection of ideal veins for access

CV — central vein; CO — cardiac output; CVC — central venous catheter; PVD — peripheral vascular disease; CAD — coronary artery disease; C/I — contraindication; CCF — congestive cardiac failure arm. This is a simple test that indicates patency of the cephalic vein. It can be performed in the consulting room and used to determine patients who require further preoperative investigation such as a colour flow Doppler ultrasound or venography. If the cephalic vein is patent, the site of the access can be planned to allow the anastomosis to the radial artery being placed as distal as possible. The anatomical "snuffbox" fistula between the tendons of extensor pollicis longus and brevis, is the most distal of the endogenous AVF's and gives the longest length of cephalic vein. Published results for snuff-box fistulas compare favourably with the two-year overall cumulative patency rates published following the preferred Cimino fistulas.

When planning the access, one should ensure that sufficient length of adequately sized vein is available after arterialisation to allow two needles to be inserted sufficiently separated to avoid recirculation of dialysed blood. The availability of a suitable vein remains the limiting factor. In up to 30% of patients, suitable native vessels will not be available and a prosthetic polytetrafluoroethylene (PTFE) "bridge" fistula will be necessary.

Fistulae that use small or inadequate veins, although technically feasible, as evidenced by a patent anastomosis, will be difficult to cannulate, fail to provide efficient dialysis and will usually occlude soon after placement due to difficult and traumatic cannulations. The use of these small veins in the hope that the increased blood flow will dilate them enough to give adequate dialysis is a mistake. The early failure rate of 8–24% is a result of two commonly encountered pitfalls:

- The use of a cephalic vein which is patent at the level of the wrist, but which is actually occluded in mid-course, usually as a result of previous venepunctures and iv cannulation. In such situations, the patent distal vein allows a successful anastomosis to be created but the local resistance is such, that the fistula invariably fails in the first 24–48 hours.
- The cephalic vein may not be well developed in the forearm although a suitable sized vein is present at the wrist. In such situations the distal cephalic vein often drains into a network of small tortuous forearm veins, which even if successfully arterialised, would not mature enough to sustain adequate HD. These veins are often thin walled, friable and easily torn when needled.

4.3.2. *Diagnostic evaluation prior to permanent access*

The gold standard investigation is venography but additional imaging techniques such as Doppler ultrasound, arteriography and magnetic resonance imaging are indicated in selected cases, especially where multiple previous accesses have been performed.

Venography prior to placement of access is indicated in patients with the following:

- Oedema in the extremity in which access is planned
- Collateral vein development in any planned access site
- Differential extremity size, if that extremity is contemplated as the access site
- Current or previous insertion of either subclavian catheter or pacemaker of any type in the venous drainage of a planned access
- Previous arm, neck or chest trauma or surgery in the region of the venous drainage for the planned access site
- Multiple previous access procedures in extremity planned as an access site

Oedema of the extremity, obvious collateral vein development, either of the arm or anterior chest wall or differential extremity size, almost certainly indicates inadequate venous drainage or central vein obstruction (2,3). Such defects should be corrected prior to access placement and this can be done either by balloon dilatation or a combination of balloon dilatation and vascular stent placement. Access should never be placed on the same side as an existing transvenous pacemaker or an existing subclavian catheter, unless all other options have been exhausted. It has been well documented that transvenous catheter placement, particularly in the subclavian vein, is associated with a higher rate of subclavian vein stenosis (3,4).

Doppler studies may be used in lieu of venography when assessing the central veins (5,6). However, this method is less accurate than venography for evaluation of central vein strictures, is highly operator dependent and requires an enthusiastic radiologist for it to be useful. The National Kidney Foundation Dialysis Outcome Quality Initiative (NKF-DOQI) workgroup concluded in their interim report that Doppler studies or magnetic resonance imaging may be preferred to venography and arteriography in patients with residual renal function in whom contrast agents should be avoided.

Arteriography is seldom required, but can be useful in patients in whom the Allen test suggests an inadequate vascular arch or in patients in whom extremity ischaemia is suspected. When access grafts are to be placed in the lower limbs, pre- and post-exercise Doppler and arteriography may be very helpful.

4.4. Selection of Permanent Vascular Access

As a general principle, all patients requiring chronic haemodialysis should be considered candidates for some form of surgically created long-term vascular access. Wrist (radiocephalic) and elbow (brachiocephalic) primary fistulae are the preferred types of access for many reasons:

- Simple to create
- Excellent patency once established and lower complication rate compared to other access types (7–9)
- Lower morbidity associated with their creation
- Improved flow over time

Poor selection of arteries and veins prior to surgery will very often result in poor extracorporeal flow and inadequate dialysis. As mentioned above, a working knowledge of the topographical anatomy of upper limb vessels is essential to selecting the appropriate type and site of access. For this reason, it is important that patients, when seen in the access clinic, are assessed by an experienced access surgeon in order to make the right access decision. Fistulae that fail to mature within four months, should be considered as primary access failures. The access clinic is an ideal setting for reassessing slowly developing fistulae as it allows early and objective prognostication making the decision to pursue other options easier.

The brachiocephalic primary AV fistula is often the second choice of initial access. There is a higher flow rate compared to a wrist fistula and unsightly aneurysmal veins are more easily covered and may offer cosmetic benefit. The major disadvantage with elbow fistulae is that they are associated with an increased incidence of steal syndrome compared to the radiocephalic fistulae. They offer a much shorter length of vein to needle. Use of the basilic vein has several disadvantages including difficult cannulation because of the deep location of the vein, higher risk of steal syndrome and arm swelling.

The non-dominant rule should apply as far as possible, but stubbornly adhering to the non-dominant rule may be detrimental to the patient. Once the two common pitfalls have been looked for and both the Allen and tourniquet tests have been satisfactorily performed, the patient can be counselled as to what type of vascular access will be placed. If the surgeon assessing the patient feels that the cephalic vein or artery at the wrist is inadequate, then they should proceed proximally to the cubital fossa. It is important to keep in mind that the proximal cubital fossa AV fistula may predispose elderly, atherosclerotic and diabetic patients to steal syndrome because of their tendency to enlarge over time, even when care is taken to keep the AV anastomosis within the recommended 5–7 mm length.

The cephalic vein is very often the limiting factor in determining both early and long-term outcome. If the venous anatomy of the dominant arm better conforms to the required specifications, that side should be used. In reality, the decision as to the type, site, side and timing of access depends on many factors:

- The anticipated time to commencement of haemodialysis
- Experience and judgement of the surgeon
- Success rate of particular access type
- The status of the venous system
- How important it is to avoid temporary access
- Possible role of CAPD in the treatment of a particular patient
- The blood flow rates required for a particular dialysis prescription

Surgeons must understand the difficulty some patients have in coming to terms with haemodialysis. Clear explanation of the decision-making process will alleviate this trauma. This is particularly important if pre-operative examination or investigation such as venograms suggest that the dominant limb is the preferable site, because patients can become very distressed by this. At the first visit, if assessment suggests that the non-dominant arm is the arm of choice, this should be preserved, as it is more likely to be free of previous trauma, and in addition would allow the most useful arm to be free during dialysis. Once the limb is chosen, this should be protected from venesection and intravenous cannula insertion prior to fistula formation. Simply asking the patient to change his/her wrist watch site from the protected arm may facilitate increased awareness. It is important to be very clear with patients as to why further investigation is required. If patients are not adequately counselled, they may not be receptive to post-operative manipulation such as subclavian vein stenting or be able to deal with early access failures, which are not uncommon.

If a primary AV fistula cannot be established, the next choice is to use synthetic bridge grafts or transposition of the basilic vein. Currently, the most satisfactory and widely used prosthetic graft is polytetrafluoroethylene (PTFE). They can be placed in a straight, looped and curved configuration. They can be used in the upper arm, forearm or the anterior aspect of the thigh. Currently in the United States the vast majority of patients who go onto haemodialysis will have an AV graft placed. The converse is true for haemodialysis patients in the United Kingdom and Europe. The advantages of PTFE grafts include:

- availability of a large surface area for cannulation (8,10)
- technical ease of needling
- short maturation time compared to AV fistulae (14 days)
- relative ease of unblocking by radiological and surgical means

The long-term patency of AV grafts is certainly inferior to autologous AV fistulae and the life expectancy ranges from 18 months to 5 years. This depends largely on where the grafts are placed. Certainly, grafts placed in the forearm using smaller vessels, tend to have lower patency rates, as compared to grafts placed in the upper arm or in sites using larger vessels. At present, there is no consensus as to the best site for an AV graft placement, but my preference is to use an upper arm curved graft. The decision to use AV graft as primary access should be taken only after careful assessments (including history, examination, tourniquet and Allen tests) show inadequate cephalic vein in the forearm and arm. If the basilic vein appears to be superficial, a brachiobasilic fistula may be attempted. The technique of basilic vein transposition or superficialisation is described but is quite difficult to perform. Complications associated with it include difficulty of cannulation and an increased thrombosis rate. Therefore a decision to perform basilic vein transposition must be based on evidence of a satisfctory basilic vein obtained by venography or Doppler ultrasound.

4.5. Preservation of Vein for Access

Arm veins, particularly the cephalic veins of the non-dominant arm should not be used for venepunctures or intravenous catheters. The dorsum of the hand should be the preferred site for intravenous lines in patients with chronic renal failure and all venepuncture sites should be rotated. Educating patients and doctors about the benefits of vein preservation will reduce the proportion of patients with unsuitable veins due to previous venipuncture.

It is often quite difficult to know when to begin to preserve the vein in patients with end stage renal disease, but a good rule is that if the creatinine is > 3 mg/dl, vessel preservation should start. If the patient has been seen in the clinic and there is adequate time, then a medical bracelet could be worn. This would inform hospital staff to avoid IV cannulation of essential veins. Subclavian vein catheterisation should be avoided at all costs for temporary access in all patients with chronic renal failure, due to the risk of central vein stenosis. This again should be pointed out to the patient, so that they can keep medical staff informed.

4.6. Subsequent Visits to the Access Clinic

Once seen at the clinic, patients are either scheduled for surgery or sent for special investigations. After surgery, it is important to follow-up on the patient and assess the access maturation. All too often, this is left to the physician or dialysis nurse and contact is only established between surgeon and patient when access malfunction or failure occurs. With timely referral, a new access can be performed at a different site, still allowing for maturation time before dialysis is started. Maturation time varies and for AV grafts, this time (approximately three weeks), is required for reduction of surgically induced swelling and allowing for graft adherence to surrounding tissue. Ideally, one should place AV grafts (PTFE) between three and six weeks prior to use, although recent data suggest that these grafts can be used within 24 hours of placement.

In summary,

- All patients requiring chronic haemodialysis should be referred to an access clinic
- Native AV fistulae will give the best 4- and 5-year patency rate, and certainly require fewer interventions compared to other access types (11,12)
- In order to increase the percentage of native AV fistulae, it is best for patients to be referred for early access assessment, while dialysis is still months away
- When patients are first seen in the clinic, surgeons should endeavour to place a permanent access rather than be forced into placing temporary access because of time constraints
- Each centre should establish a database to track the types of access created and document complication rates
- The goal of the access clinic should also be one of re-evaluation to assess primary access failure and to attempt to re-establish native AV fistulae whenever possible

4.7. Complications Seen at the Access Clinic

An essential part of the function of the access clinic is follow-up. This will allow surgeons to accurately audit their work and provide a setting for re-investigation, re-assessment and further access provision. The most common early complication of AV fistulae formation is thrombosis. In the early post-operative period, the critical question that should be answered prior to returning a patient to the operating room for exploration of a thrombosed fistula is, the level of experience of the performing surgeon and the degree of satisfaction with the initial operation. In general, a fistula with good flow and a good thrill at the end of the procedure should always be returned to the operating room for thrombectomy should thrombosis occur soon after operation. If the vein was tenuous at assessment and the best procedure was performed under the circumstances by an experienced surgeon, there is little to be achieved by re-exploration. Such a patient should be referred to the access clinic for re-assessment. In these cases, patients will very often require investigation, if this was not done prior to the first procedure.

A complication which is more commonly seen in the access clinic is late thrombosis of an established AVF. This is often associated with the surgical procedure for an unrelated problem, over-zealous haemodialysis, hypotension related to post-dialysis bleeding or myocardial infarction or excessive post dialysis compression. Provided all systemic factors have been corrected, most acute thromboses in well-established fistulae should be assessed for radiological thrombectomy.

Venous stenoses can be reduced by regular surveillance of AV fistulae. Such narrowing can be identified angiographically prior to complete occlusion and treated by percutaneous transluminal angioplasty or surgical revision. The access clinic provides an ideal setting for assessment of fistula dysfunction and initiation of action to salvage access in patients with early warning signs of access failure.

The development of *venous aneurysms* calls for regular review of the patient in the access clinic. If these aneurysms remain stable, as they can do for many years, no further treatment is required, provided dialysis staff needle the fistula intelligently, in particular, rotating needling sites. The surgeon must ascertain adequacy of skin cover over the aneurysmatic portion of the AV fistula. Should the skin form part of the wall (false aneurysm), excision is advised otherwise conservative management is advocated.

Other complications encountered in the access clinic include:

- "steal syndrome"
- ischaemic monomelic neuropathy
- troublesome venous hypertension and
- late thrombosis, secondary to neointimal hyperplasia

Steal syndrome is a rare complication of radiocephalic fistulae but it is a common complication of the brachiocephalic fistula, especially in diabetics and is related to a progressive increase in fistula size with reversal of flow in the distal radial artery. Patients present with pain, paraesthesia and muscular weakness in the affected limb. Symptoms are usually worse on dialysis. Often, the examination will reveal a cool hand and loss of the distal radial pulse. Ways of confirming steal syndrome include:

- pre- and post-compression Doppler studies
- plethosmography and
- angiography which should be used in selected patients in order to decide what procedure to perform

There are a number of operations described for ischaemia secondary to the steal syndrome, the most successful being that described by Haimov (13) in which the distal radial artery is ligated and a venous bypass graft placed from the proximal radial artery to the ligated distal stump. Distal ischaemia develops in approximately 1–9% of patients with bridge grafts. Investigation should include Doppler studies and angiography because of a high incidence of arterial stenosis.

Ischaemic monomelic neuropathy is a form of motor nerve dysfunction associated with brachiocephalic fistulae or bridge grafts. It almost always occurs in diabetics and the cause is related to pre-existing arterial disease, underlying neuropathy and an element of steal. Most patients present to the clinic with a history of pain following fistula creation. If duplex scanning before and after compression Dopplers do not confirm the diagnosis of steal syndrome, ischaemic monomelic neuropathy should be considered. Often, the only effective treatment is fistula ligation.

Venous hypertension is caused by proximal vein obstruction (see Fig. 4.1). Early occurrence of venous hypertension is due to inadequate preoperative assessment but this complication may develop over time due to sclerosis at needling sites. With reversal of flow in the veins draining the hand or forearm the affected limb becomes engorged, painful, hyperpigmented and sometimes even ulcerated. Venography or fistulography, balloon dilatation with or without stenting using a Wall stent (see Fig. 4.2), is an effective way of treating this complication.

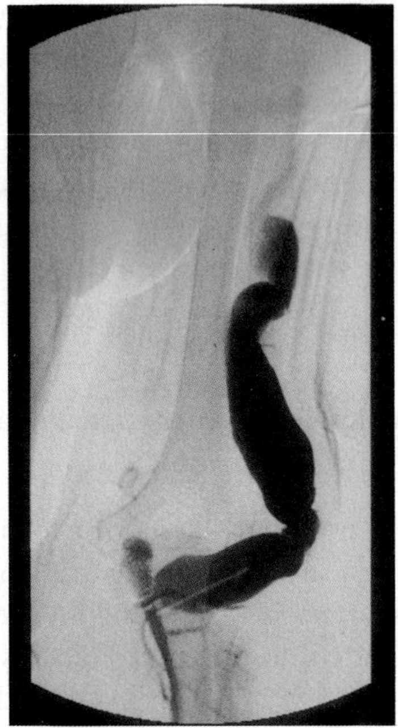

Fig. 4.1. Aneurysmal fistula and venous hypertension secondary to venous outflow stenosis.

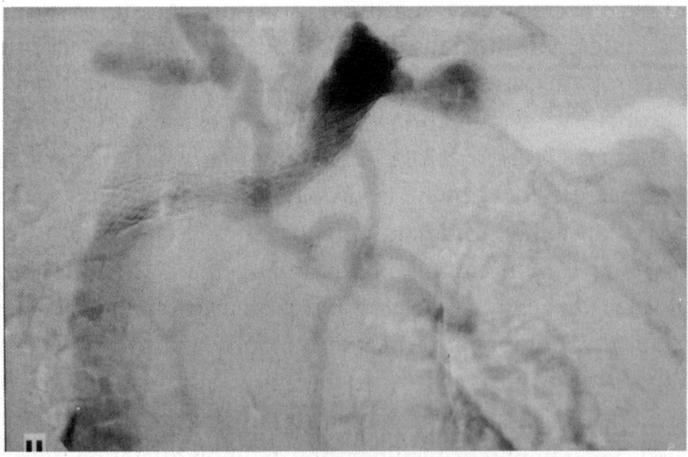

Fig. 4.2. Central venous stenosis secondary to repeated subclavian cannulation — treated by endovascular stenting.

Almost all forms of angioaccess are prone to thrombosis which in turn leads to access failure. Few access sites will escape without an episode of thrombosis. Autogenous fistulae probably last twice as long as PTFE bridge grafts. Statistical projection suggests that *all* PTFE fistulae will have failed by 7.5 years while only half of the autogenous fistulae will be non-functional over the same period of time (14).

Thrombosis is the leading cause of access dysfunction and failure as a result of neointimal hyperplasia. It has been suggested that physical examination of an access site, particularly, an access graft should be performed weekly (15,16). The NKF-DOQI guidelines suggest that a reasonable prediction of AV graft flow can be made by determining flow characteristics (pulse, thrill and bruit). For example,

- A palpable thrill at the arterial and venous segments of the graft predicts flow of greater than 450 ml/min
- A pulse suggests lower flows and
- Intensification of a bruit suggests a stricture or a stenosis (1)

Flow rates are very difficult to interpret and dialysis units in different parts of the world would accept different flow rates. In most units in the USA, blood flows of less than *600 ml/min* in access grafts are regarded as having a higher rate of access thrombosis (17–19). Because problems can develop over a period of time it is best to have the access monitored at regular intervals. It is important for dialysis units to identify patients developing access dysfunction and refer them to the access clinic for prompt action. Such patients, on a selective basis, should undergo fistulography and either angioplasty or surgical revision of the fistula. Although fistulography provides the most accurate information about vascular access patency and configuration, it is not cost-effective to carry it out on all patients with access dysfunction. Indications for fistulography include:

- graft thrombosis
- prolonged bleeding
- fistula arm oedema
- pain in the arm with dialysis

- elevated venous pressures
- elevated fistula recirculation
- unexplained decrease in delivered dialysis dose
- decreased access flow

Salvage of access before severe stenosis and thrombosis develop improves patency and decreases the incidence of thrombosis. Although insertion of endovascular stents have been advocated as a method of preventing recurrent stenosis after angioplasty (20), there is no concrete evidence of any beneficial role.

4.8. Special Problems

4.8.1. *Patients with underlying cardiac dysfunction*

Overt vascular-access-related cardiac decompensation occurs rarely, even in patients with underlying cardiac disease. However, cardiac hypertrophy is a frequent complication of long-term dialysis even in patients with optimal correction of anaemia using recombinant erythropoetin. This complication is common in patients with a primary AV fistula which has become grossly dilated and aneurysmal allowing shunting of a large percentage of the cardiac output. Often the cardiac symptoms are erroneously ascribed to the aneurysmal fistula and it is difficult to decide whether or not to ligate a vital access for the dubious benefit of reducing the risk of high output cardiac failure. It is important to be able to distinguish cardiac hypertrophy in end-stage renal disease from an access-related high output failure or a cardiomyopathy. Limiting fistula flow by banding or by tying off some of the draining veins is often associated with access thrombosis. Change of access modality where possible, by using either indwelling central venous catheter or peritoneal dialysis, may provide the less complicated option.

4.8.2. *Patients with diabetes and peripheral vascular disease*

Due to microvascular disease, these patients are considered to be at high risk for vascular access. Patients with diabetes mellitus, abnormal arterial supply, vascular anomalies, atherosclerotic disease or multiple previous vascular accesses are at risk of developing limb ischaemia post construction of AV access. Limb ischaemia distal to an AV access can occur at any time from a few hours to months following access construction. Severe ischaemia can cause irreparable injury to nerves within hours and must be considered a surgical emergency. Mild ischaemia manifested by subjective coldness, paraesthesias and objective reduction in skin temperature with no loss of sensation or motion is common and generally improves with time.

Patients at risk therefore require objective evaluation of relevant vascular anatomy before the formation of access in the limb. Most patients seen in the access clinic will not need to have their arterial tree imaged unless there is good objective evidence of underlying disease. Certainly any patient who has required recurrent access procedures or has had a proven arterial thrombosis or has previous ischaemia monomelic neuropathy should undergo arteriography prior to reconstructing any further access. It may be more prudent to avoid such extremities altogether.

Patients who develop mild ischaemia following access procedure should undergo symptomatic therapy (such as wearing of gloves), have frequent physical examinations at the access clinic where special attention should be paid to any subtle neurological changes in muscle wasting (21). Lack of improvement or deterioration of symptoms will necessitate surgical intervention.

4.8.3. *Recurrent thrombosis of grafts and fistulae*

As documented by Schwab *et al.* (22), some patients are predisposed to recurrent access failure. Prior to further access surgery such patients should be assessed for causes of hypercoagulability, including protein S, protein C and lupus anticoagulant. Once further access is obtained, they should be monitored with Doppler flow studies (23) and regular digital subtraction fistulography. My personal practice is to put these types of patients on anti-platelet therapy, preferably aspirin, and where indicated, anticoagulants are used.

4.9. Patient Education

It is important that patients understand the highly significant role angioaccess plays in ensuring trouble-free dialysis and that they play a vital role in achieving a long-use life from any access procedure. Patients therefore need to be educated about how to take care and monitor their own access — in other words, be taught what a thrill means, where to feel for a thrill and what the implications of the loss of a thrill are. For example, disappearance of a thrill halfway up their AV fistula since the last haemodialysis session constitutes a significant deterioration and must be reported to the dialysis unit. All patients should be educated regarding:

- Preservation of veins in their upper extremities
- The "art" of maintaining an infection free access
- The appropriate degree of compression required after access needling (excessive compression may result in thrombosis and too little compression may result in false aneurysm formation and bleeding)
- Prompt reporting of complications of their access

Long-term haemodialysis patients become very conversant with their access and can play a role in advising new patients regarding their access. Dialysis units throughout the world are becoming increasingly understaffed and "experienced dialysis patients" will be required to make dialysis staff aware of problems with their access. Very often, patients know a lot more about their access than they are given credit for. For example, many patients surveyed on our unit were able to tell us what the venous dialysis pressure measurements were over the previous few weeks. Similarly, many patients were able to tell us what flow rates they were achieving on dialysis. This should be exploited and patients could assist with their own surveillance.

4.10. Communication

In order for an access clinic to function successfully, there needs to be good rapport and communication with the dialysis unit(s) that it serves. The best way to do this is:

- To provide an annual report, which includes an annual audit of all access being performed in that particular unit. The audit should include number of patients passing through, primary and secondary patency rates, as well as complications and revision surgery.
- There should be regular meetings between surgeons, dialysis staff and nephrologists. These meetings could be conducted three-monthly or more frequently, depending on how each unit wants to do this and what problems they have encountered. It is also a useful forum to discuss problem patients.
- For surveillance methods to be useful in the early detection of failing fistulae and grafts, a feedback loop to the access clinic from the dialysis units must be maintained. Excellent communication can be provided by a research or access liaison nurse. Unless a formal line of communication is established, the task of maintaining a decent surveillance programme becomes burdensome and ineffective.
- Protocols about the management of common access problems need to be established and adequately publicised. This must contain a clear line of communication either through the acute dialysis sister (site member of the access team) or a specialist registrar or resident, whose job it is to maintain links between the dialysis units and the access clinic. In some units, regular meetings with members of the dialysis staff are held to discuss access problems.
- Finally, communication also involves education of medical staff and dialysis staff and there should be regular updates between medical staff and dialysis staff in the form of academic meetings to discuss new innovations.

Ensuring efficient and effective dialysis relies on good access. This requires a dedicated, experienced and enthusiastic team. The access clinic allows for well planned surgery, good post-operative monitoring and aggressive screening in order to pick up early complications. This will result in improved patient dialysis, less frequent hospitalisation for access morbidity, reduced dialysis workload and overall healthcare costs. When run by a surgeon with interest in vascular access, aided by a liaison nurse with good communication links with nephrologists and dialysis units, the *access clinic* provides an ideal setting for planning, monitoring and screening patients in order to ensure efficient and effective dialysis.

References

1. National Kidney Foundation — Dialysis Outcome Quality Initiative (1997). *Clinical Practice Guidelines for Vascular Access*. National Kidney Foundation.
2. Glanz, S., Bashist, B., Gordon, D.H. *et al.* (1988). Axillary and subclavian vein stenosis: Percutaneous angioplasty. *Radiology*, **168**, 37–373.
3. Schwab, S.J., Quarles, L.D., Middleton, J.P. *et al.* (1988). Hemodialysis-associated subclavian vein stenosis. *Kidney Int*, **33**, 1156–1159.
4. Trerotola, S.O. (1994). Interventional radiology in central venous stenosis and occlusion. *Semin Intervent Radiol*, **11**, 291–304.
5. Sands, J., Young, S. and Miranda, C. (1992). The effect of Doppler flow screening studies and elective revisions on dialysis access failure. *ASAIO J*, **38**, M524–M527.
6. Tordoir, J.H.M., Hoeneveld, H., Eikelboom, B.C. and Kitslaar P.J.E.H.M. (1990). The correlation between clinical duplex ultrasound parameters and the development of complications in arterial venous fistulae for hemodialysis. *Eur J Vasc Surg*, **4**, 179–184.
7. Palder, S.B., Kirkman, R.L., Whittemore, A.D. *et al.* (1985). Vascular access surgery for hemodialysis: Patency rates and results of revision. *Ann Surg*, **202**, 235–239.
8. Fan, P.-Y. and Schwab, S.J. (1992). Vascular access – Concepts for the 1990s. *J Am Soc Nephrol*, **3**, 1–11.
9. Kinnaert, P., Vereerstraeten, P., Toussaint, C., Van Geertruyden, J. (1977). Nine years experience with internal arteriovenous fistulas for hemodialysis: Study of some factors influencing results. *Br J Surg*, **64**, 242–246.
10. Owens, M.L., Stabile, B.E., Gahr, J.A. and Wilson, S.E. (1979). Vascular grafts for hemodialysis: Evaluation of sites and materials. *Dial Transplant*, **8**, 521–530.
11. Churchill, D.N., Taylor, D.W., Cook, R.J. *et al.* (1992). Canadian hemodialysis morbidity study. *Am J Kidney Dis*, **19**, 214–234.
12. Mehta, S. (1991). Statistical summary of clinical results of vascular access procedures for hemodialysis. In Vascular Access for Hemodialysis — II. (eds. B.G. Sommer and M.L. Henry), Chap. 11, pp. 145–147, W.L. Gore & Associates.
13. Schanzer, H., Schwartz, M., Harrington, E. *et al.* (1988). Treatment of ischaemia due to "steal" by arteriovenous fistulae with distal artery ligation and revascularization. *J Vasc Surg*, **7**, 770–773.

14. Kjerlakian, G.M., Roedersheimer, L.R., Arbaugh, J.J. *et al.* (1986). Comparison of autologous fistulae vs PTFE graft fistulae for angioaccess in haemodialysis. *Am J Surg*, **152**, 238–243.
15. Beathard, G.A. (1996). Physical examination of AV grafts. *Sem Dial*, **5**, 74.
16. Trerotola, S.O., Scheel, P.J., Powe, N.R. *et al.* (1996). Screening for access graft malfunction: Comparison of physical examination with US. *J Vasc Interv Radiol*, **7**, 15–20.
17. Schackelton, C.R., Taylor, D.C., Buckley, A.R. *et al.* (1987). Predicting failure in polytetrafluoroethylene vascular access grafts for hemodialysis: A pilot study. *Can J Surg*, **30**, 442–444.
18. Strauch, B.S., O'Connell, R.S., Geoly, K.L. *et al.* (1992). Forecasting thrombosis of vascular access with Doppler colour flow imaging. *Am J Kidney Dis*, **19**, 554–557.
19. Depner, T.A. and Reasons, A. (1996). Longevity of peripheral of A-V grafts and fistulae for hemodialysis is related to access blood flow. *J Am Soc Nephrol*, **7**, 1405–1405.
20. Beathard, G.A. (1993). Gianturco self expanding stents in the treatment of stenosis in dialysis access grafts. *Kidney Int*, **43**, 872–877.
21. Mattson, W.J. (1987). Recognition and treatment of vascular steal secondary to hemodialysis prostheses. *Am J Surg*, **154**, 198–201.
22. Schwab, S.J., Raymond, J.R., Saeed, M. *et al.* (1989). Prevention of haemodialysis fistulae thrombosis and early detection of venous stenosis. *Kidney Int*, **36**, 707–711.
23. Sands, J., Young, S. and Miranda, G. (1992). The effect of Doppler flow screening studies and elective revisions in dialysis access failure. *ASAIOJ*, **38**, M524–M527.

CHAPTER 5

RECIRCULATION AND DIALYSIS ACCESS

Paul W Chamney BEng(Hons), PhD, AMIEE
Department of Electrical and Electronic Engineering
University of Hertfordshire
Hatfield, Herts AL10 9AB, UK

5.1. Introduction

Recirculation is a collective term describing any combination of factors causing a proportion of dialysed blood returning from the dialysis machine to bypass the patient's systemic circulation and to be re-dialysed. Recirculation, which may be manifested in different ways, can lead to significant under-dialysis, particularly in centres practising short dialysis. Knowledge of the different forms of recirculation has important implications in the interpretation of recirculation data.

Any form of arteriovenous access in particular, will always present some degree of recirculation known as *cardiopulmonary recirculation.* The important management issue for the health care provider is to ensure that the effects of recirculation are minimised to within acceptable limits. In addition, where recirculation cannot be eliminated but its effect may be predicted, it is appropriate to introduce a suitable time correction for the rational planning of treatment. Quantitative and accurate measurement of recirculation, easily performed on a regular basis, is therefore essential in the clinical setting.

The availability of recirculation measurement devices has been of considerable assistance in the diagnosis of vascular access problems. Measurement of the access flow rate is also possible with recirculation monitors by reversing the arterial and venous blood lines to the access. Under these conditions, recirculation is deliberately induced and its magnitude depends on the available access flow rate. While there is little doubt regarding the utility of recirculation and access flow rate measurements, a solid grasp of the principles is mandatory in order that the correct interpretation may be drawn from the information presented. Unfortunately, the different forms of recirculation and the plethora of mathematical theories presented in the literature has led to considerable confusion in the application of recirculation monitors. Therefore, it is the aim of this chapter to address the following issues:

- An account of how the various forms of recirculation can occur in different types of vascular access
- The appropriate theory which underpins the principles of recirculation
- The typical values of recirculation measurements which are likely to be encountered during routine practice
- The utility of induced recirculation in reversed lines for measurement of access flow
- A method to correct dialysis treatment times for the effects of recirculation
- A discussion of the limitations of traditional urea based methods for the quantification of recirculation
- An overview of the principles of automated recirculation measurement
- A review of the devices available on the market for recirculation monitoring
- A suggested set of practical guidelines for the application of recirculation measurements in the routine clinical setting

5.2. Recirculation in the Arteriovenous Access

Access recirculation (AR) can occur in any method of vascular access. The most basic form of AR can arise in the arteriovenous access (fistula or graft) and is best illustrated with a simple example. In Fig. 5.1(a), the fistula is supplied with an access flow rate (Q_A) of 300 ml/min. Blood is drawn from the arterial needle and returned to the fistula downstream via the venous needle. The term *shunt* will be used throughout this chapter to describe the

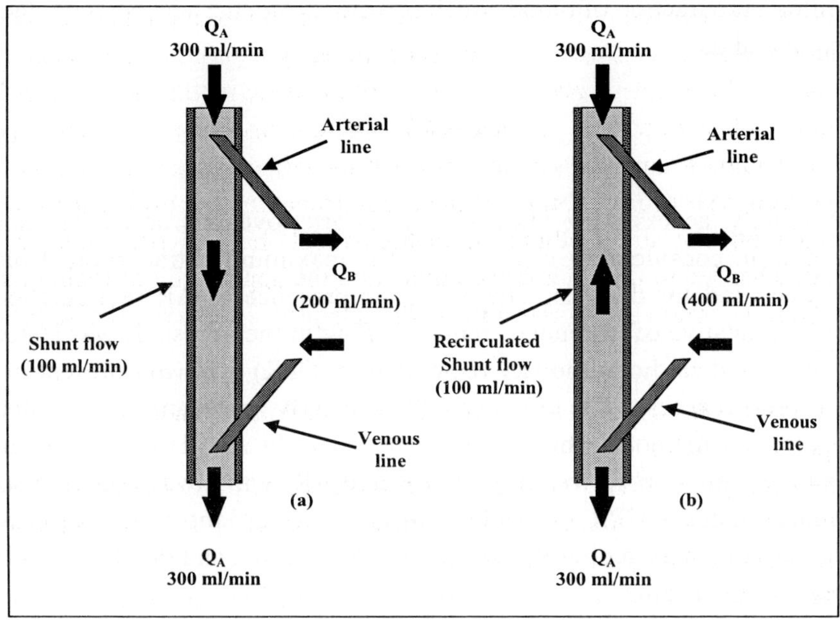

Fig. 5.1. Shunt conditions in an arteriovenous access. (a) In the absence of access recirculation, blood flows in the "forward" direction through the shunt. (b) If blood flow rate exceeds the available access flow rate, flow through the shunt is reversed and access recirculation occurs.

section of blood vessel between the arterial and venous needles (or catheter tips in the case of central venous lines). Providing the blood flow rate in the extracorporeal circuit (Q_B) is less than Q_A, then blood flows in the "forward" direction through the shunt. Therefore, the shunt flow rate (Q_S) is simply:

$$Q_S = Q_A - Q_B = 100 \text{ ml/min}. \tag{5.1}$$

If the blood pump demands more flow than can be supplied by the access, i.e. $Q_B > Q_A$ then the direction of flow in the shunt is reversed. Again, the shunt flow must be the difference between Q_A and Q_B, i.e. 100 ml/min. In other words, a proportion of the blood flow rate from the venous needle is drawn back to the arterial needle causing recirculation.

Therefore, the fraction of blood which is recirculated, the access recirculation, is expressed as:

$$\mathrm{AR} = \frac{Q_\mathrm{B} - Q_\mathrm{A}}{Q_\mathrm{B}}. \tag{5.2}$$

Normally, access flow rates, Q_A, in an arteriovenous access are above 700 ml/min, considerably in excess of the maximum extracorporeal blood flow rate used for dialysis. Therefore, the presence of AR recirculation is usually indicative of a stenosis, reducing flow in the access. It was stated at the outset that cardiopulmonary recirculation (CPR) is *always* present in the arteriovenous access. The effect of CPR is clearly irrelevant in recirculation measurement methods which specifically quantify AR. However, methods which measure *total recirculation* (CPR and AR) will always observe some recirculation due to CPR, even when AR is completely absent. The discussion of cardiopulmonary recirculation is more complicated and deferred to a later section in this chapter.

5.2.1. *Needle location and orientation*

The degree of recirculation that is measured is often influenced by the fistula needle location and orientation. It is virtually impossible to quantify all possible effects and indeed unpredictable measurements can arise. Optimal needle orientation is indicated in Fig. 5.1 with the arterial needle facing upstream and the venous needle returning blood downstream in the access. If the fistula needles are positioned correctly, then the theory is instructive and the correct conclusions may be drawn from the data. It should be appreciated that the presence of a stenosis can significantly affect flow conditions in the access. Furthermore, the location of needles relative to a stenosis is an important consideration. If blood cannot be supplied to the arterial needle at the rate demanded by the blood pump, either as a result of insufficient supply from the access or by recirculation of venous blood, the vessel may collapse under negative pressure. This generally causes a catastrophic fall in arterial pressure resulting in an arterial pressure alarm (1), and any recirculation measurement is immediately terminated.

It is important to realise the consequence of locating the venous needle upstream. This can easily occur particularly in the case of PTFE grafts where the direction of flow may not be evident. The same effect may be observed by inadvertent reversal of the blood lines connected to the fistula needles. Either way, dialysed blood is returned upstream. Inevitably, a proportion of this dialysed blood will be drawn up by the downstream arterial needle resulting in recirculation. Clearly, this is a *user-induced* form of recirculation which is an artefact in the context of the diagnosis of a failing fistula. It is exactly this principle of reversed lines or needles which, if applied correctly, may be used to determine the access flow rate.

The distance between fistula needles is an important consideration. It has been shown by Krivitski *et al.* that in the case of access flow measurement, fistula needle separation of the order of several centimetres can introduce errors at blood flow rates below 150 ml/min. Under these conditions, needle orientation relative to the vessel wall can have a substantial impact on the measured access flow (2). This implies that standard access recirculation measurements are subject to the same phenomena although there is little supporting evidence in the literature.

5.3. Recirculation in the Central Venous Catheter Access

There is little doubt that AR usually occurs in central venous catheters (3–7). The data shown below (Table 5.1) is typical of recirculation measurements which may be observed during routine practice.

Leblanc *et al.* have also reported data indicating that AR increases with blood flow rate (5). By contrast, other data suggest AR is independent of blood flow rate in central venous catheters. However, a significant influence of blood flow rate could be observed in femoral catheters (3).

In the case of central venous access under normal conditions, the flow available in the access is considerably in excess of the flow rate required for dialysis. Difficulties in achieving flow for dialysis are more likely to be attributable to clots or occlusion of the arterial lumen. Therefore, assuming adequate flows have been established for dialysis, *reversal* of the bulk shunt flow rate (as described in the previous section) is an unlikely possibility.

Table 5.1. Typical recirculation values to be encountered in central venous catheters at three blood flow rates. Data measured using the Fresenius Blood Temperature monitor, courtesy of Dr. M. Kroker, Kreiskrankenhaus Bad Hersfeld, Germany.

	Blood flow rate (ml/min)		
	200	300	400
No. of patients	27	26	9
Minimum AR	2	1.1	1.8
Maximum AR	32	25.7	33.2
Mean AR	9	10.2	11.7
SD	6.7	6.8	10

Since AR is observed in practice, it is obvious that there must be other effects which predispose to AR. The magnitude of AR measurements is unpredictable and clearly no correlations with other measured variables are indicated. This has thwarted any attempts to propose quantitative theories.

5.3.1. *Causes of recirculation in central venous lines*

In many cases, the overall effects of recirculation may only be quantified by consideration of *flow velocities* in a blood vessel rather than simply *bulk flow rate*. Rigorous analysis of flow profiles in large blood vessels is a complex subject, well beyond the scope of this book, but discussion of some basic ideas is justified. During application of the extracorporeal blood flow rate, for example, it may be demonstrated experimentally that the catheter tip migrates to one side of the blood vessel resulting in a complex flow velocity profile. Any such effect which results in non-uniform flow of blood through a vessel could potentially lead to recirculation.

The venous tip of a central venous catheter is normally located in the vicinity of the right atrium. The right atrium serves as a mixing chamber for the blood returning from the superior and inferior venae cava resulting in considerable turbulence. If the separation between the venous and arterial tip of the catheter is of the order of a few centimetres, then it is possible that turbulent blood flow could cause dialysed blood to be recirculated to the arterial tip. In addition, the cardiac cycle induces pulsatile flow. During contraction of the right ventricle, flow to the right atrium ceases momentarily, providing an opportunity for dialysed venous blood to be drawn towards the arterial tip. In view of these possible effects, it follows that maximising the separation between the venous and arterial tip would certainly help to minimise these forms of access recirculation. Indeed, the recently introduced Tesio twin catheters are associated with less access recirculation (7). This may be attributable to the increased separation of the arterial and venous tips afforded by this type of access which is generally in excess of 6 cm.

It is often necessary during routine practice to reverse the lumens of the catheter to obtain the required blood flow for dialysis. In similarity with AV fistula access, reversal of the lumens induces recirculation (5,8). Although this practice should be avoided in well-functioning catheters, it has been suggested by Twardowski *et al.* that this may be acceptable in patients in whom adequate blood flows is difficult to achieve (6).

5.4. Cardiopulmonary Recirculation

Access recirculation has been long established, and in fact represents the only form of recirculation that can occur in central venous access. Only recently, it was recognised that cardiopulmonary recirculation was an inevitable consequence following the creation of an arteriovenous fistula (9).

Figure 5.2 below shows a simplified schematic of the circulatory system, an arteriovenous fistula and an extracorporeal circuit. This example illustrates the condition when the extracorporeal blood flow rate (Q_B) is *less* than the access flow rate (Q_A). The bold figures represent the blood flow rates in the respective limbs of the circuit and the italic figures indicate the flow rates of dialysed blood. For the sake of simplicity, it is assumed that all of the blood returning from the venous needle has been completely cleared of solute. The ventricles of the heart and the pulmonary vessels of the lungs may be regarded

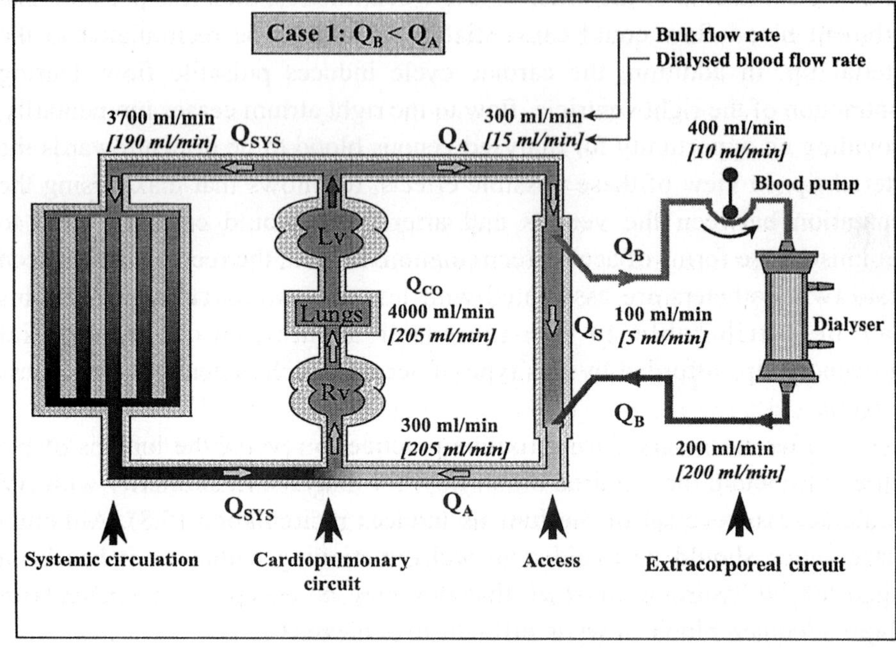

Fig. 5.2. Cardiopulmonary recirculation in the arteriovenous access.

collectively as the *cardiopulmonary circulation*. In a normal subject, blood returning from the systemic circulation passes through the cardiopulmonary circulation in a single path.

The creation of a fistula provides a second path through which blood may be returned to the cardiopulmonary circulation. The cardiac output Q_{CO} is then divided into the systemic flow rate Q_{SYS} and the access flow rate Q_A. Therefore, a component of the cardiac output containing dialysed blood is diverted to the arterial end of the fistula to be re-dialysed. Provided the blood flow rate Q_B in the extracorporeal circuit is less than the available Q_A, then only cardiopulmonary recirculation (CPR) occurs. In this case, a small proportion of the dialysed blood in the arterial side of the access passes directly through the shunt (the path between the arterial and venous needles), returning to the cardiopulmonary circulation. Consequently, there is always a fraction of dialysed blood which continuously circulates through the fistula and cardiopulmonary circuit, not entering either the systemic circulation or the extracorporeal circuit.

The fraction of dialysed blood flowing through the shunt is continually mixed with the dialysed blood returning from the venous needle of the extracorporeal circuit. During each pass of the dialysed blood in the venous side of the access, mixing occurs with the output from the systemic circulation. After several passes, the concentrations of solute in the various paths achieve a steady state within a few minutes with respect to the mean solute concentration of the patient. In this context, the mean solute concentration represents the aggregate output of the systemic circulation immediately before mixing with dialysed blood. Analysis of the recirculation in steady state is straightforward. The problem is to find the proportion of the venous blood which recirculates back to the arterial needle through the cardiopulmonary circulation. Therefore, the solute concentration or proportion of cleared blood at the output of the venous needle is irrelevant.

Let Q_{BCP} represent the *aggregate* flow rate of venous blood in the cardiopulmonary circuit. Q_{BCP} must be higher than Q_B at all times since some of Q_B will have recirculated through the shunt during the previous pass.

Hence, in steady state

$$\frac{Q_{BCP}}{Q_{CO}} Q_S + Q_B = Q_{BCP} \tag{5.3}$$

and

$$Q_S = Q_A - Q_B, \tag{5.4}$$

where Q_{CO} is the cardiac output, and Q_S is the shunt flow rate and Q_A is the access flow rate. Solving for Q_{BCP} yields:

$$Q_{BCP} = \frac{Q_B Q_{CO}}{Q_{CO} - Q_S}. \tag{5.5}$$

Therefore, the cardiopulmonary recirculation (CPR) is simply Q_{BCP} expressed as a proportion of the cardiac output, so

$$CPR = \frac{Q_B}{Q_{CO} - Q_S}. \tag{5.6}$$

CPR is by definition the fraction of the original venous blood flow rate Q_B which is recirculated to the arterial needle. An identical fraction is presented to the arterial needle since no further mixing occurs between the output of the cardiopulmonary circuit and the access. Inspection of Eq. 5.6 reveals that the degree of CPR increases with blood flow rate. Substituting the flow rates shown in the example of Fig. 5.2, then

$$CPR = \frac{Q_B}{Q_{CO} - Q_S} = \frac{200}{4000 - 100} = 5.1\%. \tag{5.7}$$

If the shunt flow Q_S is relatively small compared with the cardiac output Q_{CO}, then the expression for CPR may be approximated to:

$$CPR \approx \frac{Q_B}{Q_{CO}} = \frac{200}{4000} = 5\%. \tag{5.8}$$

This example of the approximate expression yields a CPR of 5%, slightly less than the exact expression. Clearly, such an approximation is particularly useful in the clinical setting and sufficiently accurate for most practical purposes. In addition, the need for access flow measurement (Q_A) is eliminated. Finally, it should be clear that CPR is always present in the arteriovenous access and cannot be avoided. During routine practice, it is very unlikely that CPR will exceed a value of 12 to 14% at the highest blood flow rates used for dialysis.

5.5. Combined Cardiopulmonary and Access Recirculation

In the case of the arteriovenous access, CPR is always present. Therefore, the presence of AR must occur in combination with CPR. In the context of dialysis efficiency, it is the total recirculation which is relevant, i.e. the sum

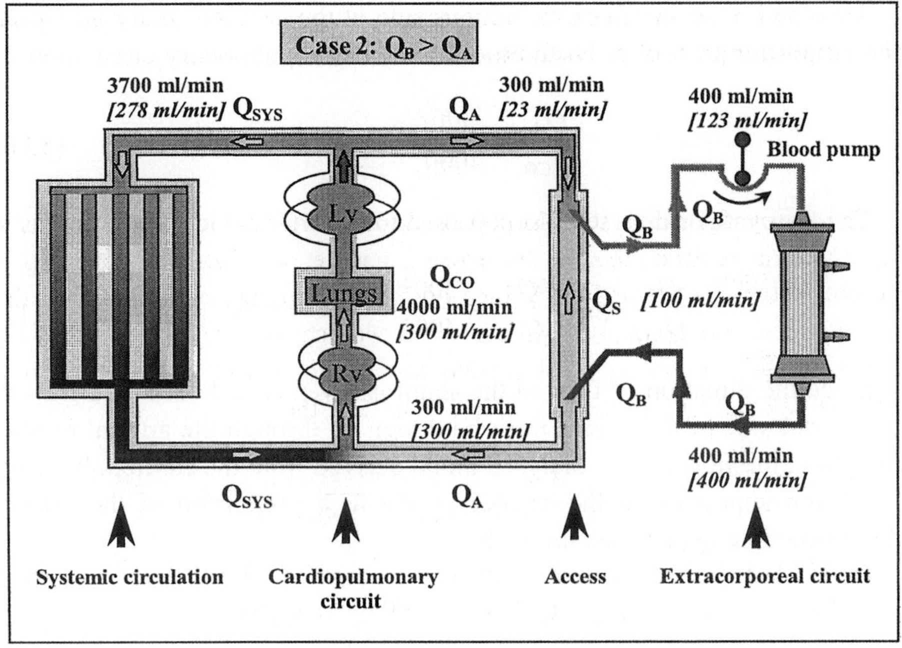

Fig. 5.3. Combined cardiopulmonary and access recirculation.

of AR and CPR. If patency of the fistula is the subject of investigation, then CPR is irrelevant and AR must be identified separately.

In the example shown in Fig. 5.3, the blood flow rate in the extracorporeal circuit (Q_B) exceeds the available access flow rate (Q_A). This causes the flow in the shunt to be reversed leading to AR. Therefore:

$$\mathrm{AR} = \frac{Q_B - Q_A}{Q_B} = \frac{400 - 300}{400} = 25\%. \tag{5.9}$$

The presence of AR influences the CPR and therefore a new expression for CPR must be derived. Again for simplicity, it may be assumed that all blood returning from the venous needle has been completely cleared of blood. (Although this does not happen in practice, it makes no difference to the analysis.)

All blood returning from the venous side of the access is dialysed blood. The proportion of this dialysed blood in the cardiopulmonary circulation is

$$\frac{Q_{\mathrm{A}}}{Q_{\mathrm{CO}}}=\frac{300}{4000}=7.5\%. \tag{5.10}$$

The flow rate of dialysed blood passed to the arterial side of the access is

$$\frac{Q_{\mathrm{A}}}{Q_{\mathrm{CO}}}\cdot Q_{\mathrm{A}}=\frac{Q_{\mathrm{A}}^{2}}{Q_{\mathrm{CO}}}=\frac{300^{2}}{4000}=22.5\ \mathrm{ml/min}. \tag{5.11}$$

Since the direction of flow in the shunt is reversed, all the blood flow in the arterial side of the access must be drawn up through the arterial needle. Therefore, the flow rate of dialysed blood derived from the cardiopulmonary circulation appearing in the arterial needle as a proportion of the arterial blood flow rate Q_{B} defines the CPR:

$$\mathrm{CPR}=\frac{Q_{\mathrm{A}}^{2}}{Q_{\mathrm{CO}}Q_{\mathrm{B}}}=\frac{300^{2}}{4000\times 400}=5.6\%. \tag{5.12}$$

The total recirculation (TR) may then be determined:

$$\mathrm{TR}=\frac{Q_{\mathrm{B}}-Q_{\mathrm{A}}}{Q_{\mathrm{B}}}+\frac{Q_{\mathrm{A}}^{2}}{Q_{\mathrm{CO}}Q_{\mathrm{B}}}. \tag{5.13}$$

In this example shown in Fig. 5.3, the TR is 30.6%.

5.6. Verification of the Recirculation Theory

5.6.1. *In vitro experiment*

Figure 5.4 shows the results of an *in vitro* study which demonstrates the principles of recirculation in an arteriovenous access (10). Blood flow rate (Q_{B}) has been increased sequentially and a measurement of the total recirculation has been obtained at each increment of Q_{B}. The theoretical model of total recirculation is described by Eqs. 5.7 and 5.13, which quantifies

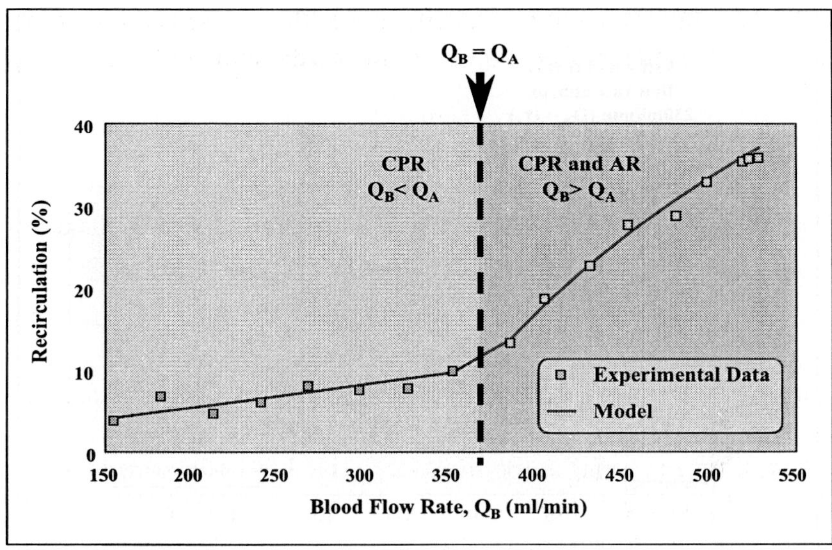

Fig. 5.4. *In vitro* verification of recirculation theory.

CPR (if $Q_B < Q_A$) and combines CPR and AR (if $Q_B > Q_A$), respectively. A correlation coefficient of 0.964 ($p < 0.001$) resulted when this model was fitted to the experimental data. The magnitude of recirculation at different blood flow rates is clearly demonstrated. When Q_B is less than Q_A, only cardiopulmonary recirculation (CPR) is present and the degree of CPR is approximately proportional to the applied blood flow rate. At the onset of access recirculation where $Q_B > Q_A$, the total recirculation rises rapidly with increasing flow rate. Under these conditions, both cardiopulmonary and access recirculations are present.

5.6.2. *Case study, in vivo recirculation investigation*

A similar study to that presented in Sec. 5.6.1 has been performed *in vivo* in a patient exhibiting high rates of recirculation (10). In this patient, the available access flow for dialysis was clearly restricted to approximately 230 ml/min. Again, for blood flow rates in excess of the access flow rate, a

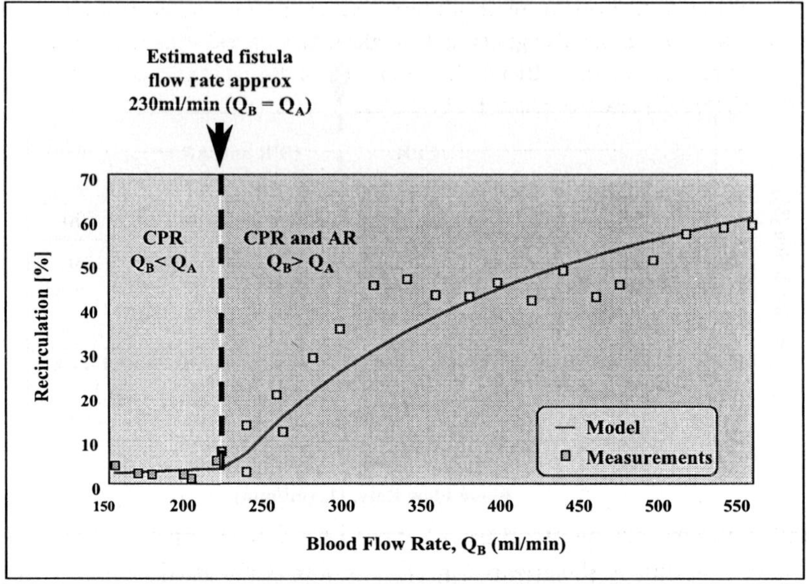

Fig. 5.5. *In vivo* verification of recirculation theory.

significant rise in total recirculation results due to the presence of access recirculation. The model was fitted to the data with a correlation coefficient of 0.651 ($p < 0.001$). These results also suggest agreement between the theory and measured data. However, there are some obvious departures of the measured data from the model. Since ten minutes was required for the acquisition of each recirculation measurement, a 4-hour study was necessary to obtain the data set. It cannot be assumed therefore that the cardiac output or access flow rate remained constant during this period of time.

5.6.3. *Recirculation data in the arteriovenous access*

The data shown in Table 5.2 below represents typical total recirculation measurements which may be encountered in the arteriovenous access during routine practice. Several points should be highlighted. Firstly, the mean total recirculation (TR) increases with the blood flow rate, ($p < 0.001$ between

Table 5.2. Typical recirculation values to be encountered in arteriovenous fistulae/grafts at four blood flow rates. Data measured using the Fresenius Blood Temperature monitor, courtesy of Dr. M. Kroker, Kreiskrankenhaus Bad Hersfeld, Germany.

	Blood flow rate (ml/min)			
	200	300	400	500
No. of patients studied	110	119	102	29
No. of TR >14% (AR)	1	5	5	7
Minimum TR	0	1.5	0	5.3
Maximum TR	40.8	51.9	53.8	34.0
Mean TR	5.8	7.4	8.9	11.6
SD	4.3	5.4	6.4	6.2

recirculation measurements at all adjacent blood flow rates), consistent with the theory. Secondly, the mean TR is less than 14% at all blood flow rates indicating the presence of CPR only in the majority of cases. Finally, the number of patients developing access recirculation (AR) increases with blood flow rate.

5.7. Determination of the Access Flow Rate (Method of Line Reversal)

It was highlighted earlier that if the arterial and venous needles were reversed, this effect could be usefully employed for the determination of access flow rate. It must be stressed that in this technique, recirculation is deliberately induced. The same principles for analysis of recirculation may then be applied as detailed in the following sections.

5.7.1. *Access flow rate in central venous lines*

Reversing the lines clearly increases the flow between the lumen of the catheter. This net interlumen flow rate (Q_{IL}) is simply

$$Q_{IL} = Q_B + Q_A. \tag{5.14}$$

Therefore, the flow rate of dialysed blood as a proportion of Q_{IL} will define the induced access recirculation (AR_I). So,

$$AR_I = \frac{Q_B}{Q_B + Q_A}. \tag{5.15}$$

Rearranging Eq. 5.15 leads to the Krivitski expression for calculation of the access flow (2):

$$Q_A = Q_B \left(\frac{1 - AR_I}{AR_I} \right) \tag{5.16}$$

Currently, there is an absence of data in the literature pertaining to the range of access flow rates to be expected during routine practice. Further studies are necessary in order to evaluate the utility of Eq. 5.16 in the case of central venous catheters.

5.7.2. *Determination of the access flow rate in the arteriovenous access*

It was noted earlier that cardiopulmonary recirculation is an inevitable consequence of the arteriovenous (AV) access. Indeed, this situation remains unchanged despite reversal of the blood lines. Therefore, in accordance with the theory developed previously for recirculation in the AV access, it is necessary to consider both the access and cardiopulmonary components of recirculation, i.e. the total recirculation. Although the total recirculation is irrelevant in this context for calculating dialysis efficiency, it does have important consequences when different access flow measurement methods are applied.

Figure 5.6 shows the combination effects of cardiopulmonary recirculation (CPR) and induced access recirculation (AR_I). While the flows in the arterial and venous needles have been reversed, the location of the needles are unchanged. The derivation of AR_I is identical to that of the central venous lines, the resulting expression given by Eq. 5.15.

The derivation of CPR is more complicated since there are a number of mixing and flow division effects to consider. The blood flowing through the

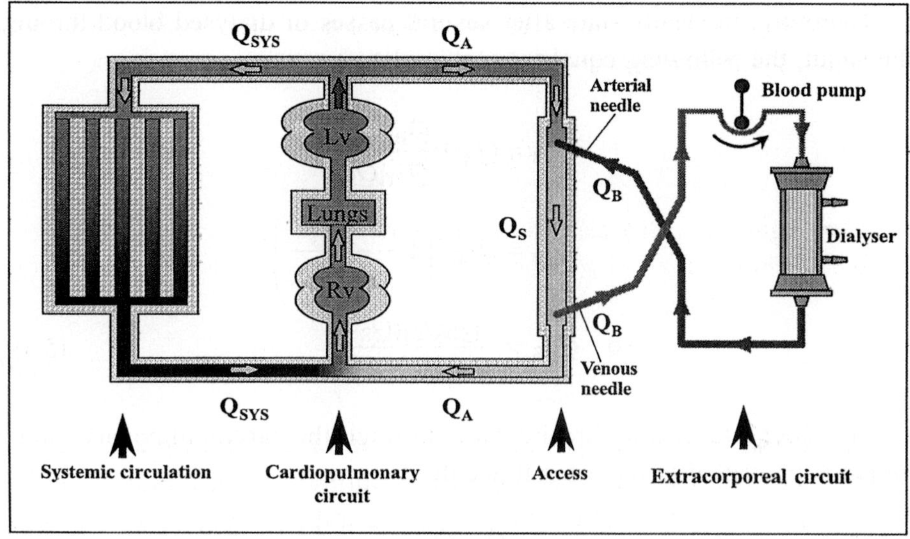

Fig. 5.6. Measurement of access flow rate by method of reversed blood line.

shunt will contain dialysed blood returning directly from the arterial needle (Q_B), in addition to a fraction which is recirculated through the cardiopulmonary circulation. Steady-state conditions result after several passes of dialysed blood through the cardiopulmonary circuit. If this aggregate flow of dialysed blood in the shunt is denoted by Q_{BS}, then the following analysis may be applied.

At the venous needle, the shunt flow (Q_S) is divided into the venous blood flow rate (Q_B) and the access blood flow rate (Q_A). Therefore, the proportion of Q_{BS} flowing through the venous side of the access is

$$Q_{BS} \cdot \frac{Q_A}{Q_S}, \tag{5.17}$$

and the proportion of dialysed blood flowing to the arterial needle from the cardiopulmonary circulation is

$$\frac{Q_{BS} Q_A^2}{Q_{CO} Q_S}. \tag{5.18}$$

Therefore, in steady state after several passes of dialysed blood through the shunt, the following equality must apply

$$Q_{BS} = Q_B + \frac{Q_{BS}Q_A^2}{Q_{CO}Q_S}$$

$$Q_B = Q_{BS} \cdot \left(1 - \frac{Q_A^2}{Q_{CO}Q_S}\right)$$

$$\text{so} \quad Q_{BS} = \frac{Q_B Q_{CO} Q_S}{Q_{CO}Q_S - Q_A^2}. \tag{5.19}$$

The flow rate of blood recirculated through the cardiopulmonary circuit which is presented to the arterial needle is

$$\frac{Q_{BS}Q_A^2}{Q_{CO}}. \tag{5.20}$$

Therefore, the fraction of this flow rate in the shunt may be written as:

$$\frac{Q_{BS}Q_A^2}{Q_{CO}Q_S}. \tag{5.21}$$

Hence, the flow rate into the venous needle must be

$$\frac{Q_{BS}Q_A^2}{Q_{CO}Q_S} \cdot \frac{Q_B}{Q_S}. \tag{5.22}$$

The flow rate into the venous needle expressed as a proportion of the arterial blood flow rate Q_B thus defines the CPR:

$$\text{CPR} = \frac{Q_{BS}Q_A^2}{Q_{CO}Q_S^2} \tag{5.23}$$

and substituting the expression for Q_{BS} in Eq. 5.23 yields

$$\text{CPR} = \frac{Q_A^2}{Q_{CO}Q_S} \cdot \frac{Q_B Q_{CO} Q_S}{Q_{CO}Q_S - Q_A^2} = \frac{Q_B Q_A^2}{Q_{CO}Q_S^2 - Q_A^2 Q_S}. \tag{5.24}$$

If total recirculation is measured, AR_I must be isolated from CPR. Access flow Q_A may then be determined as before via Eq. 5.16. Depner *et al.* has reported values for Q_A ranging from 125 to 2860 ml/min during *in vivo* studies (11).

Table 5.3. Summary of the recirculation expressions.

Configuration	Total recirculation	Access recirculation	Cardiopulmonary recirculation
Fistula/graft normal line $Q_B <= Q_A$	$\frac{Q_B}{Q_{CO} - (Q_A - Q_B)}$	None	$\frac{Q_B}{Q_{CO} - (Q_A - Q_B)}$
Fistula/graft normal line $Q_B > Q_A$	$\frac{Q_A^2}{Q_B Q_{CO}} + \frac{Q_B - Q_A}{Q_B}$	$\frac{Q_B - Q_A}{Q_B}$	$\frac{Q_A^2}{Q_B Q_{CO}}$
Fistula/graft lines reversed $Q_B <= Q_A$	$\frac{Q_B \cdot (Q_B + Q_A)}{(Q_B + Q_A) \cdot Q_{CO} - Q_A^2} + \frac{Q_B}{Q_B + Q_A}$	$\frac{Q_B}{Q_B + Q_F}$	$\frac{Q_B Q_A}{(Q_B + Q_A) \cdot Q_{CO} - Q_A^2}$
Fistula/graft lines reversed $Q_B > Q_A$	$\frac{Q_B \cdot (Q_B + Q_A)}{(Q_B + Q_A) \cdot Q_{CO} - Q_A^2} + \frac{Q_B}{Q_B + Q_A}$	$\frac{Q_B}{Q_B + Q_A}$	$\frac{Q_B Q_A}{(Q_B + Q_A) \cdot Q_{CO} - Q_A^2}$
Central venous access normal	No analytical expression available	No analytical expression available	None
Central venous lines reversed	$\frac{Q_B}{Q_B + Q_{CV}}$	$\frac{Q_B}{Q_B + Q_{CV}}$	None

Key to variables:

Q_B Extracorporeal blood flow rate [ml/min]

Q_{CO} Cardiac output [ml/min]

Q_A Arteriovenous access flow rate [ml/min]

Q_{CV} Central venous access flow rate [ml/min]

5.8. Summary of the Recirculation Expressions

The key points central to the analysis of recirculation are the identification of the different forms of recirculation (i.e. AR and CPR), the flow conditions under which these occur, consideration of the access type and the connection of the blood lines.

Table 5.3 provides a summary of all the recirculation expressions discussed in the previous sections.

5.9. The Influence of Recirculation on Dialysis Efficiency

5.9.1. *Calculation of effective clearance*

In the presence of recirculation, the performance of the dialyser remains unchanged, i.e. recirculation *does not* reduce dialyser clearance since K_d is determined by flow and the membrane product area and is *independent* of the input solute concentrations. Nevertheless, it is obvious that the rate of solute cleared from the patient must be reduced during recirculation. Therefore, it is convenient to specify an *effective* dialyser (K_e) clearance. This is nothing other than clearance of the dialyser as seen by the patient taking into account the recirculation effects. For the purposes of calculating effective clearance, the access may be considered as a "black box" as shown in Fig. 5.7.

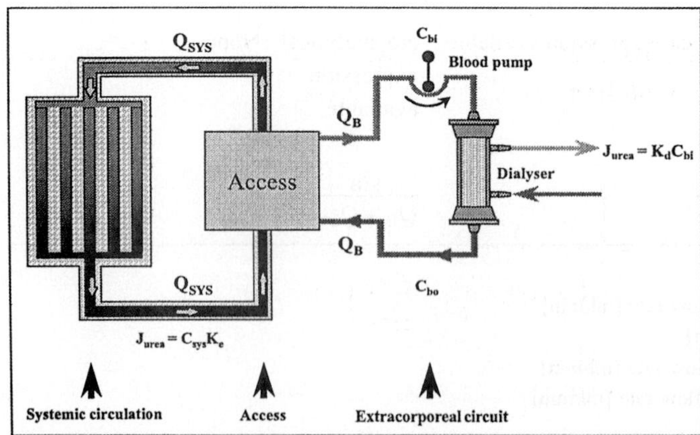

Fig. 5.7. The black-box approach for calculation of effective clearance.

By mass balance, the flux J_{urea}, of solute (urea in this example) cleared by the dialyser must be equal to the flux at the output of the systemic circulation. Hence

$$C_{SYS}K_e = C_{bi}K_d \tag{5.25}$$

therefore

$$\frac{C_{bi}}{C_{SYS}} = \frac{K_e}{K_d}. \tag{5.26}$$

The flux of blood flowing into the arterial needle may be expressed as

$$Q_B C_{bi} = R_{TOT} Q_B C_{bo} + (1 - R_{TOT}) Q_B C_{SYS}, \tag{5.27}$$

where TR is the total recirculation. Since

$$C_{bo} = \frac{C_{bi}(Q_B - K_d)}{Q_B} \tag{5.28}$$

then Eq. 5.26 may be rewritten as:

$$\frac{C_{bi}}{C_{SYS}} = \frac{1 - R_{TOT}}{1 - R_{TOT}\left(1 - \dfrac{K_d}{Q_B}\right)}. \tag{5.29}$$

Hence, combining Eqs. 5.29 and 5.26, an expression for the effective clearance reported by Gotch *et al.* (12) may be derived

$$K_e = \left(\frac{1 - R_{TOT}}{1 - R_{TOT} \cdot \left(1 - \dfrac{K_d}{Q_B}\right)}\right) K_d. \tag{5.30}$$

5.9.2. *Interpretation of effective clearance*

The theory of recirculation presented in the earlier sections of this chapter could be regarded as a somewhat academic exercise, particularly in the face of dedicated recirculation monitoring equipment. However, the applicability of rigorous analysis becomes immediately apparent when considering the effective clearance of solute from the patient in the presence of recirculation. Figure 5.8 shows clearance curves under different conditions. If there were no limits on the rate of diffusion in the dialyser (infinite clearance), all blood returning from the dialyser could be completely cleared of solute, irrespective of the applied blood flow rate (Q_B). Clearly, the dialyser cannot clear blood at a rate faster than the applied Q_B. If this concept is extended to the condition when the blood flow rate exceeds the access flow rate (in the case of AR), the dialyser cannot clear blood at a rate greater than the access flow rate. Thus, further increases in Q_B after the onset of access recirculation *do not* improve clearance.

The clearance curve of a standard dialyser (K_d) in the absence of any recirculation, is indicated in Fig. 5.8. In Fig. 5.8, CPR causes K_e to deviate from K_d. At the onset of AR, Q_B exceeds Q_A and K_e becomes progressively worse with increasing Q_B. The clinical significance is immediately apparent since dialysis efficiency can only be maximised by matching Q_B to Q_A.

5.9.3. *Correcting treatment time for recirculation*

The expression derived for effective clearance (Eq. 5.30) is valid for all forms of recirculation. Dialysis dose is usually prescribed using the expression

$$\frac{K_d T_d}{V}. \tag{5.31}$$

Let the efficiency factor be denoted by σ. So

$$\sigma = \frac{1-R}{1-R\left(1-\dfrac{K_d}{Q_B}\right)} \tag{5.32}$$

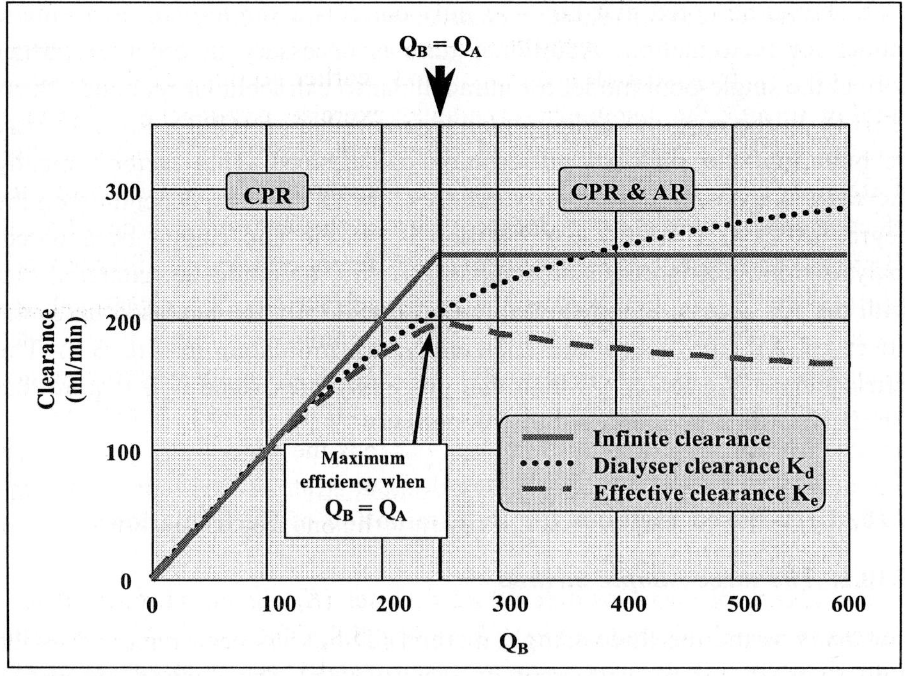

Fig. 5.8. Dialyser clearance K_d increases with blood flow rate. Recirculation modifies K_d leading to an effective clearance K_e.

then $K_e = \sigma K_d$, and from Eq. 5.31, it is clear that the dialysis time must be increased by the factor $\frac{1}{\sigma}$. $\frac{1}{\sigma} T_d = T_d + \delta T_d$ where δT_d is the additional time which must be added to T_d in order to compensate for recirculation, and solving for δT_d yields

$$\delta T_d = T_d \left(\frac{1}{\sigma} - 1 \right).$$

Combining Eq. 5.32, it may be shown that

$$\delta T_d = \frac{R \cdot K_d \cdot T_d}{Q_B (1 - R)}. \tag{5.33}$$

It should be noted that Eq. 5.33 only corrects a *single-pool* urea kinetic model for recirculation. Additional time is necessary in order to further correct the single-pool model for intracellular to extracellular rebound effects (13,14). In the case of cardiopulmonary recirculation (CPR), Eq. 5.33 may be built into the standard prescription calculations. This factor typically increases dialysis time by approximately 10 to 12 minutes, depending on the degree of CPR. If access recirculation is present and cannot be avoided, dialysis time will increase substantially. Again, it should be reiterated that with the AV access, blood flow rate Q_B should only be increased up to the onset of AR for maximum efficiency, as indicated in Fig. 5.8. The consequence of a lower Q_B than that originally prescribed will require that the dialysis time be increased to compensate.

5.10. Urea-Based Methods for Determination of Recirculation

5.10.1. *The three-sample method*

For many years, the three-sample method (TSM) has been regarded as the gold standard for measurement of recirculation. The method requires a peripheral venous blood sample for measurement of urea concentration. In addition, a pair of blood samples must be taken simultaneously at the arterial and venous sides of the dialyser (15). It was thought that the peripheral venous sample would provide the reference against which the arterial and venous blood samples could be compared. Recirculation was calculated using the expression:

$$R_{TSM} = \frac{C_P - C_A}{C_P - C_V} \tag{5.34}$$

where C_P, C_A and C_V represent the concentrations of urea in the peripheral, arterial and venous blood samples, respectively. Consequently, if the arterial and peripheral blood samples contained the same urea content, then the TSM would measure zero recirculation. By the same token, 100% recirculation would be observed if the arterial and venous blood concentrations were equal. Practical application has since revealed that the method always returns recirculation values of approximately 12% even when access recirculation is absent (16). The reasons for this observation have been elucidated by consideration of some elementary circulatory principles. Figure 5.9 is a simplified schematic of the circulation. The central compartment is perfused with the majority of the cardiac output. By contrast, the peripheral circulation is perfused relatively poorly with blood flow rates rarely exceeding a few hundred millimetres per minute. As a result, the clearance of urea from the peripheral circulation is relatively slow. Therefore, at any time after the start of dialysis, the solute concentration in the peripheral circulation will be higher than that of the central circulation. This gives rise to the venous disequilibrium, first highlighted by Depner (17) and has since been acknowledged by others (18,19). Tattersall *et al.* later demonstrated the significance of this effect in the context of a two pool urea kinetic model (16), the results of which are shown in Fig. 5.10.

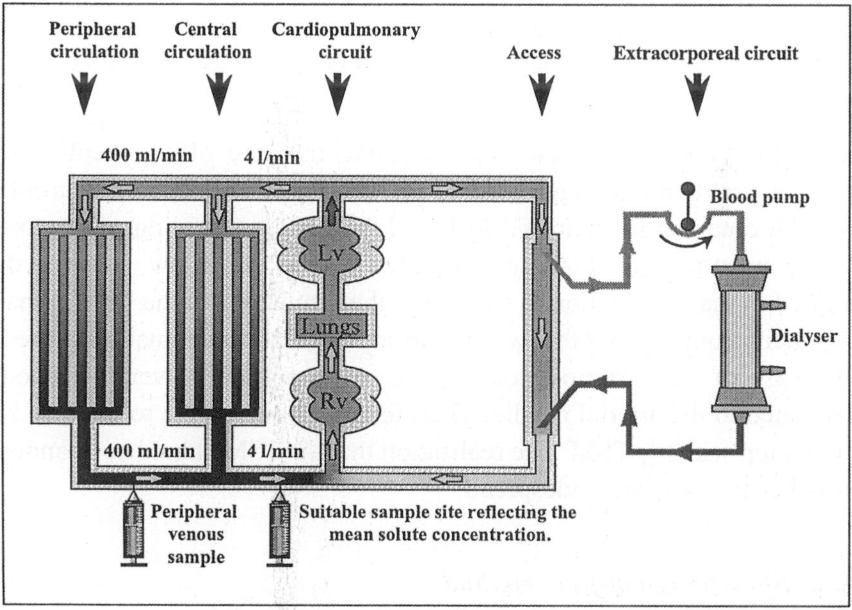

Fig. 5.9. Schematic of the circulatory system. The peripheral circulation is perfused relatively poorly by comparison with the central circulation, leading to venous disequilibrium.

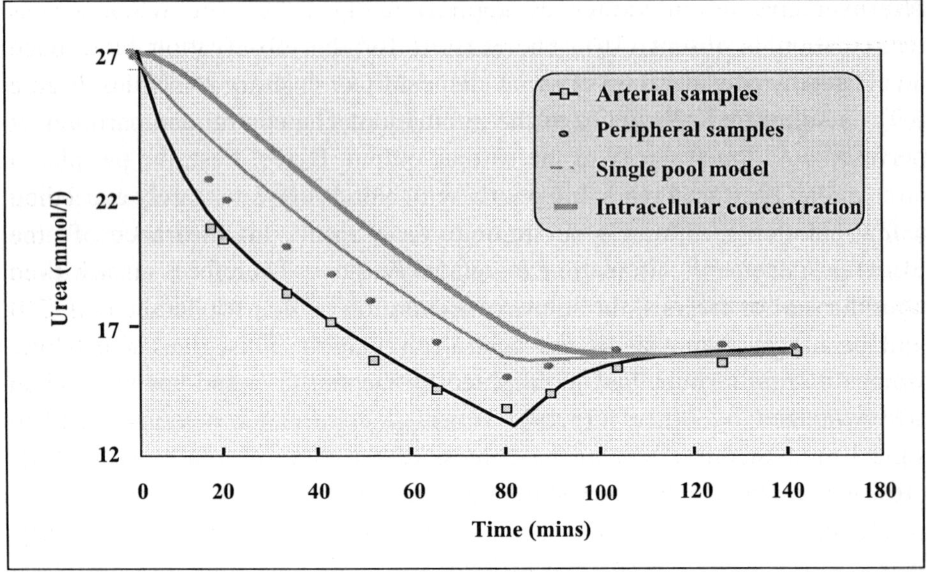

Fig. 5.10. The concentration of urea in arterial blood samples is always lower than that of the peripheral venous samples during dialysis. [Data courtesy of Tattersall *et al.* (16).]

In order to obtain a more representative measure of a patient's mean solute concentration, an aggregate of the central and venous compartments should be obtained as indicated in Fig. 5.9. It is the sampling of this mean solute concentration which is attempted in the stop/slow flow method described in the next section to overcome the limitations of the TSM. Finally, it is evident from Fig. 5.9 that where cardiopulmonary recirculation is present in the case of the arteriovenous access, a proportion of venous blood is recirculated to the arterial needle. Therefore, it is impossible to observe zero recirculation with the TSM. The realisation that the TSM should be abandoned is now becoming more widespread.

5.10.2. *Slow-flow/stop-flow method*

The *slow-flow* method is an improvement over the TSM, substituting the peripheral sample with a "slow-flow" sample. All other samples' required and associated calculations are identical to the TSM. Providing care is exercised in the timing of blood samples, the effects of CPR which introduce artefacts in the TSM may be overcome. The slow-flow method is applicable to all types of vascular access but may only be used for the detection of *access recirculation*.

The key to the successful application of this method is to obtain a blood sample at the arterial inlet to the dialyser which is representative of the blood urea concentration in the cardiopulmonary circulation. This is achieved by reducing the blood flow to approximately 50 ml/min for 30 seconds (20) and then withdrawing a blood sample from the arterial line port, the "slow-flow" sample. A variation of this technique is the stop-flow method in which the blood flow rate is reduced to zero. It has been demonstrated in studies involving the slow-flow and stop-flow techniques that the two methods are comparable (19). Nevertheless, regardless of the method, accurate timing of the slow-flow sample is essential. If the sample is withdrawn too early, it is likely that there will be insufficient time to allow the dialysed blood to clear the arteriovenous access. Similarly, adequate time is necessary for blood from the cardiopulmonary circulation to be drawn up to the arterial sample port. Consequently, the slow-flow samples and arterial sample will be similar and the method will return a value of zero recirculation, irrespective of the presence or absence of AR. In the case of late sampling, the dialysed component of blood in the cardiopulmonary circulation will be cleared. Furthermore, the effects of post dialysis rebound caused by solute equilibration between compartments will also contribute to a higher slow-flow concentration resulting in an overestimation of recirculation (13,18). Indeed, validation studies against a reference method have highlighted a systematic over-estimation of recirculation (21).

5.11. Automated Methods for Measurement of Recirculation

5.11.1. *General principles of recirculation monitoring*

The majority of methods for automated recirculation measurement require the injection of a marker (usually hypertonic saline) into the venous side

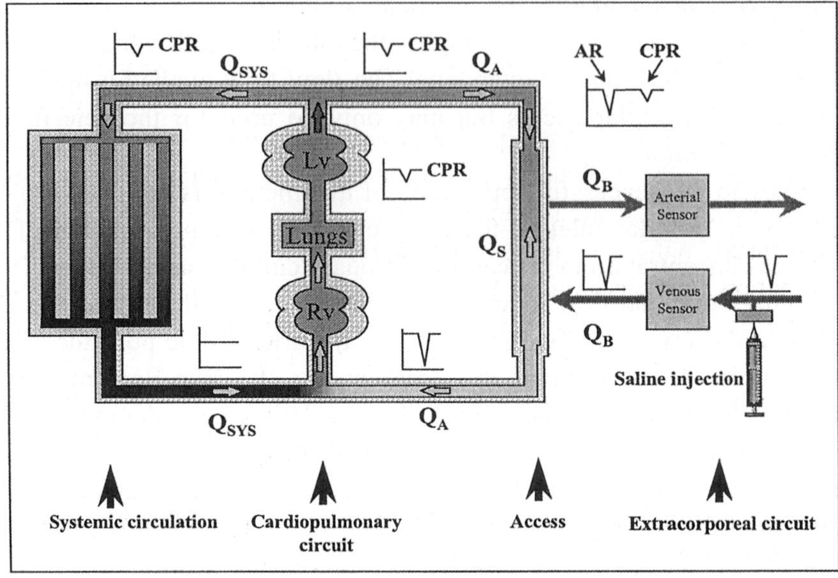

Fig. 5.11. Passage of a saline bolus (recirculation marker) through the circulation.

of the extracorporeal circuit. A venous sensor typically measures a property of the blood composition as the bolus of saline passes into the patient as shown in Fig. 5.11. The bolus of saline is then attenuated by the mixing of blood throughout the circulation. Any proportion of the bolus which is recirculated back to the extracorporeal circuit is detected by an arterial sensor. The ratio of the venous and the arterial bolus indicates the degree of recirculation. Strictly speaking, the detection of AR by the arterial head precedes the detection of CPR as AR represents the shorter transit time for the saline bolus. The first recirculation equipment exploiting these principles was developed over a decade ago by Aldridge *et al.* (22,23). Since this pioneering work, a number of commercially available recirculation monitors which are variations of this technique have been developed, described in the following sections.

5.11.2. *Transonics system*

The Transonics HD01 haemodialysis monitor (Transonics Systems Incorporated) provides a sensor which clips onto standard blood tubing on the arterial side of the extracorporeal circuit. The sensor monitors both the flow and the dilution of blood with saline, the information from which is processed by the associated Transonics HD01 haemodialysis monitor. These data are further analysed via dedicated software running on a laptop PC for determination of recirculation. Injection of a 4 to 5% bolus of hypertonic saline into a venous sample port is necessary for recirculation measurements. The arterial sensor determines the recirculated fraction by monitoring the velocity of ultrasound through blood which is modified by the dilutional effect of the saline. Ultrasound velocity in blood is largely determined by the protein content (24) and since this increases during dialysis as a result of ultrafiltration, calibration of the sensor must be performed prior to measurement of recirculation. Calibration is readily achieved by injection of saline into an arterial line port. Use of a second sensor located on the venous side of the extracorporeal circuit eliminates the need for two saline injections since both calibration and measurement may be performed at the same time. In addition, the use of dual sensors allows the effects of cardiopulmonary and access recirculation to be distinguished and quantified with increased accuracy (25). *In vitro* experiments have demonstrated that access flows can be measured with an error of only 5%, more than that acceptable for routine clinical application (6). Similar errors have been reported from *in vivo* studies (26). The reproducibility of the HD01 has been assessed in patients by Lindsay *et al.* (27). In the study, reproducibility was calculated by normalising the standard deviation between three successive measurements in the same patient. Results indicated that AR could be determined with a reproducibility of 9.1%, i.e. if the HD01 measures 25% AR in a patient, a repeated measurement may yield a value between 22.7 and 27.3%. In the case of access flow measurements, the reproducibility was found to be 13%. Consequently, the transonics system is well established and has been used as a reference against which other methods have been compared (27).

5.11.3. *Gambro haemodialysis monitor (HDM)*

The haemodynamic monitor (HDM) (Gambro Healthcare Incorporated) employs clip-on arterial and venous sensors which determine changes in blood conductivity non-invasively by electromagnetic induction. In order to perform a recirculation measurement, 1 ml of hypertonic saline is injected into a venous sample port generating a "conductivity bolus". The differential change in conductivity between the arterial and venous sensors enables the magnitude of the recirculation to be calculated. This method monitors access recirculation (AR) only. In a study involving over 250 investigations undertaken by Lindsay *et al.* to assess the performance of the HDM (28), a high correlation (0.94, slope 0.95) was found between the HDM and the slow-flow method (described above) in patients with AR ranging from 5 to 60%. The reproducibility of the HDM over consecutive dialyses in the same patients was found to be ±1.8%. Finally, Lindsay devised a method to verify the accuracy of the HDM in patients with arteriovenous fistulae/grafts. This was achieved by simulating exactly 10% AR resulting from injection of known volume of hypertonic saline in both the arterial and venous sides of the extracorporeal circuit. The mean AR returned by the HDM in 93 patients was 9.96% ±1.59 SD, confirming the precision of this method. Subsequent studies by Lindsay have confirmed that the performance of the HDM yields a reproducibility of 7.5% for AR and 10.6% for access flow measurement, comparable with the Transonics HD01 system (27).

5.11.4. *Crit-Line*

Crit-Line monitor (In-Line Diagnostics), is well established as a device for blood volume monitoring (29). The utility of the Crit-Line has been extended recently for the purposes of recirculation monitoring, available in the model III version. A dedicated cuvette must be inserted in the extracorporeal circuit at arterial inlet of the dialyser into which an optical sensor is engaged. This sensor is sensitive to changes in the optical density of blood, the properties of which depend on the blood hydration content. Recirculation measurement requires two saline injections both of which must be infused over a period of 10 seconds. The first saline bolus (10 ml) is injected into the arterial line upstream of the sensor for calibration purposes. This has the effect of diluting the blood which decreases the optical density monitored by the sensor. A second 10 ml bolus is injected into the venous sample port. The proportion of this saline bolus which is recirculated through the access then detected by the arterial sensor. Recirculation measurements below 4% are not quantified by the Crit-Line. The performance of this device has been evaluated monitor by Lindsay *et al.* (30). These studies have reported an excellent AR reproducibility of ±7.82%. The Crit-Line has been compared against the Transonics HD01 yielding a high degree of correlation ($R = 0.95$). However, a systematic error has been observed — at AR values below 20%, the Crit-Line over-estimates the AR observed by the HD01 and vice versa.

5.11.5. *Fresenius blood temperature monitor (thermal dilution)*

The Fresenius blood temperature monitor (BTM) (Fresenius Medical Care) employs the principle of thermal dilution (31) as the marker for recirculation measurement. Two temperature heads which clip over standard blood tubing measure the blood temperature in the arterial and venous sides of the extracorporeal circuit. Since the BTM is integrated with the dialysis machine, it is the only method available which is fully automated, requiring a single button press to initiate a recirculation measurement. A bolus of "cold" blood is generated for several minutes by cooling the dialysis fluid temperature by several degrees. This temperature bolus is monitored by the venous head before the bolus passes into the patient. Fractions of the temperature bolus which are recirculated through the cardiopulmonary circuit and through the shunt (if access recirculation is present) are monitored by the arterial temperature for a period of 5 to 8 minutes. Since the original temperature bolus makes several passes through the cardiopulmonary circulation and shunt, it is important to appreciate that the BTM measures *total recirculation* (TR). TR is the sum of the cardiopulmonary and access recirculations (if present). TR is particularly useful for direct correction of dialysis treatment times. The reproducibility of the BTM *in vitro* is ±3 to 5% but this has yet to be confirmed *in vivo*. Small errors may arise in the case of central venous catheters since at low blood flow rates, partial temperature equilibration may take place across the lumen of the catheter. Clearly, this is dependent on the type of catheter, especially lumen wall thickness.

5.12. Summary of Methods for Recirculation and Access Flow Monitoring

Table 5.4 below summarises the important aspects of the main methods available for recirculation measurements. Note that measurement of access flow rate is possible with all automated recirculation monitors by blood line reversal to induce recirculation. However, the implementation of this method varies among manufacturers. Some devices are specifically designed to calculate the access flow rate directly (e.g. Transonics), while others simply provide the induced recirculation value. In cases where only the recirculation value is available, the access flow rate may be obtained by application of Eq. 5.16. Care should be exercised when applying this expression since only the induced access recirculation AR_I should be used — measurements of cardiopulmonary recirculation or total recirculation (CPR and AR_I) will lead to incorrect results.

5.13. Practical Application of Recirculation Methods

The following points should be considered during routine use of recirculation monitoring methods:

- Needle location and orientation can significantly affect recirculation measurements.
- The position of central venous catheters can affect the magnitude of AR.
- In general, the greater the separation between the arterial and venous lumens in a central venous catheter, the less the likelihood of AR.
- Irrespective of the method used for measurement of recirculation, it is important to be aware of exactly which information is obtained. Each form of recirculation leads to a specific interpretation.
- Ensure that the correct expressions for recirculation have been applied (see Table 5.3).

Table 5.4. Overview of recirculation measurement methods.

	Measurement				Reproducibility			
Method	AR	CPR	TR	Access flow	Recirculation	Access flow	Advantages	Disadvantages
Three-sample method	✓	✓	✓	×	Very poor	N/A	None, should be abandoned!	Many, due to venous disequilibrium.
Stop-flow method	✓	✓	✓	×	Poor	N/A	Inexpensive, no dedicated equipment requirement.	Timing of sampling is critical and the possibility of errors cannot be eliminated.
Crit-Line* (In-Line Diagnostics) — Saline dilution	✓	×	×	✓	7.8%	7.8%	A single integrated unit which may be used with any dialysis machine.	Requires a special purpose cuvette. Relatively complex procedure for handling. CPR is not measured.
HD01** haemodialysis monitor (Transonics Systems Inc.) — Saline dilution	✓	✓	×	✓	9.1%	13%	High accuracy. In addition to recirculation measurements, extracorporeal blood flow rates can be measured with high precision.	Relatively expensive.

Table 5.4. (*Continued*)

Method	Measurement				Reproducibility		Advantages	Disadvantages
	AR	CPR	TR	Access flow	Recirculation	Access flow		
BTM (Fresenius Medical Care) — Thermal dilution	✓	✓	✓	✓	5%[#]	Unconfirmed	Simplest method available, integrated with dialysis machine. Measures total recirculation which is useful for calculation of effective clearance.	Slow, can only be used in machine fitted with BTM.
Haemodynamic monitor (Gambro) — Saline dilution	✓	×	×	✓	7.5%	10.6%	High accuracy.	CPR is not measured.

*Calibration required; **Calibration not required if dual sensor system employed; [#]*In vitro* data only; AR = Access recirculation, CPR = Cardiopulmonary recirculation, TR = Total recirculation.

- In order to correct dialysis times for the inefficiency introduced by recirculation, the total recirculation must be determined. In the case of the arteriovenous access, the effect of CPR must be taken into account. If a direct measurement of CPR is not available, a worst case value of 14% can be assumed.
- Before taking any corrective action, a repeat measurement of recirculation is always advisable. Recheck that the correct expressions for recirculation have been applied.
- Ultrafiltration affects the measured recirculation in some methods. Since recirculation measurements can be performed quickly, it is recommended that the ultrafiltration rate be reduced to zero during recirculation tests.
- It is often worth verifying the effective blood flow rate prior to analysis. The effective blood flow rate may differ from the displayed blood flow rate depending on the particular manufacturer of dialysis machine (5).
- Access flow rate may be determined by reversal of blood lines. If a method measures total recirculation, it is possible to determine the access flow rate, thus *avoiding* the necessity to reverse the blood lines *provided* the access flow rate is lower than the maximum blood flow rate required for dialysis. This method is achieved by measurement of total recirculation over a range of blood flow rates (see Fig. 5.5).

5.13.1. *Recirculation protocols*

In order to maximise the utility of recirculation data in the clinical setting, two protocols are suggested for arteriovenous and central venous accesses, Figs. 5.12 and 5.13, respectively. For the purposes of general application, the arteriovenous access protocol considers total recirculation (TR) which includes the effects of CPR and AR. However, arteriovenous access patency can be assessed directly if the recirculation method observes AR only.

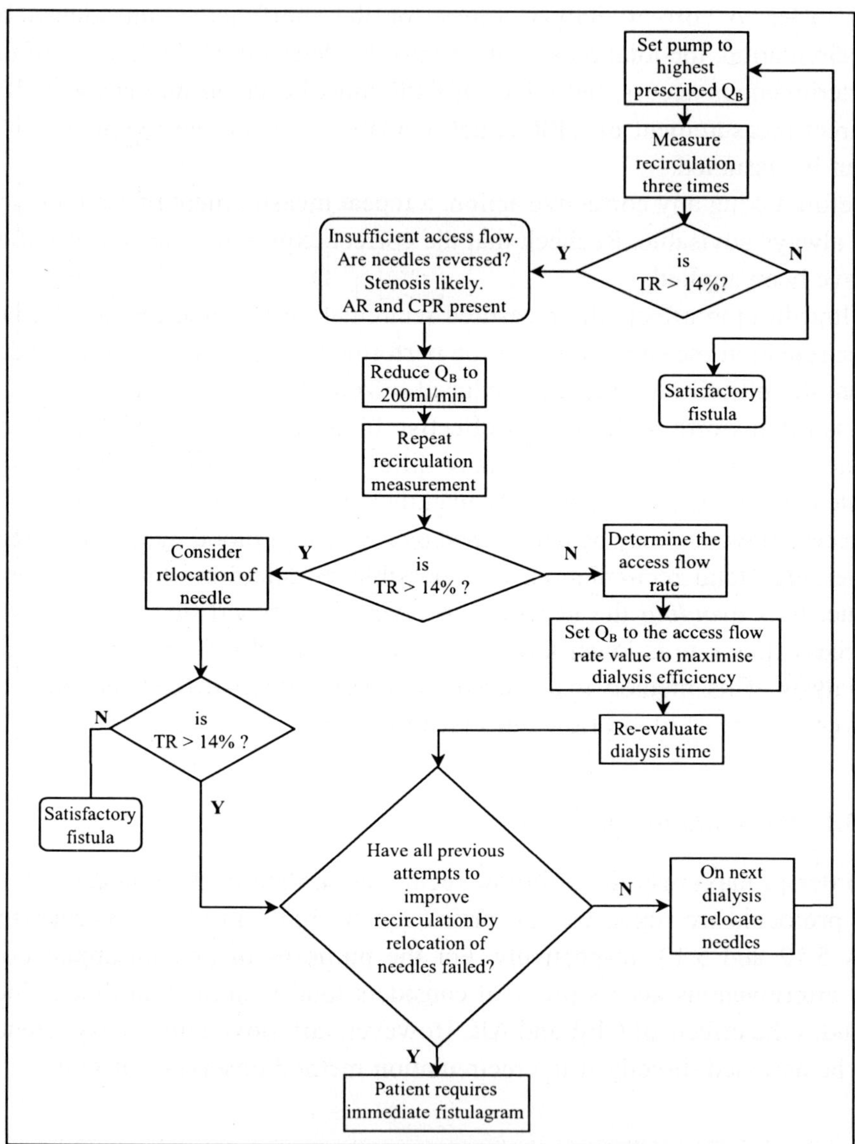

Fig. 5.12. Protocol for the application of recirculation measurements in the arteriovenous access.

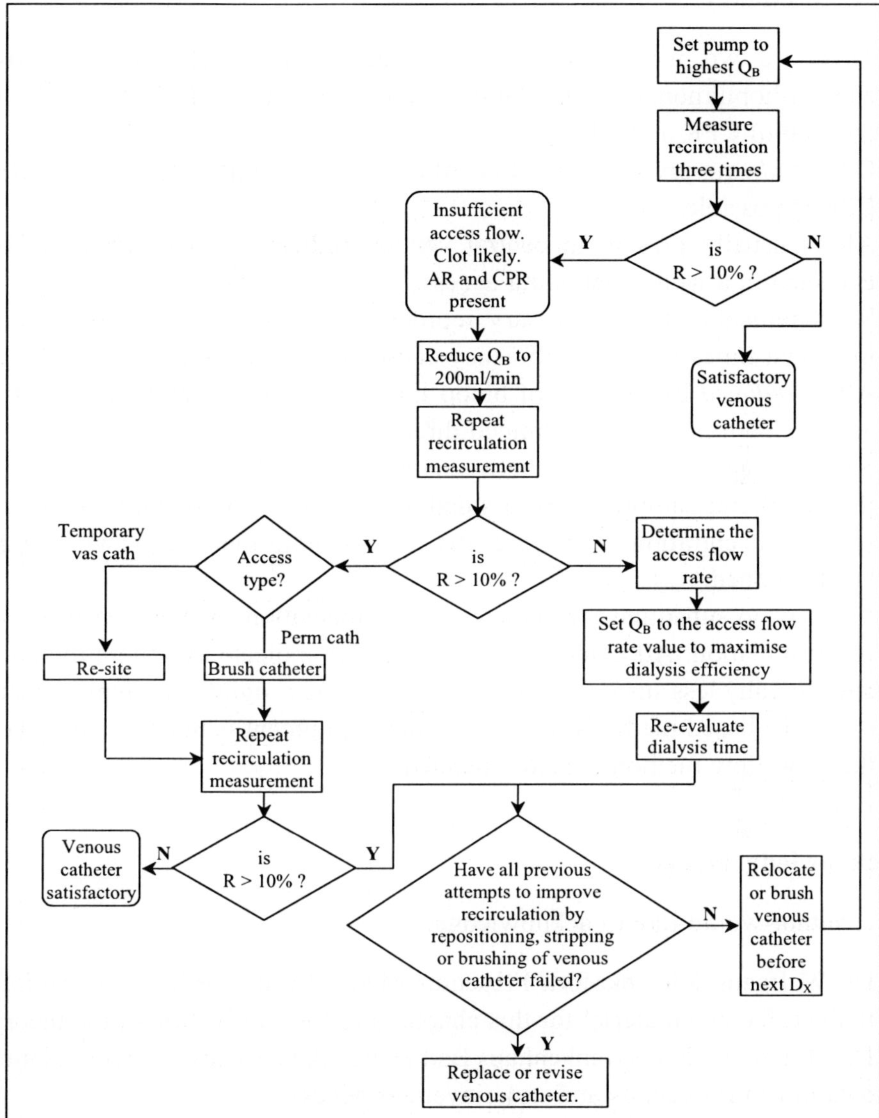

Fig. 5.13. Protocol for the application of recirculation measurements in the venous catheter access.

5.14. Summary

- There are two primary forms of recirculation: access recirculation (AR) and cardiopulmonary recirculation (CPR). The total recirculation (TR) is the sum of CPR and AR.
- CPR is *always* present in the case of an arteriovenous access. Up to 14% CPR is possible.
- AR is usually present in central venous catheters. Up to 10% can be expected in a well-functioning access.
- Measurement of TR is necessary in order to compensate prescribed dialysis treatment times for the inefficiency caused by recirculation.
- AR is *induced* by reversal of blood lines. This configuration enables the access flow rate to be determined directly.
- Since disequilibrium occurs between peripheral and central compartments, the peripheral sample is not a suitable reference representative of mean solute concentration. Therefore, the three-sample method (TSM) should be abandoned.
- There is no substitute for dedicated instrumentation for the measurement of recirculation. All evidence in the literature indicate that these methods are generally less susceptible to artefacts and are a significant improvement over urea based methods. If recirculation equipment is not available, then the slow-flow method is recommended.

Acknowledgements

The author would like to acknowledge:

- Dr. Wolfgang Kleinekofort, Fresenius Medical Care for his contributions to the reference material for this chapter and for verification of the theory.
- Dr. M. Kroker, Kreiskrankenhaus Bad Hersfeld, Germany, for recirculation data in central venous and arteriovenous accesses.
- S. Byrne and C. Hodson for their contributions regarding the practical application of recirculation monitoring.
- Lister Hospital, Stevenage, UK, for the patient study data.
- S.J. Atkinson for edits to the illustrations.

References

1. Byrne, S., Tomlinson, C.D., Bowser, M. *et al.* (1994). A new proactive approach to the managament of arteriovenous fistulae. *EDTNA*, **4**, 10–15.
2. Krivitski, N.M. (1995). Theory and validation of access flow measurement by dilution technique during hemodialysis. *Kidney Int*, **48**, 244–250.
3. Kelber, J., Delmez, J.A. and Windus, D.W. (1993). Factors affecting delivery of high-efficiency dialysis using temporary vascular access. *Am J Kidney Dis*, **22**, 24–29.
4. Sombolos, K. *et al.* (1993). Efficacy of dual lumen jugular venous catheter hemodialysis when venous lumen is used as arterial lumen. *Nephron*, **65**, 147–149.
5. Leblanc, M. *et al.* (1998). Effective blood flow and recirculation rates in internal jugular vein twin catheters: Measurement by ultrasound velocity dilution. *Am J Kidney Dis*, **31**, 87–92.
6. Twardowski, Z.J. *et al.* (1993). Blood recirculation in intravenous catheters for hemodialysis. *J Am Soc Nephrol*, **3**, 1978–1981.
7. Prabhu, P.N. *et al.* (1997). Long-term performance and complications of the Tesio twin catheter system for hemodialysis access. *Am J Kidney Dis*, **30**(2), 213–218.
8. Depner, T.A. (1997). Does monitoring have the potential for preventing vascular access failure in dialysis patients? *Contemporary Dialysis & Nephrology*, **Sept**, 22–24.
9. Schneditz, D. and Polaschegg, H.D. (1992). Cardiopulmonary recirculation in dialysis: An under-recognised phenomenon. *ASAIO J*, **38**(3), M194–M196.
10. Chamney, P.W. *et al.* (1994). Modelling cardio-pulmonary and fistula recirculation. *EDTNA*, **20**, 17–21.
11. Depner, T.A. and Krivitski, N.M. (1995). Clinical measurement of blood flow in hemodialysis access fistulae and grafts by ultrasound dilution. *ASAIO J*, **41**, M745–M749.
12. Gotch, F.A. (1984). Models to predict recirculation and its effect on treatment time in single-needle dialysis. In *First International Symposium on Single-Needle Dialysis*. (eds. S. Ringior, R. Van-Holder and P. Ivanovich, p. 305) ASAO Press: Cleveland.
13. Tattersall, J.E. *et al.* (1999). Recirculation and the post-dialysis rebound. *Nephrol Dial Transplant*, **11**, 75–80.

14. Tattersall, J.E. *et al.* (1996). The post-dialysis rebound: Predicting and quantifying its effect on *Kt/V*. *Kidney Int*, **50**, 2094–2102.
15. Gibson, S.M., Von Albertini, B. and Bosch, J.P. (1990). Reproducible measurement of recirculation without peripheral venipuncture. *Kidney Int*, **37**, 297 (Abstract).
16. Tattersall, J.E. *et al.* (1993). Haemodialysis recirculation detected by the three sample method is an artefact. *Nephrol Dial Transplant*, **8**, 60–63.
17. Depner, T.A. *et al.* (1991). High venous urea concentrations in the opposite arm: A consequence of hemodialysis-induced compartment disequilibrium. *ASAIO J*, **37**, 141–143.
18. Sherman, R.A. and Kapoian, T. (1997). Recirculation, urea disequilbrium, and dialysis efficiency: Peripheral arteriovenous versus central venovenous vascular access. *Am J Kidney Dis*, **29**, 479–489.
19. Sherman, R.A. *et al.* (1994). Recirculation reassessed: The impact of blood flow rate and the low-flow method reevaluated. *Am J Kidney Dis*, **23**(6), 846–848.
20. Sherman, R.A. and Levy, S.S. (1991). Assessment of a two-needle technique for the measurement of recirculation during hemodialysis. *Am J Kidney Dis*, **18**, 80–83.
21. Kapoian, T., Steward, C.A. and Sherman, R.A. (1997). Validation of a revised slow-stop flow recirculation method. *Kidney Int*, **52**(3), 839–842.
22. Aldridge, C. *et al.* (1984). The assessment of arteriovenous fistulae created for haemodialysis from pressure and thermal dilution measurements. *J Med Engin Tech*, **8**(3), 118–124.
23. Aldridge, C. *et al.* (1999). Instrument design for the bedside assessment of recirculation during hemodialysis. *Proc EDTNA-ERCA*, **14**, 255–260.
24. Schneditz, D. and Kenner, T. (1989). A sound-speed sensor for the measurement of total protein concentration is disposable, blood perfused tubes. *J Acoust Soc Am*, **86**, 2073–2081.
25. Krivitski, N.M. *et al.* (1998). Accuracy of dilution techniques for access flow measurement during hemodialysis. *Am J Kidney Dis*, **31**(3), 502–508.
26. Depner, T.A., Krivitski, N.M. and MacGibbon, D. (1995). Hemodialysis access recirculation measured by ultrasound dilution. *ASAIO J*, **41**(3), M749–M753.
27. Lindsay, R.M. *et al.* (1998). A comparison of methods for the measurement of hemodialysis access recirculation and access blood flow rate. *ASAIO J*, **44**, 62–67.

28. Lindsay, R.M. *et al.* (1998). Accuracy and precision of access recirculation measurements by the hemodynamic recirculation monitor. *Am J Kidney Dis*, **31**(2), 242–249.
29. Steuer, R.R., Harris, D.H. and Conis, J.M. (1993). A new optical technique for monitoring hematocrit and circulating blood volume: Its application in renal dialysis. *Dial Transplant*, **22**, 260–265.
30. Lindsay, R.M., Rothera, C. and Blake, B.G. (1998). A comparison of methods for the measurement of hemodialysis access recirculation. *ASAIO J*, **44**, 191–193.
31. Kraemer, M. and Polaschegg, H.D. (1993). Automated measurement of recirculation. *Proc EDTNA/ERCA J*, **2**, 6–9.

CHAPTER 6

THE VALUE OF ULTRASONIC IMAGING IN DEFINING THE ANATOMY FOR VASCULAR ACCESS

Adil Eltayar MSc, FRCS (Ireland) and Andrew Nicolaides MS, FRCS
Irvine Laboratory for Cardiovascular Investigation and Research
St Mary's Hospital, London, W2 1NY, UK

Gabriel Szendro MD, FRCS
Soroka Medical Centre and Faculty of Health Sciences
Ben-Gurion University, Beer-Shera, Israel

6.1. General

Although the results of renal transplantation have improved over the last 15 years, and chronic ambulatory peritoneal dialysis (CAPD) has become an alternative form of treatment, most patients with end-stage renal failure (ESRF) still spend an appreciable amount of time on haemodialysis either waiting for transplants or as a definitive treatment. The number of potential access sites for haemodialysis per subject is limited and careful planning of procedures is essential. Early detection, localisation and characterisation of lesions that might compromise haemodialysis is extremely important because they may allow correction before failure of the access site (1). Extending the life of an existing fistula or graft is of great benefit.

A percutaneously inserted central venous line provides immediate vascular access in patients with ESRF in urgent need of dialysis. However, it is now recognised that these lines are associated with a high incidence of subclavian and superior vena caval thrombosis, and the creation of an AV fistula later on in such cases will lead to venous engorgement, swelling, considerable pain and graft failure. The Cimino–Brescia radiocephalic fistula is the method of choice for the establishment of vascular access in patients with ESRF.

However, about 30% of AV fistulas fail early, within three months of surgery (2). The cause of early failure is often unclear, although the quality of the vessels is thought to play an important role.

Small-sized, stenosed or partially-thrombosed cephalic veins being repeatedly cannulated for infusion of fluids or drugs make a successful outcome less likely. Wong *et al.* (3) showed that sites of vein stenosis following cannulation can be detected prior to surgery, thus avoiding the use of such veins or tackling the lesion during surgery. Anastomotic neointimal hyperplasia is yet another important reason for early failure. Assessment of the vessels prior to access surgery consists usually of confirmation of the presence of a good radial pulse and evaluating the degree to which the cephalic vein dilates after application of a tourniquet. In patients who are obese or have deep-lying veins, this is difficult to accomplish. The deep veins of the upper limb tend to follow the arteries deep within the limb and are impossible to palpate, whereas the superficial veins lie within the superficial fascia forming a variable network. Demonstration of the veins varies considerably and is dependent on sex, age, ethnic origin and obesity. Sometimes, they are extremely difficult to palpate as they are normally thin walled, variable in position, concealed within the superficial fascia and have a low internal pressure. Gentle brushing of the skin surface may be all that is required to indicate the course of superficial veins.

With the advance of high-resolution ultrasound scanners, it is now possible to obtain non-invasive qualitative and quantitative data on vessel size and blood flow characteristics prior to and following surgery. Sonography is non-invasive and offers several advantages compared with those of venography. Examinations can be done rapidly and bilaterally, can be extended to include imaging of the jugular vein along with the subclavian veins and avoids all angiographic complications. Development of sonographic techniques for assessing the deep veins draining into the superior vena cava has lagged behind those for the deep venous system in the lower extremity;

this is partly due to the anatomical constraints imposed by the acoustic shadow of the clavicles and the sternum limiting the sonographic imaging capability (5).

6.2. Grey-Scale Image Characteristics of Normal Veins

The lumen of a normal vein is echo-free, and the wall is thin and smooth. Functioning valve leaflets can be seen. Thickening of the wall indicates pathology.

The vein is held open by the low pressure of the blood within the lumen, thus, even a small amount of external pressure will obliterate the lumen. This simple observation of compressibility is of great diagnostic importance. Vein compressibility is best assessed on cross-sectional image (Figs. 6.1 and 6.2). Acoustic medium pads (Gel block) need to be used in scanning the very superficial veins if high-resolution transducer is not available.

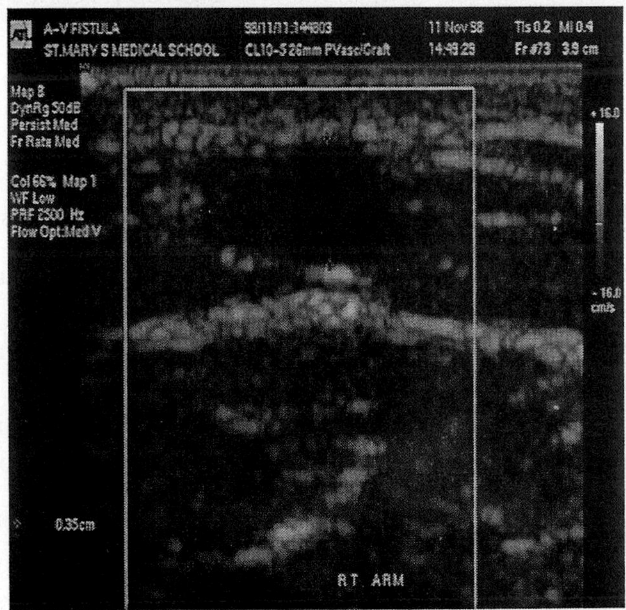

Fig. 6.1. Transverse section of the patent cephalic vein.

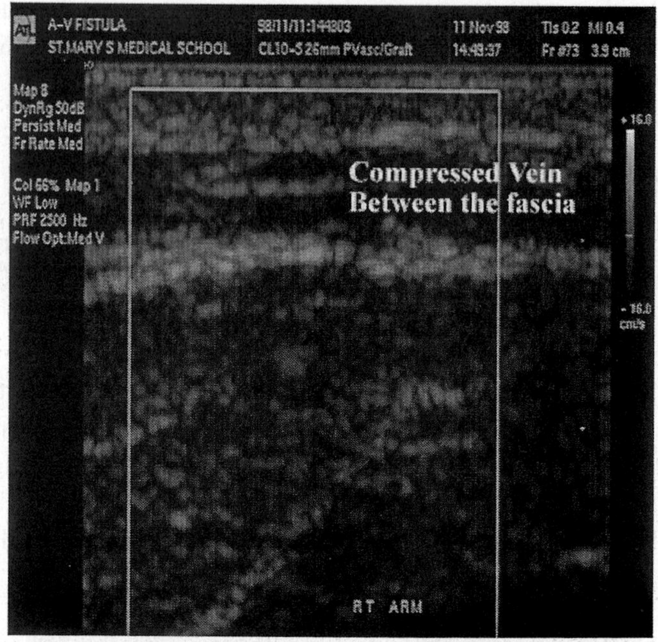

Fig. 6.2. Transverse section of the compressed cephalic vein.

6.3. Doppler Flow Characteristics of Normal Veins

Blood flow in normal veins has five important Doppler features:

- Spontaneous flow — absence of spontaneous flow is abnormal.
- Phasic flow — the flow changes with respiration.
  Absent phasic flow is abnormal. Under certain circumstances, an abnormal flow pattern can be described as a continuous flow.
- Cessation of flow by the Valsalva manoeuvre is normal. This important response demonstrates the patency of the veins proximal to the site of insonation.
- The flow is unidirectional, towards the heart
- Augmentation of flow by compression of the distal part of the extremity is an additional sign of normal venous flow. Delayed or loss of augmented flow indicates an obstruction distal to the site of examination.

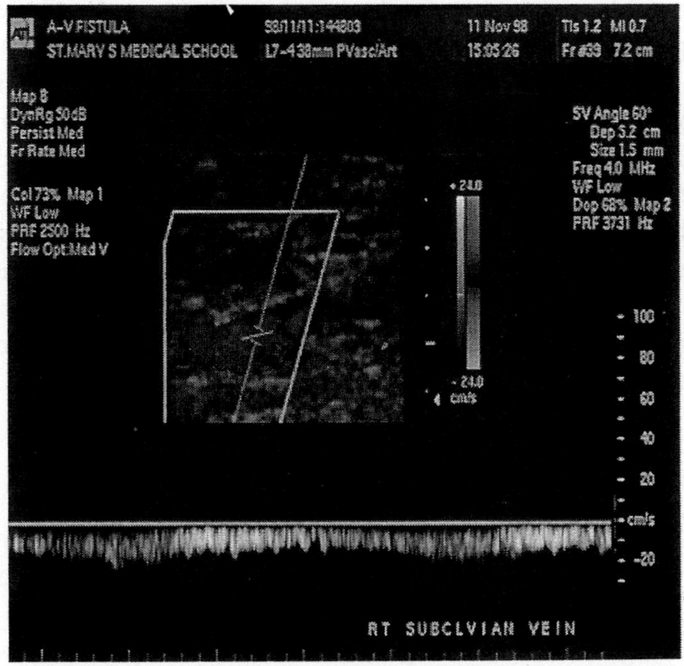

Fig. 6.3. Longitudinal section of the subclavian vein showing normal venous flow.

6.4. Use of Duplex Ultrasound and Colour Doppler with Regard to Vascular Access

Duplex scanning is the combination of both B-mode real-time imaging and pulsed Doppler. Real-time imaging allows assessment of the vascular anatomy in grey scale while the haemodynamics are evaluated by using pulsed Doppler and colour flow. The resolution of ultrasound equipment has greatly improved over the last decade, thus allowing a much more detailed visualisation of the vessels while colour Doppler helps to assess patency and stenoses more rapidly. Ultrasound is a highly reproducible, objective, non-invasive method of investigation, with no inconvenience to the patients either in terms of pain or ionising radiation. The sensitivity and specificity of the test are however operator dependent. Ultrasound gives precise information on vessel size, intravascular extension of a thrombus and perivascular tissue changes. These capabilities suggest that ultrasound should be the first imaging procedure to be employed in the study of the vascular anatomy for planned vascular access in haemodialysis patients. Angiography can then be done when ultrasound findings are uncertain or when a more detailed vascular map is needed for surgery. Measurement of flows and velocities through PTFE dialysis grafts is a valuable monitoring technique for assessing their integrity and for early detection of failing shunts and stenoses in asymptomatic patients with still-patent graft (5,6). Routine pre-operative and post-operative colour Doppler ultrasound scans are therefore recommended for every patient in whom a vascular access graft is planned or has been implanted.

6.5. Duplex Ultrasound in Arteriovenous Fistulas and Grafts

In normally functioning grafts and fistulas, waveforms of flow in the supplying arteries proximal to the take-off of the graft and throughout are monophasic, with peak systolic velocities of 100–400 cm/sec and end-diastolic velocities of 60–200 cm/sec. The draining veins have arterial pulsation waveforms with peak systolic velocities of 30–100 cm/sec. These veins have none of the previously mentioned venous flow characteristics. Arterial and venous stenoses, graft thrombosis, aneurysm and pseudoaneurysm formations, and arterial steal are abnormalities that can threaten or destroy graft function or jeopardise the patients' limbs and, in many cases, can be diagnosed sonographically (7). Although abnormal haemodynamics in access sites are usually detected by the flows during haemodialysis, sonographic evaluation at the time of an initial dysfunction may reveal an underlying correctable abnormality, and specific therapy may be instituted before further deterioration. Duplex ultrasound is a reliable method for assessing anatomical features and graft function. The best cut-off points for distinguishing a successful outcome with good functioning from failing grafts are cross-sectional areas of 8.5 mm^2 and a blood flow rate of 425 ml/min (8).

6.6. Indications for Ultrasound in AV Fistula Surgery

6.6.1. *Pre-operative scanning*

1. To check arterial inflow.
2. To assess venous anatomy and patency prior to AV fistula formation or graft insertion.
3. To measure recipient and donor vessel size pre-operatively.
4. To identify the presence of cephalic vein branches needing ligation to prevent diversion of flow from the fistula.
5. To study subclavian vein patency as stenosis there leads to early failure of the fistula.

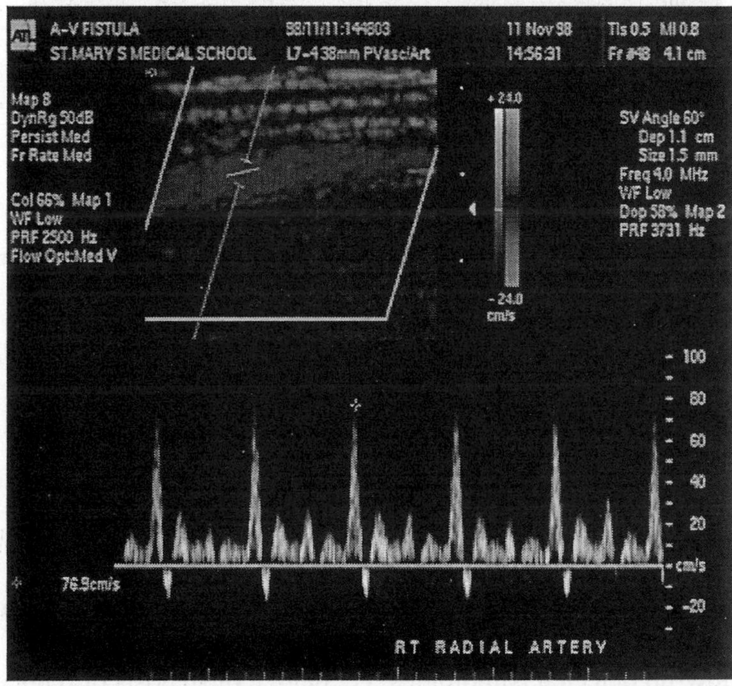

Fig. 6.4. Longitudinal section of the radial artery showing normal triphasic spectral waveform.

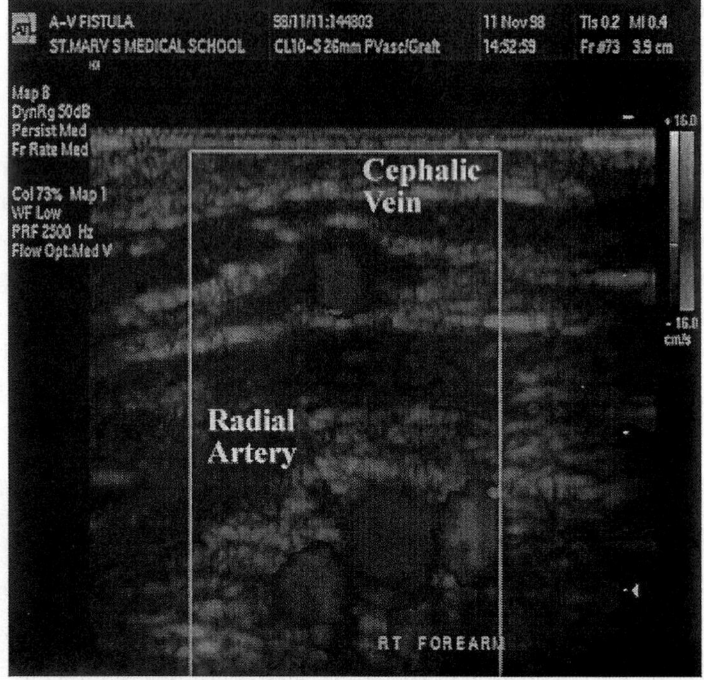

Fig. 6.5. Transverse section of the radial artery and the corresponding vein.

6.6.2. *Post-operative scanning*

1. To study flow velocities and volumes post-operatively.
2. To assess puncture site stenosis.
3. To detect anastomotic strictures.
4. To study subclavian vein patency as stenosis leads to thrombosis and early failure of the fistula.
5. To detect perivascular abnormalities such as haematoma and aneurysmal or pseudoaneurysmal dilatations.
6. To follow-up detected pathologies; i.e. degree of stenosis or reduced flow.

6.7. Anatomy of the Venous System Used in Vascular Access

6.7.1. *Anatomy of the venous system used in vascular access*

The jugular vein receives blood from the brain, face and the neck. It begins at the jugular foramen as a continuation of the sigmoid sinus after which it descends in the neck in the carotid sheath and unites with the subclavian vein behind the medial end of the clavicle. Immediately below the skull, the internal carotid artery and the lower four cranial nerves lie antero-medial to the vein which then runs downwards medial to the internal carotid and common carotid artery.

The sternomastoid muscle overlaps the vein above and covers it below. Just below the transverse process of the atlas, the vein is crossed by the accessory nerve at its middle part by the inferior roots of the ansa cervicalis

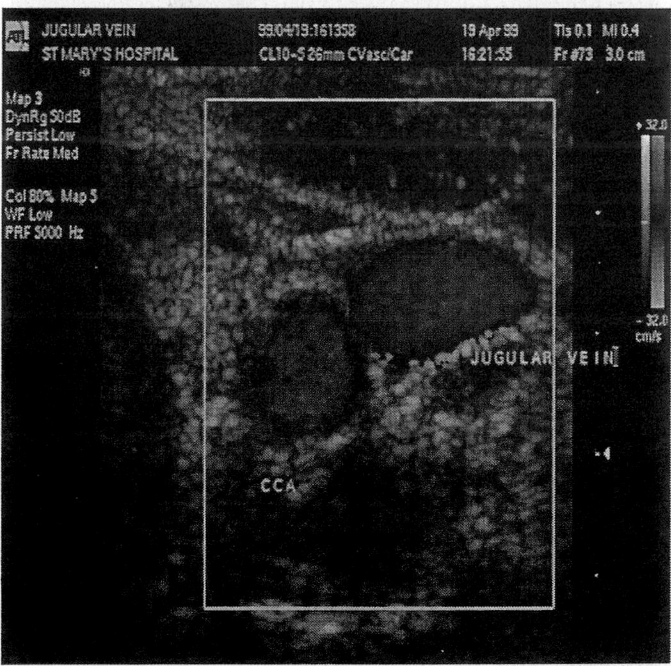

Fig. 6.6(A). Transverse view of the jugular vein and common carotid artery.

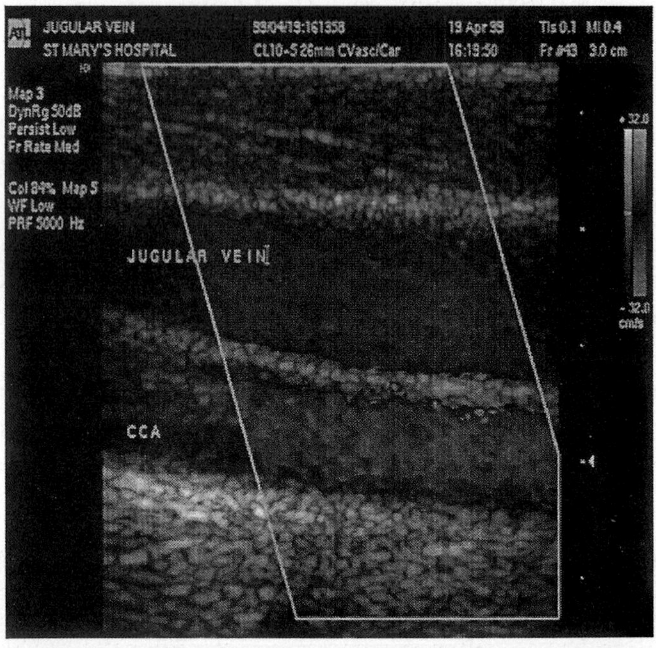

Fig. 6.6(B). Longitudinal view of the jugular vein and common carotid artery.

and near its lower end by the anterior jugular vein. For practical purposes, the surface anatomy of the internal jugular vein is of great importance. A line drawn from the medial end of the clavicle to the interval between the ramus of the mandible and mastoid process delineates the course of the vein. The external anatomical landmarks for cannulation of the internal jugular vein are numerous but are somewhat unreliable; therefore, an ultrasound survey of its anatomy is recommended for selecting the most appropriate puncture sites, thus reducing the risks and complications associated with percutaneous insertion of dialysis catheters [Figs. 6.6(A) and 6.6(B)].

The axillary vein lies medial to the axillary artery at the outer border of the first rib where it becomes the subclavian vein. The subclavian vein is deep to the medial end of the clavicle, both ventral and inferior to the subclavian artery. At its proximal end, the subclavian vein joins the internal jugular vein to form the brachiocephalic vein, which continues to the superior vena cava. Both the jugular and the subclavian veins are used for central venous access; however, the deep veins of the upper limb are hardly ever used in venous access surgery.

The radial and ulnar veins are paired vessels of the venae comitantes; each travels alongside its corresponding artery. They usually join at the elbow to form a single radial and ulnar vein. The radial and ulnar veins join above the cubital fossa to form the brachial vein which can be single or venae comitantes. At the lower border of the teres major muscle it receives the basilic vein to form the axillary vein. Very rarely, the brachial vein is used as a site for the venous anastomosis of dialysis grafts. Moreover, due to its deep location and small size, it is not used for the creation of AVF.

6.7.2. *The superficial veins of the upper limb*

In terms of haemodialysis access, the veins of interest are cephalic, basilic, median cubital and ante-brachial veins. The cephalic vein is formed at the anatomical snuff box. It then curves proximally from the radial end of the dorsal plexus round the radial side of the forearm to its ventral side. It then ascends in front of the elbow, superficial to a groove between the brachioradialis and biceps muscles. It crosses superficial to the lateral cutaneous nerve of the forearm and ascends lateral to the biceps to pierce the clavipectoral fascia, after which it crosses the axillary artery to join the axillary vein.

Anatomical variations in the course of the cephalic vein do occur and include the following:

- An accessory vein that joins the proximal to the distal end crossing the cubital fossa.
- Big median cubital veins transferring most of the blood from the cephalic to the basilic vein. In this case, the cephalic will be absent higher up.
- It may be connected to the external jugular vein by an anterior branch crossing the clavicle.

The basilic begins medially in the dorsal plexus of the hand. It ascends posteromedially in the forearm inclining forward to the anterior aspect distal to the elbow. It then ascends superficial to the biceps muscle to perforate the deep fascia midway in the arm. Higher up, it continues medial to the brachial artery up to the lower border of the teres major muscle. Finally, it joins the venae comitantes of the brachial artery to form the axillary vein. The median vein of the forearm drains the superficial palmar veins. Then, it ascends to the forearm to join the basilic or median cubital vein.

6.7.3. *Veins of the lower limbs*

Seventy-five percent of the venous return from the lower extremities is channelled through the deep system and only 25% through the superficial veins (the reverse is the case in the upper limbs). The deep system in the legs is therefore of greater importance than the superficial system. As in the hands, all deep veins in the lower extremities accompany their corresponding arteries. Communications exist between the deep and superficial systems throughout the limb through the perforating veins that, under normal circumstances, maintain flow in one direction only — from superficial to deep.

The superficial and deep femoral veins unite about 1–2 cm below the inguinal ligament to form the common femoral vein which drains the long saphenous vein and becomes the external iliac vein above the inguinal ligament.

In the groin, the femoral vein lies medial to the femoral artery at the base of the femoral triangle, both enclosed in the femoral sheath. As a landmark for cannulation purposes, one should find the femoral artery pulse at the mid-inguinal point.

6.8. Technique of Ultrasound Examination

6.8.1. *Equipment*

The examination is carried out using high-resolution colour duplex Doppler ultrasound with a high-frequency linear array transducer of the order of

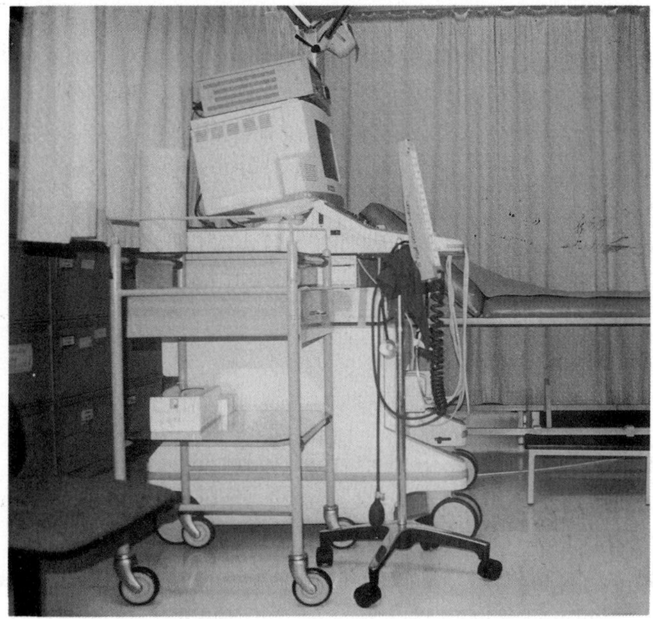

Fig. 6.7. ATL 3000 colour flow duplex scanning machine.

10 MHz for the superficial structures and 7 MHz for the deeper veins (Fig. 6.7). The anatomy under examination by ultrasound is checked in both transverse and longitudinal sections.

6.8.2. *Protocol in AV-fistula anatomy identification*

The examination is performed with the patient sitting on the edge of a couch with feet on the step or lying flat on the couch, depending on the planned access site. The operator sits on a stool facing the patient. The limb under examination is extended and placed in a relaxed position over a pillow (Figs. 6.8 and 6.9). In an upper limb study, coupling gel is applied from the upper arm to the wrist along the line of the cephalic vein which is checked using B-mode and colour flow Doppler. B-mode is used for measuring vein diameter (antero-posteriorly), and checking for wall thickness, thrombosis,

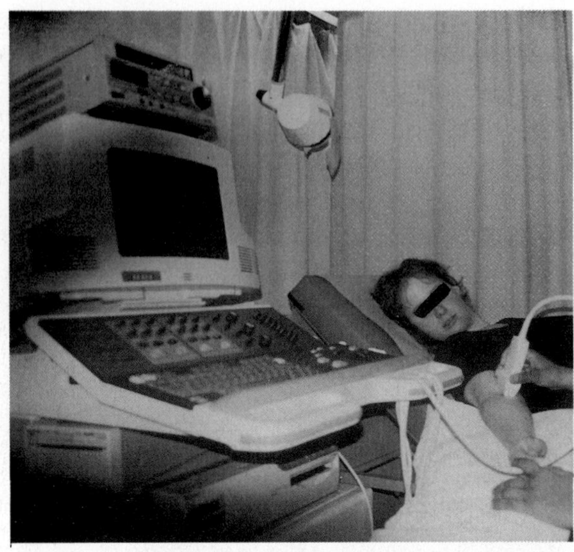

Fig. 6.8. Examination of the right elbow blood vessels using L7-4 MHz linear array transducer.

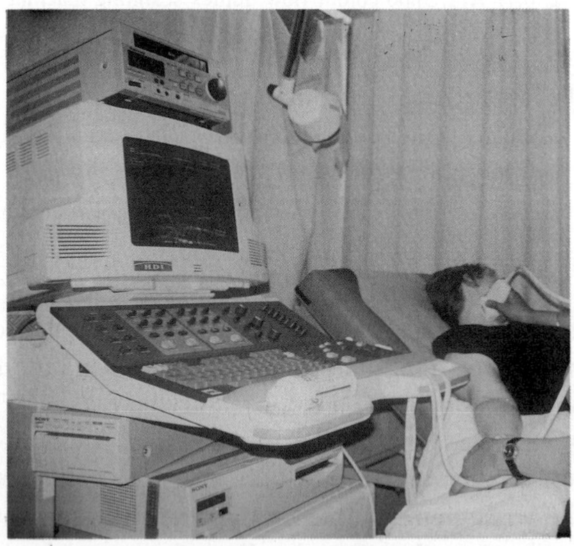

Fig. 6.9. Examination of the right internal jugular vein using L7-4 MHz linear array transducer.

branches, other intraluminal pathologies and compressibility of the vein. Colour flow and Doppler are used for assessing vessel patency and venous flow disturbances. Arterial flow (subclavian, brachial and radial) is evaluated using waveform analysis. In the case of a lower extremity study, the same examinations are done in the femoral artery and vein.

6.8.3. *Protocol in central venous access identification*

The routine examination of the veins in the upper limb to exclude thrombosis and outflow obstruction includes examination of the internal jugular, brachiocephalic, subclavian and axillary veins. Both sides are studied even when the disease is suspected to be unilateral. The positions of the sternum and clavicle make sonography of the superior vena cava, the brachiocephalic veins and proximal portion of the subclavian veins difficult. Colour Doppler imaging is used primarily to aid in rapid localisation of the vessels, to differentiate the artery from the adjacent vein, to determine the presence of thrombosis and to screen for focal narrowing. The patient is examined supine with no elevation of the head, which is rotated away from the side of examination. The internal jugular vein is located by scanning transversely. It is examined with compression from the cephalic to the caudal extremes in order to exclude occlusive or non-occlusive thrombosis. In the lower neck, it is difficult or even impossible to achieve easy closure under compression due to the bony attachment of the sternomastoid muscle. The vessels are then examined in a longitudinal orientation using colour Doppler. The brachiocephalic vein is examined whenever possible, as also is the subclavian vein. It is important to realise that these veins cannot be collapsed by external compression if bony anatomy is in the way. It is also important to recognise the normal anatomical relationship of the subclavian vein to the subclavian artery, so that a large venous collateral is not mistaken for the former. All the previously mentioned characteristics of normal venous flow should be looked for. In addition, a transmitted arterial type waveform should be detected in the subclavian vein. This sign is considered to be a reliable though indirect sign of patency.

Scanning is often used to direct a "blind" venous puncture for a central vein access. Ultrasound-assisted vascular puncture is considered to be a fast and safe alternative for puncturing these deep and pulseless vessels and for puncturing under difficult conditions. The ultrasound-guided cannulation of the internal jugular vein is associated with increased rates of success and a decreased complication rate, especially in high-risk patients.

Acknowledgement

The authors wish to thank Ms P Khodabakhsh from the Irvine Laboratory for Vascular Research, St Mary's Hospital, London for her contribution to this chapter.

References

1. Middleton, W.D. *et al.* (1989). Color Doppler sonography of haemodialysis vascular access. *AJR*, **152**, 633–639.
2. Palder, S.B. *et al.* (1985). Vascular access for haemodialysis; patency rates and results of revision. *Ann Surg*, **202**, 235–239.
3. Wong, V. *et al.* (1996). Factors associated with early failure of arteriovenous fistulae for haemodialysis access. *Eur J Vasc Endovasc Surg*, **12**, 207–213.
4. Longley, D.G. *et al.* (1993). Sonography of the upper and jugular veins. *AJR*, **160**, 957–962.
5. de-Preciout, V. *et al.* (1994). Comparison of different monitoring techniques for vascular access in chronic renal failures. *Nephrologie*, **15**, 87–90.
6. Pieterman, H. and Tordior, J.H. (1986). Non-invasive evaluation of prosthetic dialysis shunts in asymptomatic patients. *ROFO-Fortschr-Geb-Rontgenstr-Nuklearmed*, **145**, 541–546.
7. Finlay, D.E. *et al.* (1993). Duplex and colour Doppler sonography of haemodialysis arteriovenous fistulas and grafts. *Radiographics*, **13**, 983–989.
8. Lin, S.L. *et al.* (1997). Predicting the outcome of haemodialysis arteriovenous fistulas using duplex ultrasound. *J Formos Med Assoc*, **96**, 864–868.

CHAPTER 7

ANAESTHETIC MANAGEMENT

Philip Korsah MB, ChB, DA, FFARCSI
Clinical Shock Study Group
Western Infirmary
Glasgow, G11 6NT, UK

Nick Pace MB, ChB, FRCA, MRCP, MPhil
Department of Anaesthesia
Western Infirmary
Glasgow, G11 6NT, UK

7.1. General Principles of Anaesthesia for Dialysis Access

Anaesthesia for dialysis access is aimed at providing the patient with good analgesia and possibly sedation and amnesia, taking into account the patient's safety and comfort. The choice between a local/regional anaesthetic technique and a general anaesthetic has many influences and depends on considerations like patient's choice, anaesthetist's preference and the surgical procedure to be undertaken.

The insertion of a peritoneal dialysis catheter or central venous catheter usually requires a general anaesthetic while the formation of arteriovenous fistulae or other vascular access may allow the use of local/regional anaesthetic techniques.

The surgeon requires a relaxed, co-operative patient who can keep quite still for the duration of the procedure. There is also a distinct advantage to the surgeon and patient if the vessels to be used for fistula formation are dilated.

7.2. Chronic Renal Failure — Associated Diseases

The causes and effects of end stage renal failure are of great importance in anaesthesia but will not be discussed in great detail here.

Diabetes — the complications of diabetes include ischaemic heart disease, both autonomic and peripheral neuropathies, and frequent or chronic infections. These may present intra-operative problems for the anaesthetist in addition to the necessity for monitoring and controlling blood glucose.

Ischaemic heart disease presents as angina or myocardial infarction, with the severity of the problem being elucidated from past medical history.

Peripheral neuropathies tend to mainly affect the extremities, especially the lower leg. It is therefore a very important consideration when positioning patients on the operating table and, in particular, that precautions are taken to protect their heels.

The presence of autonomic neuropathies is more difficult to establish pre-operatively because they may not be obvious until the patient is stressed by an operative procedure. Postural hypotension may occur due to the loss of sympathetic tone to peripheral arterioles.

Vagal neuropathy may lead to a tachycardia at rest and loss of sinus arrhythmia. Cardiovascular reflexes like the Valsalva manoeuvre may be impaired and may cause arrhythmias if high airway pressures are required to ventilate the patient. Gastroparesis may occur due to vagal nerve damage to the stomach and may present an aspiration risk during general anaesthesia.

Hypertension is very common and may be the cause or result of chronic renal failure. There are two mechanisms resulting in hypertension in chronic renal failure:

a) The activation of the renin-angiotensin–aldosterone pathway, and
b) The retention of electrolytes (especially sodium and potassium) and water with impaired excretory function leading to an increase in blood volume and blood pressure.

The second of these two mechanisms becomes more important as renal function deteriorates.

Connective tissue disorders and some of the vasculitides may lead to chronic renal failure. The patient may be on long-term steroid therapy and thus will require intra- and post-operative steroid replacements.

Patients with SLE may have other problems including arthritis, pleurisy or pleural effusions (up to 66%), pneumonitis, atelectasis and eventually may develop a restrictive lung defect. They may also develop Raynaud's phenomenon, pericarditis, aortic valve lesions, myopathy or secondary Sjogren's syndrome.

Anaemia is common and a haemoglobin (Hb) concentration of 6–9 g/dl is not unusual. Anaemia may have been caused by losses during haemodialysis, due to uraemia or haemolysis. The most important cause of anaemia in chronic renal failure is failure of production due to decreased levels of erythropoetin.

This anaemia is usually well tolerated as an increased cardiac output and low systemic vascular resistance compensate for it. Furthermore, the Hb–oxygen dissociation curve shifts to the right, encouraging off-loading of oxygen. Blood loss in surgery for dialysis access is seldom of significance and it is rare for patients to require transfusion intra- or post-operatively.

Side effects associated with recombinant human erythropoetin include hypertension, usually occurring in the first four months, myalgia and influenza-like symptoms.

Bleeding disorders also occur. The main mechanism for this is an alteration in platelet–vessel wall interaction, as well as platelet–platelet interaction. The platelet count is usually normal. The best indicator of a bleeding tendency in uraemic patients is the bleeding time.

Hypovolaemia — the patients may already be on chronic dialysis (peritoneal or haemodialysis). Often, they will be dialysed immediately prior to theatre and can be markedly hypovolaemic. This may manifest itself, if uncorrected, as hypotension at induction of anaesthesia. It is important to be aware of the state of fluid balance and, if possible, arrange for a less aggressive dialysis regime pre-operatively.

7.3. The Effect of Renal Failure on Anaesthetic Drugs

Renal failure produces significant changes to normal physiology, which can affect and alter the effects of drug used in anaesthesia. In summary:

- The changes in the apparent volume of distribution of the drug depend on plasma protein binding, tissue binding and total body water. A decrease in serum protein concentration will lead to an increase in the bioavailability of drugs that are highly protein-bound. This is because the free drug (and therefore active) concentration is much higher. Thiopentone, etomidate and propofol are all anaesthetic induction agents that are highly protein-bound, and have to be titrated to the desired effect. A pre-determined dose calculated on the patient's weight will usually lead to myocardial depression and significant hypotension.
- Acidosis slows down the metabolic degradation of non-depolarising muscle relaxants. It may also result in a higher plasma concentration of non-ionised drugs, e.g. lignocaine, increasing the fraction of a given dose that is able to penetrate lipid membranes.
- Suxamethonium, a depolarising muscle relaxant, is the agent of choice in "rapid sequence inductions", and can cause a significant rise in serum potassium. A rapid sequence induction is used when patients are at high risk of aspiration, for example, with significant reflux due to a hiatus hernia or gastroparesis.
- Renal failure will have a significant effect on both renal and non-renal mechanisms of drug metabolism. Many lipid-soluble drugs undergo hepatic biotransformation to water-soluble metabolites, e.g. morphine, which is broken down to an even more potent active metabolite (morphine-6-glucuronide). This is a potent respiratory depressant and may be responsible for post-operative respiratory problems. The dose of morphine used in the patient-controlled analgesia (PCA) technique, for example, will therefore have to be decreased. The standard regime in our hospital for patients with renal failure is a 1-mg bolus with a 15-min lockout period. The patient will have to be well analgesed before PCA is started.
- Uraemia may also be associated with central nervous system depression, causing a reduction of up to 50% in the dose of drug required to achieve

adequate sedation. This should be taken into consideration when pre-medication is prescribed and when sedating patients for regional/local blocks.

7.4. Anaesthetic Management

7.4.1. *Pre-operative evaluation*

This includes an assessment of the cardiovascular, respiratory, neurological and renal systems, and any appropriate investigations that the clinical history and examination require. The anaesthetist should visit the patient pre-operatively to obtain a clinical history from the patient, to check the availability of appropriate results, to order further investigations if necessary and discuss the anaesthetic management with the patient. A sedative may be prescribed as a pre-medication agent, most commonly an oral short-acting benzodiazepine such as temazepam.

A full knowledge of the patient's current fluid balance, drug history, blood biochemistry and haematology results and an assessment of the patient's acid-base status are essential. An electrocardiograph and a chest X-ray must be obtained where indicated, e.g. hypertension, angina, and pulmonary oedema. There should be time to normalise blood biochemistry pre-operatively by dialysis.

All the patient's normal medications should be reviewed. The patient may be on a variety of anti-hypertensive drugs which may alter the patients vasomotor control under anaesthesia, e.g. angiotensinogen converting enzyme inhibitors (ACEIs), beta-blockers. It is usual for the patient's normal anti-hypertensive and anti-angina drugs to be administered on the morning of the operation, and the normal anti-hypertensive and anti-angina drugs should be prescribed as well. It is our practice, however, to omit these when the patient is undergoing creation of a fistula.

The patient should ideally be fasted for six hours, and if required, a prokinetic agent like metoclopramide and a H_2 antagonist or a proton pump inhibitor like omeprazole may be prescribed with the pre-medication. Uraemia and haemodialysis are known to delay gastric emptying.

The choice of an anaesthetic technique should be discussed thoroughly with the patient and surgeon, and if a local or regional technique were chosen, the possibility of changing to a general anaesthetic is discussed. With good selection and preparation, this should not happen very often.

The anaesthetic technique chosen will depend on the procedure to be done and the patient's clinical and psychological conditions, as well as the need to create the ideal conditions for successful surgery. The American Society of Anaesthetists (ASA) grading system (Table 7.1) is used to assess the risks of morbidity and mortality in relation to anaesthesia, and is assessed using the severity of the patient's disease and the difference between elective and emergency surgeries.

Table 7.1. The ASA physiological status scale.

ASA Rating	Definition of Grade	Mortality Rate %
I	A normally healthy patient	0.06–0.8
II	A patient with mild systemic disease	0.27–0.4
III	A patient with severe systemic disease that is not incapacitating.	1.8–4.3
IV	A patient with incapacitating systemic disease that is a constant threat to life.	7.8–23
V	A moribund patient who is not expected to survive 24 hours with or without an operation.	9.4–51

Add "E" as a suffix in emergency operations. The mortality rate is increased in emergency cases.

7.4.2. *General anaesthesia*

The choice of general anaesthetic is one that must be clearly considered. The pharmacokinetics and pharmacodynamics of drugs used to administer a general anaesthetic are very often affected by the altered physiology of the patient, as discussed above. A balanced general anaesthetic provides the patient with profound sedation, analgesia and if required muscle paralysis, while keeping the airway controlled and protected. The aim is to keep the patient well oxygenated with a heart rate and blood pressure within approximately 20% of baseline measurements.

Once the patient is in the anaesthetic room, routine monitoring should be instituted. This includes continuous ECG, pulse oximetry and non-invasive blood pressure unless invasive pressure monitoring is indicated for cardiovascular instability or arterial blood sampling. Intravenous access should be established with a large bore cannula, avoiding any limb where there is an arteriovenous fistula or any vein which could potentially be used for formation of an arteriovenous shunt in the future, such as the forearm veins. The veins on the dorsum of the hand are the best choice. Unfortunately, patients with chronic renal failure often have poor peripheral veins, which may decrease the choice considerably.

The patient should be pre-oxygenated and then induction commenced. Induction of anaesthesia is usually achieved with an induction agent such as propofol, thiopentone or etomidate titrated to effect. If there is any possibility of regurgitation, a rapid sequence induction is used, though it should be noted that suxamethonium may cause a significant increase in plasma potassium. If this is a problem, it is possible to perform a rapid intubation using a moderate dose of the synthetic opiate alfentanil in combination with propofol. Alfentanil is given initially in a dose of 1–2 mg intravenously, followed by a bolus of propofol in normal induction doses of 2–2.5 mg/kg. This technique produces reasonable intubating conditions though the patient invariably coughs on intubation. A non-depolarising muscle relaxant such as atracurium can then be given.

If the patient has been adequately fasted, he/she may be induced with an induction agent like propofol and, once asleep, a muscle relaxant is then administered. Atracurium is commonly used as it is predominantly metabolised by Hoffman degradation and therefore does not accumulate in renal failure. The other muscle relaxant sometimes used is vecuronium, though about 20% of its elimination is by the renal route. It has weakly active metabolites though accumulation and "re-curarisation" very rarely occurs. Intubation is then performed and anaesthesia maintained with a volatile inhalational agent or an infusion of propofol. Isoflurane is the volatile agent of choice as only 0.2% is metabolised to potentially nephrotoxic free fluoride ions.

The insertion of a peritoneal dialysis catheter often requires a general anaesthetic, even though it can quite safely and efficiently be done under local anaesthetic. In our opinion, however, it is quite stressful to the patient. Thus, in our practice, patients who have significant pre-existing illness (e.g. recent myocardial infarction) would normally be given a general anaesthetic as we believe it to be less stressful to the patient than a local technique.

The need for paralysing the patient once anaesthetised (and therefore mechanically ventilating them) will depend on the degree of surgical technical difficulty and the patient's cardiovascular and respiratory status. Most patients, and certainly obese patients, with a body mass index (BMI) over 30, will usually require intubation after induction, and mechanical ventilation. We have however paralysed and ventilated selected patients using a laryngeal mask airway (LMA) once they are anaesthetised, for insertion of a Tenckhoff catheter, without incident. It would be prudent to be prepared to change this to an oral endotracheal tube at any time during the procedure.

Local anaesthetic infiltration of the insertion site while under general anaesthesia provides good post-operative analgesia.

The anaesthetic agents described may all cause a degree of hypotension and the effect of a drop in mean arterial pressure on an arteriovenous shunt should be monitored. Using vasoconstrictor or inotropic drugs to maintain a good mean blood pressure, as measured using a non-invasive or even invasive methods, does not necessarily provide good flow to the arteriovenous fistula formed and may even be counter-productive. In this situation, inotropic drugs such as dopamine are a better choice.

The insertion of temporary central venous dialysis lines by the anaesthetist should not be done if the continuous monitoring of the patient will be compromised. It should be noted that central dialysis lines recently capped off after haemodialysis would have been "heplocked" with about 2 ml of 1000 iu/ml heparin. If these lines are used at induction, this heparin must be aspirated before anything is injected into the lines. If used safely, there should be no problems with heparinisation or infection. The temporary dialysis line must be heplocked at the end of the procedure.

The insertion of permanent central venous catheters or haemocath creates a potential problem for the anaesthetist as the surgeon is working very close to the airway and, furthermore, it may not be accessible under surgical drapes. It is therefore our recommendation that patients undergoing haemocath insertion are intubated and ventilated.

Analgesia is a potential problem in renal patients. After the insertion of peritoneal dialysis catheters, some local anaesthetic should be infiltrated around the wound. Intravenous opioids may be required intra- and post-operatively, though simple analgesics are often adequate for post-operative analgesia for these procedures. Most opioids are metabolised in the liver and excreted by the kidney. Morphine is metabolised to morphine-6-glucuronide, an active metabolite that is renally excreted, so once good analgesia is achieved, a decreased frequency and dose of analgesic is enough to maintain good analgesia.

7.4.3. *Regional/local anaesthesia*

This is most suited to arteriovenous shunt formation in the upper limb. It may also be appropriate for percutaneous insertion of a central veno-venous dialysis line. Typical regional blocks are axillary, supraclavicular and interscalene brachial plexus blocks. These are described in more detail later.

It is also possible to block the individual nerves supplying the area to be operated on. This would particularly apply to the forearm.

Regional blockade will also block the sympathetic innervation to the upper arm and therefore cause vasodilatation of the blood vessels in the arm. This makes the vessels easier to operate on and improves flow in the shunt. The patient will however not be able to protect his/her arm (due to loss of sensation) in the immediate post-operative period and arrangements must be made to keep the arm protected.

Regional/local blockade is desirable even if a general anaesthetic is administered because it provides the patient with good post-operative analgesia, thereby reducing their requirements for opiates and other simpler analgesics.

The pre-operative preparation of a patient for a procedure under local anaesthetic should be just as thorough as it would be for a general anaesthetic. Monitoring intra-operatively should also be of the same standard as for general anaesthesia, including continuous ECG monitoring, pulse oximetry and regular non-invasive blood pressure measurement. There should also be the capacity to supply oxygen by face mask, nasal cannulae or some other simple device. The procedure should not start without the availability of resuscitation equipment and a doctor who can use them (apart from the surgeon operating because he/she may not be in a position to do so).

7.4.4. *Brachial plexus blocks*

There are various techniques described for the injection of local anaesthetic around the brachial plexus. These approaches are (from proximal to distal): interscalene approach, subclavian perivascular approach, supraclavicular approach and axillary approach.

The supraclavicular and subclavian perivascular approaches produce the most complete limb blockade. The interscalene block provides a good block of the shoulder, but often misses the ulnar nerve distribution in the arm. The axillary nerve block is easier to perform but may miss out the musculocutaneous nerve, though this is easily blocked separately. The axillary nerve block is probably the easiest to perform and is favoured by the occasional user.

The correct placement of the regional block needle is confirmed by either eliciting paraesthesiae in the distribution of the nerves to be blocked or by using a peripheral nerve stimulator to stimulate the motor nerves to the area required to be blocked.

The peripheral nerve stimulator (PNS) should only be used once the operator has anatomically located the nerve to be blocked. The aim is to get the needle as close to the nerve without touching or penetrating it. The peripheral nerve stimulator provides a low electrical current to the tip of the regional block needle. The lower current means that the needle will have to be very close to the nerve to stimulate it and cause contraction of the muscles it supplies.

The anode (+ve) of the peripheral nerve stimulator is connected to an electrode on the patient's body, away from the site of needle insertion. The current to be delivered is set to about 2.5–3 mA and the cathode (–ve) attached to the hub of the needle. Having identified the relevant anatomical landmarks, the regional/local anaesthetic block is then undertaken as follows. As the intended target is approached, the peripheral nerve stimulator is switched on. The nerve will have to be approximately 1 cm from the tip of the needle for the current at this level to have any effect. The current delivered is then reduced as the needle gets closer to the nerve, usually to between 0.1–0.5 mA. The needle is then immobilised at this point and injection of the local anaesthetic is started. There should not be any resistance or pain on injection. Otherwise, this may suggest an intraneural or perineural injection causing direct nerve damage or increased pressure on the nerve due to a large volume in a confined space. The muscular twitch should disappear on injection of the local anaesthetic.

The onset of analgesia and a successful block should be monitored and be effective after about 30 minutes, depending on the agents used for performing the block. The extent of neurological blockade can be tested using a 21G needle to gently pinprick the skin to map out the blocked dermatomes but, more importantly, to make sure that there is no sensation in the area to be operated on.

The local anaesthetics used most commonly are bupivacaine, prilocaine, lignocaine, and ropivacaine (Table 7.2). They are available in various

Table 7.2. The pharmacological properties of the commonly used local anaesthetics.

Local Anaesthetic	Speed of Block Onset	Duration	Max. safe dose (Single Injection)	Relative Potency
Lignocaine 2%	Rapid (10–15 mins)	1.5–3 hours	300 mg (500mg with adrenaline)	1
Prilocaine 2%	Rapid (10–15 mins)	1.5–3 hours	400 mg (600 mg with adrenaline)	1
Bupivacaine 0.5%	Slow (30–45 mins)	4–12 hours	150 mg (with/without adrenaline)	4
Ropivacaine 2%	Slow (30–45 mins)	4–8 hours	150–200 mg (with/without adrenaline)	4

Based on a 75 kg patient.

concentrations, and with or without added adrenaline. Adrenaline may be added to decrease absorption of the local anaesthetic into the general circulation due to vasoconstriction, thereby decreasing the peak serum local anaesthetic concentration. In the shorter acting local anaesthetics, it may also prolong the duration of the block. In the longer acting local anaesthetics however, the duration of block is not affected. If adrenaline is added to the local anaesthetic, it should not be used to block peripheral areas like the digits or any area supplied by end arteries. The concentration of adrenaline should not be more than a 1 in 200 000 solution.

Perivascular supraclavicular brachial plexus block

- Position the patient supine with a pillow under their head and neck, turn the head away from the side to be blocked and push the shoulder down to depress the clavicle (Fig. 7.1).
- Draw a line laterally from the cricoid cartilage to the sternomastoid muscle, crossing it at its midpoint. The muscle may be accentuated by asking the patient to lift his/her head off the pillow slightly while turned to the side.
- The interscalene groove should be identified behind the posterior border of sternomastoid, at its midpoint, and followed distally.
- The pulsation of the subclavian artery can be palpated about 1 cm above the midpoint of the clavicle, in the interscalene groove (Fig. 7.2).
- Inject 2–3 mls of 2% lignocaine into the skin just above the pulsation and then introduce a 22 G short-bevelled 3.5 cm regional block needle caudally in a horizontal plane, parallel to the neck. It should enter the fascial sheath about 1–2 cm deep into the skin. A click may be felt with penetration of the fascial sheath.
- The correct placement of the tip of the needle within the fascial sheath surrounding the brachial plexus should be confirmed either with paraesthesiae or a peripheral nerve stimulator. Hold the needle in position and aspirate carefully. If blood is aspirated, pull the needle back gently until it stops. The tip should still be within the sheath.
- Inject the local anaesthetic into the sheath. A volume of 25–40 ml of 0.375% bupivacaine (Table 7.3), depending on body weight, can be injected

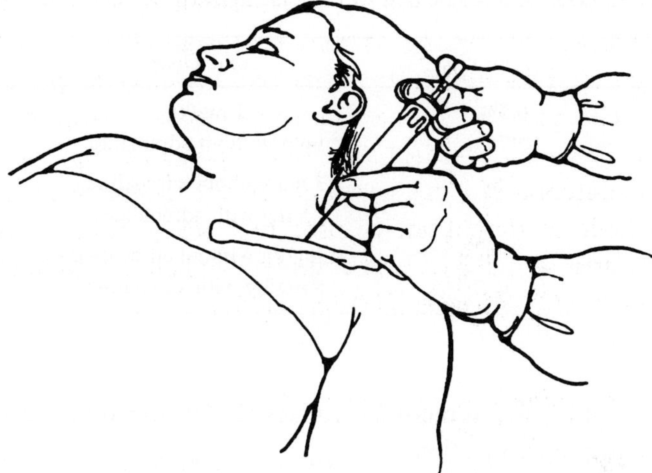

Fig. 7.1. Perivascular supraclavicular brachial plexus block: patient's position and needle direction.

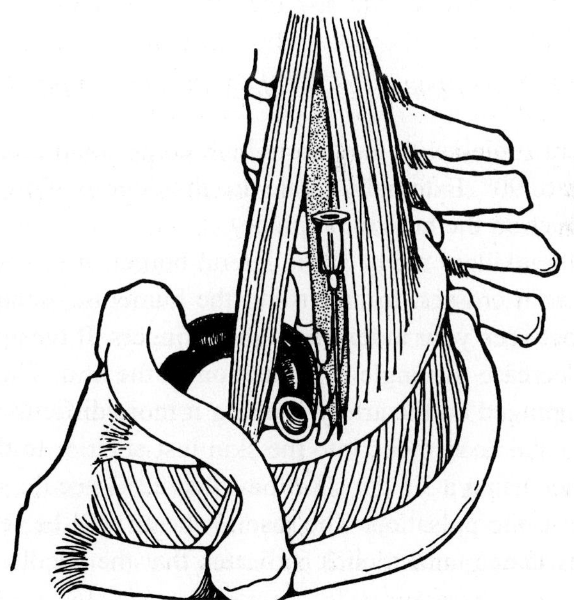

Fig. 7.2. The subclavian artery in the interscalene groove.

Table 7.3. Local anaesthetic doses for supraclavicular brachial plexus blocks.

Agent	Concentration	Toxic Dose	Volume
Bupivacaine	0.5% 0.375%	2 mg/kg with/without adrenaline	30 ml max. 40 ml max.
Prilocaine	2%	3 mg without adrenaline 5 mg with adrenaline	30 ml max.
Lignocaine	1%	4 mg/kg without adrenaline 5 mg/kg with adrenaline	40 ml max.

into the sheath. This sometimes causes distension quite distinguishable from subcutaneous injection.

- Proximal or distal spread may be encouraged by applying digital pressure below or above the injection site.
- When the patient is well analgesed and comfortable, the surgery may begin.

Axillary approach — brachial plexus block (Fig. 7.3, Table 7.3)

- Lie the patient supine with the arm to be blocked abducted to 90° and the elbow flexed to 90°. Externally rotate the shoulder to allow the forearm to lie on the couch in the horizontal plane.
- Palpate for the axillary artery at the lateral border of the pectoralis major muscle, where it crosses, to insert into the humerus. Identify its position by fixing it between your index and middle fingers. If the artery is difficult to palpate, decrease the angle of abduction of the arm. The humeral head may have impinged on the artery, making it more difficult to feel.
- Insert some local anaesthetic into the skin just anterior to the pulsation of the artery, then using a 3.5 cm short bevelled 22 G needle, insert it aiming just anterior to the pulsation. The fascial sheath will be felt as a slightly increased resistance until a click indicates that the needle has penetrated the sheath.

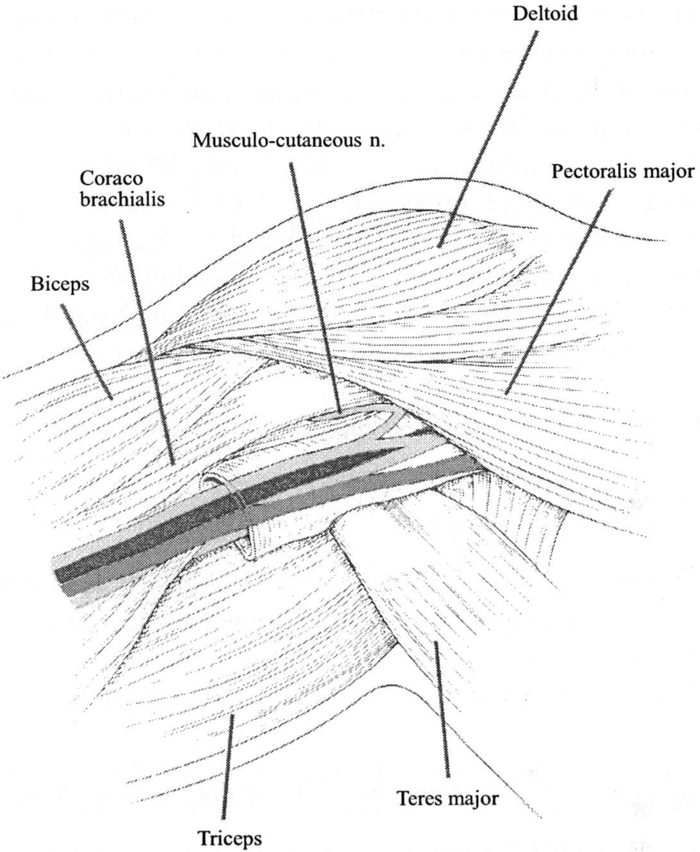

Fig. 7.3. Brachial plexus block: axillary approach.

- Correct placement of the needle can be confirmed with the appearance of paraesthesiae in the distribution of the block intended and by using a peripheral nerve stimulator to cause twitching of muscles in the same distribution.
- There should be no blood on gentle aspiration. The local anaesthetic should then be injected with firm digital pressure on the axillary sheath just distal to the injection site, to encourage more proximal flow of local anaesthetic.

- If blood is aspirated, it indicates that the needle is in the sheath within the axillary artery. The needle should be withdrawn slowly until it has just come out of the artery and into the sheath. Confirm that there is no more blood on aspiration before injecting the local anaesthetic.
- The musculo-cutaneous nerve is often missed with this approach because it branches off more proximally. It can be blocked separately by withdrawing the needle to just skin-deep, and re-directing it more anteriorly into the coraco-brachialis muscle and injecting about 5 ml of local anaesthetic into it. This should block the sensation from the distribution of the lateral cutaneous nerve of the forearm. This nerve can also be blocked at the elbow.
- Again, test the arm for sensation and movement before surgery commences.

7.4.5. *Peripheral nerve blocks*

The ulnar, median and radial nerves can be blocked individually at the elbow to provide muscle paralysis of the forearm and loss of sensation in the hand. The medial, lateral and posterior cutaneous nerves of the forearm must be blocked to provide loss of sensation to the forearm.

Median nerve and medial cutaneous nerve of the forearm blocks (Fig. 7.4)

- Abduct and extend the arm to be blocked, with the patient in a supine position.
- Identify the groove between the biceps muscle and its tendon, and the head of the pronator teres muscle on the anteromedial aspect of the forearm, just proximal to the crease of the antecubital fossa. Identify the brachial artery.
- Raise a weal of local anaesthetic in the skin just medial to the brachial artery, approximately 1–2 cm proximal to the antecubital fossa crease. Insert a short-bevelled needle at an angle of approximately 45° to the skin and gently advance until a "click" is felt as the tip of the needle penetrates the deep fascia, beneath which the median nerve lies.

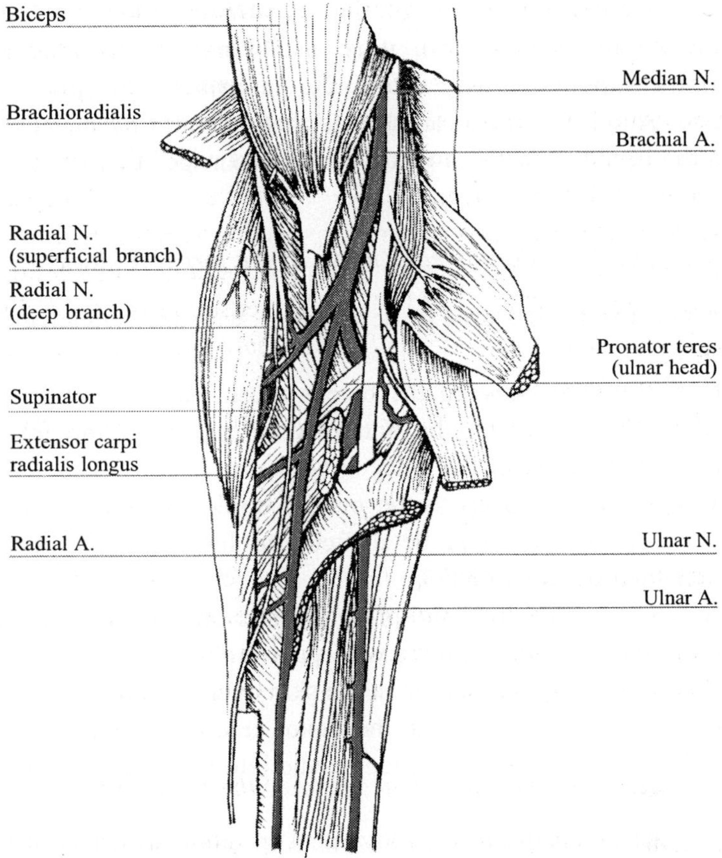

Fig. 7.4. Peripheral nerve block: relevant anatomy at the elbow.

- Paraesthesiae in the distribution of the median nerve indicate close proximity to the nerve. The peripheral nerve stimulator may also be used, though if enough local anaesthetic is placed in the right fascial compartment, the block should work.
- When the nerve is located, hold the needle still and inject 5 ml of the local anaesthetic slowly. Injection should be quite easy, with no pain or resistance on injection.

- The medial cutaneous nerve of the forearm can then be blocked by withdrawing the needle until it is just skin-deep, then advancing it proximally subcutaneously along the intermuscular groove already identified. Another 5 ml of local anaesthetic should be injected into the area as the needle is withdrawn to leave a "sausage" of local anaesthetic.

Radial nerve and lateral cutaneous nerve of the forearm block

- Abduct and extend the arm to be blocked, with the patient in a supine position.
- Identify the groove between the biceps muscle and its tendon, and the brachioradialis muscle on the anterolateral aspect of the forearm just proximal to the crease of the antecubital fossa.
- The radial nerve lies deep to the medial border of brachioradialis as it crosses the crease and can be difficult to locate. A peripheral nerve stimulator may be of some help.
- Identify the lateral condyle and place a finger on it to act as a guide, and palpate the intermuscular groove with your thumb.
- Raise a weal of local anaesthetic in the skin over the intermuscular groove approximately 2 cm proximal to the flexor crease at the anticubital fossa. Insert a 3.5 cm 22 G short-bevelled needle into the groove, aiming for the lateral condyle.
- Expect to get paraesthesiae or movement in the distribution of the nerve with peripheral nerve stimulation before making contact with bone.
- Inject the local anaesthetic (Table 7.4) into this area, always aspirating carefully first.

Table 7.4. Local anaesthetic doses for forearm nerve blocks.

Agent	Concentration	Volume	Duration
Bupivacaine	0.375%	7–10 ml	4–12 hours
Prilocaine	1%	7–10 ml	4 hours

- Withdraw the needle until it is just skin-deep, then advance the needle along the groove adjacent to the biceps tendon.
- Inject another 5 ml of local anaesthetic into this area as you withdraw the needle to block the lateral cutaneous nerve of the forearm.

Ulnar nerve block (Fig. 7.5)

- With the patient lying supine, abduct the arm and flex the elbow to about 90°.

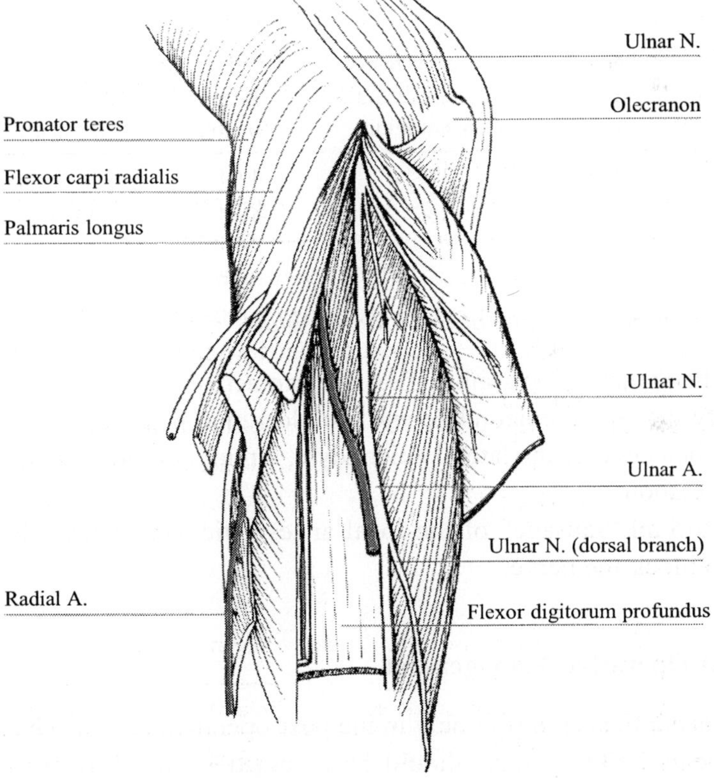

Fig. 7.5. Ulnar nerve block.

- Identify the medial epicondyle of the humerus. The ulnar nerve passes in a groove postero-lateral to the epicondyle where it can be palpated quite superficially. The nerve is in an enclosed space at this point and injection of a local anaesthetic here may cause nerve damage by increasing the pressure in the canal.
- Raise a skin weal of local anaesthetic approximately 2 cm proximal to the epicondyle to avoid the canal.
- Insert a 22 G short-bevelled needle into the skin. The tip should only be less than 1 cm deep.
- Inject the local anaesthetic (Table 7.5) slowly. There should be no pain or paraesthesiae on injection.

Table 7.5. Local anaesthetic doses for ulnar nerve block.

Agent	Concentration	Volume	Duration
Bupivacaine	0.375%	4 ml	4–12 hours
Prilocaine	1%	4 ml	4 hours

Posterior cutaneous nerve of the forearm block

- With the patient lying supine, flex the arm across the chest to about 90°.
- Identify the lateral epicondyle and the olecranon process.
- Insert the needle at the lateral epicondyle, aiming inferio-medially towards the olecranon.
- Inject a 5 ml "sausage" of the local anaesthetic subcutaneously into this area to block the nerve.

7.5. Post-Operative Management

Oxygen is administered routinely in the post-operative period. The analgesic requirements of the patient should be assessed in the recovery room and further analgesia given to make sure that the patient is comfortable before returning to the ward.

The care of the newly formed arteriovenous fistula must be meticulous. It is important to keep it well protected, especially on an arm that may still have no sensation or movement for a few hours, having been neurologically blocked.

Once the patient is stable, he/she is returned to a specialist renal ward. The patient's urea, creatinine and serum electrolytes, especially potassium, should be checked post-operatively. The patients should be reviewed post-operatively on the ward by the anaesthetist and surgeon who were involved in the operative procedure.

7.6. Conclusion

The patient with renal failure presenting for surgery for dialysis access will usually have a variety of problems that affect most systems. These patients therefore require meticulous pre-operative assessment. The variety of causes of renal failure as well as the effect of renal failure on the patients' physiology and pharmacology can make their anaesthetic management quite a challenge, though not insurmountable.

Acknowledgement

The authors would like to thank Mr. M. Mechan, Medical Illustration Dept., Western Infirmary, Glasgow.

Further Reading

Wildsmith, J.A.W. and Armitage, E.N. (1987). *Principles and Practice of Regional Anaesthesia*. Churchill Livingstone: Edinburgh.

CHAPTER 8

ARTERIOVENOUS FISTULAS

Nadey S Hakim MD, PhD, FRCS, FACS, FICS
St Mary's Hospital, London W2 1NY, UK

Endogenous arteriovenous fistula (AVF) first described in 1966 (1), remains the optimal vascular access for chronic dialysis. An internal AVF avoids the disadvantages of an external appliance, such as frequent septic and clotting problems. In spite of surgical innovations, such as autogenous vein interpositional grafts, bovine heterografts, expanded polytetrafluoroethylene (ePTFE) grafts and permanent indwelling silastic central venous catheters, the Brescia–Cimino fistula has remained the best access for maintenance haemodialysis because of its low incidence of complications and high long-term patency rate. The one- and three-year cumulative patency rates are 85–90% and 60–85%, respectively.

The major limitation to the use of an internal AVF is the absence of suitable vessels. Although construction of a distal fistula in a child weighing less than 10 kg is feasible, the results are often unsatisfactory (2). It also requires a mean maturation period of four months. The other disadvantage of an internal AVF is the need for repeated needle punctures, which may not be tolerated by some patients, and may also result in stenosis from repeated punctures around the same site.

8.1. General Technique Considerations

Pre-operative assessment, investigations and the choice of access type are covered in other chapters and will not be duplicated here. Haemodynamic studies show that flow through the fistula reaches its peak when the size of the fistula is greater than 75% of the diameter of the artery (3). Further increase in fistula flow is dependent on enlargement of the proximal artery, collateral arteries and venous outflow tracts with diminished venous resistance. Since thrombosis is still the most common cause of early fistula failure, a large fistula with a 1–1.2 cm opening is preferred. A smaller opening, 5–7 mm, is used in proximal internal fistulas where excessive shunting of blood may subsequently result from progressive enlargement of the vessels, causing the classical steal syndrome.

End-to-side anastomosis is preferred as it produces a higher fistula flow compared to the side-to-side configuration. For construction of a distal internal fistula in a child weighing less than 20 kg, the anastomosis is performed with 8/0 or 9/0 prolene sutures. Whatever technique is used, the vessels must be handled gently to minimise damage and prevent vasospasm.

The distal radial-cephalic anastomosis just above the wrist, is still the best site for an internal AVF. This provides a relatively long and straight cephalic vein for catheter insertion. It also leaves more proximal sites for future use should the radial-cephalic fistula fail. An AVF within the anatomic snuffbox has a high incidence of early failure and requires a longer maturation time (4). The proximal elbow fistula predisposes to ischaemic complications and can lead to congestive heart failure as a result of increasing flow through a chronic fistula that is made too large.

8.2. Technique for Radial-Cephalic Fistula

This procedure can be performed under local anaesthetic (without adrenaline) on a day-case basis. The patient lies in the supine position with his arm outstretched on an arm board. After aseptic preparation, the operative field is draped. The forearm is maintained in a neutral position. The radial artery pulsation is palpated, a tourniquet applied and the course of the cephalic vein outlined. After local infiltration with 1% lignocaine or 0.5% bupivacaine, a curved transverse incision is made along a line crossing the radial artery and the cephalic vein.

The cephalic vein is mobilised first to assess its calibre and dissected free from the deep subcutaneous tissues. A vessel loop is placed around the vessel and mobilisation is completed with sharp dissection and minimal manipulation to avoid vasospasm.

The wrist is then supinated to allow dissection of the radial artery. The radial artery, flanked by its two venae comitantes, is exposed when the fibrous flexor retinaculum that overlies the artery separates it easily from the venae comitantes, although an occasional branch crosses over the artery and mobilisation of the vessel is completed with ligation and division of its fine side branches.

Failure to divide these small branches often results in meddlesome bleeding. The cutaneous branch of the radial nerve to the thenar area should be preserved.

Following exposure of the artery, the vein is divided distally and the proximal end is flushed with heparinised saline. The anastomosis is best performed under magnification. Loops provide a magnification of 2.5×. An arteriotomy is made by a No. 11 blade and extended with a small straight scissors to a length of 1.0 cm. Two anchoring sutures (6/0) are used to retract the lateral walls of the artery. A polyethylene tube connected to a syringe with heparinised saline is passed proximally into each vessel to check for proximal obstruction. Both the proximal and distal vessels are irrigated with heparinised saline.

The anastomosis is started with a double-armed 6/0 or 7/0 prolene suture. The needles are passed from the inside to the outside of the vessels at a point 1 mm beyond the distal end of the arteriotomy or venotomy. A similar suture is placed at the proximal end. Both sutures are tied with the knots outside the vessels. The anastomosis should be started from the proximal end because this is functionally the more important end and a site where technical error cannot be tolerated. Continuous full-thickness sutures are placed 1 mm apart beginning 1 mm from the original knot. After the first two stitches have been completed, it is prudent to check the proximal anastomosis with a probe to make sure that the lumens have not been compromised.

Prior to completing the anastomosis, each vessel is irrigated through the tubing with heparinised saline. This manoeuvre helps to ensure patency of the vessel and to flush out any clots formed during the period of vascular occlusion. The vessel loops are released after tying the last stitch. A little bleeding may occur through the suture holes but this usually stops after a few minutes of gentle pressure.

The subcutaneous layer is approximated with interrupted 3/0 Vicryl sutures and the skin is closed with a running subcuticular 4/0 Vicryl. A good thrill should be felt over the fistula and over the vein above the level of the fistula.

8.3. Alternative Sites for AVF

8.3.1. *Anatomic snuffbox fistula*

Interest in the anatomic snuffbox fistula has recently been revived (4). The distal cephalic vein must be ligated to prevent venous hypertension. This site is rarely used now because the fistula is in a conspicuous location and the early patency rate is poor.

8.3.2. *High radial-cephalic forearm fistula*

A higher or mid-forearm radial-cephalic fistula is created if the distal vessels are unsuitable, or after failure of a distal radial-cephalic fistula. Once again, pre-operative assessment by clinical examination and Duplex ultrasound must be performed to ensure the presence of adequate vessels and to avoid unnecessary surgical exploration.

The anastomosis may be fashioned in a side-to-side manner. When performed in this fashion, the distal venous limb should be ligated to alleviate the development of venous hypertension in the hand (5).

8.3.3. *Elbow arteriovenous fistula*

When inadequate vessels or previous failure of AVFs make it impossible to create a forearm fistula, rather than proceeding with the use of synthetic graft as was the custom in the past, it is now common practice to perform an elbow AVF.

Various configurations have been described including: (i) anastomosis between the brachial artery and forearm vein; (ii) anastomosis between the brachial artery and the elbow perforating vein; (iii) anastomosis between the brachial artery and the cephalic vein; and (iv) anastomosis between the brachial artery and a superficial lysed basilic vein (6).

The brachial artery to cephalic vein configuration is the most commonly performed and certainly the easier of the options available at the elbow. Both short-term and long-term patency rates are comparable to the radial-cephalic fistula but the high flow rates associated with the elbow AVF predisposes to distal ischaemic complications. A further concern of elbow fistula is the possible higher incidence of cardiac failure and steal syndrome. The abduction of the anterior side anastomotic configuration has led to a reduction in this problem (6). Techniques used to minimise these complications include a smaller arteriotomy (5–7 mm in length), ligation of the distal vein or anastomosis of the brachial artery to the end of a perforating branch of the cephalic or median antecubital vein as described by Gracz *et al.* (7).

8.4. Transposed Basilic Vein AVF (TBAVF)

If the cephalic vein is unsuitable in the upper arm, the choice of access is between the prosthetic graft and transposed basilic vein. The basilic vein is generally of large calibre and lies deep to the fascia and is thereby protected by cannulation or repeated venepuncture. Although brachial basilic atrial-venous fistula was first described in 1976 (8), its place in the provision of long-term vascular access remains uncertain. This is because its use requires extensive dissection for exposing the deep-laying vessel and therefore the operation and the post-operative recovery phase are more prolonged and complicated. However, due to the problems associated with the use of synthetic grafts, namely infection, stenosis, and poor long-term patency rates, attention is being redirected to TBAVF. Hakaim and co-workers (1998) reported an 18-month cumulative primary patency of 79% for 26 TBAVFs placed in diabetic patients (9). In another report, TBAVF was used in preference to prosthetic graft on 31 occasions. There were no technical failures and 28 of these 31 fistulas matured and were used for dialysis (10).

8.4.1. *Technical considerations*

It is important pre-operatively to ascertain the calibre and patency of the basilic vein and brachial veins, because after TBAVF formation, the brachial veins would serve as the primary venous drainage of the forearm.

The procedure is generally carried out under general anaesthesia or regional anaesthesia by means of brachial plexus block. In a recent innovation, post-operative regional analgesia was improved by using an epidural catheter sited during the operation to allow delivery of bupivacaine (11). TBAVF can be fashioned using the technique described by Dagher *et al.* (8). Essentially, a transverse incision is made in the antecubital fossa and into the deep fascia to expose the basilic vein. This incision is then extended up on the medial aspect of the arm ("hockey-stick" incision) and the whole length of the basilic vein is mobilised taking care to preserve the medial cutaneous nerve of the forearm. Communicating tributaries of the vein are divided between ligatures. The basilic vein is then transposed to a subcutaneous tunnel on the anterior surface of the arm and anastomosis performed on the anterior side to the brachial artery and antecubital fossa using 7/0 prolene sutures. A suction drain inserted beneath the deep fascia in the original site of the basilic vein will help to prevent post-operative haematoma formation. The wound is closed in two layers.

8.5. Stapling in Vascular Access Surgery

8.5.1. *Alternative anastomotic technique*

The use of VCS (staples made from titanium), a relatively new technique in vascular surgery, has previously shown encouraging results in a variety of different vascular anastomoses in animal and human models (12,13). In these preliminary studies, the technique has been correlated with less anastomotic bleeding, decreased anastomotic and operative times, and reduced early thrombotic complications. Histological studies in animals (14) demonstrated earlier endothelialisation and less intimal hyperplasia in anastomoses created with the VCS compared to sutured anastomoses. The distinct advantage of the VCS is that they do not penetrate the vessel, disrupt the endothelium or have an intraluminal component. In addition, the interrupted anastomosis allows for dilatation and growth of the vessel.

The approximation of the artery and vein is done with 6/0 prolene stay-sutures. The arterial and venous walls are approximated and everted symmetrically by using a tissue approximation forceps and the titanium clips are applied by using the disposable clip applier (VCS autosuture) [Fig. 8.1, (A)–(C)]. Large staples (span between legs before closure 2 mm) are used in brachial AVFs and medium staples (span between legs before closure 1.4 mm) in radial AVFs. The time necessary to complete the anastomosis is approximately five minutes.

In our experience, the use of the VCS correlated with creating an excellent anastomosis, and minimising blood loss and operator time. There were no primary failures and the long-term results are very encouraging since all AVFs are currently used for dialysis.

8.5.2. *Refashioning of aneurysmatic AVF using Multifire surgical stapler*

Some of the endogenous AVFs become aneurysmatic after being used for a long time. This can lead to steal syndrome, bleeding after insertion or removal of the dialysis needles, difficulty for the patient using his arm or forearm and to an unsightly fistula. The usual way to deal with such a problem is ligation of the aneurysmatic AVF and creation of a new one. However, the proportion of haemodialysis patients who have exhausted all sites of peripheral vascular access is increasing and ingenious ways of extending access use-life must be explored. Hence, we described the technique for refashioning aneurysmatic AVF by using the Multifire GIA 60 surgical stapler (9).

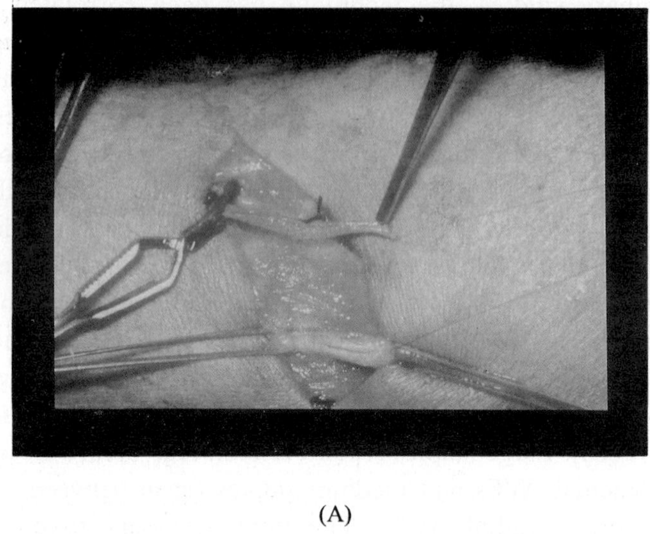

(A)

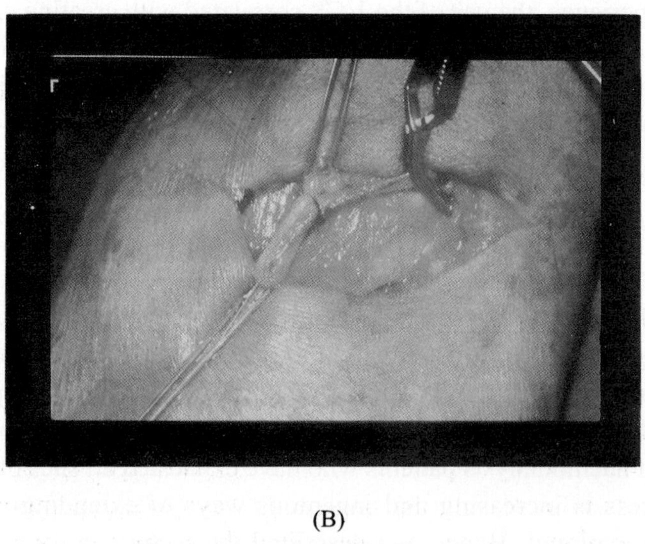

(B)

Fig. 8.1. Creation of an AVF using VCS. (A) and (B) Approximation of artery and vein with prolene stay sutures. (C) Arteriovenous anastomosis after application of VCS.

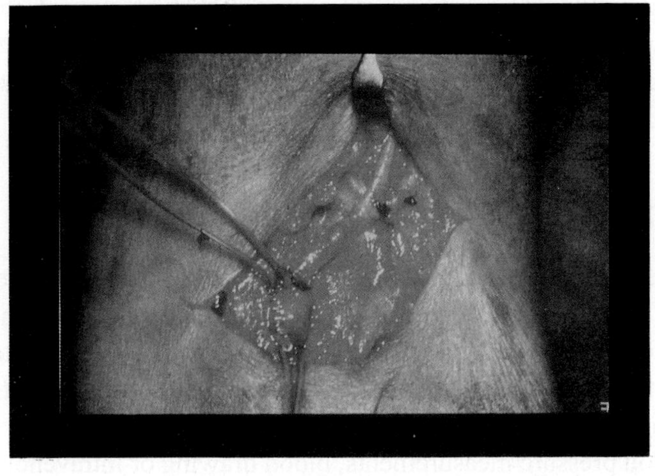

(C)

Fig. 8.1. (*Continued*)

The incision is made along the whole length of the aneurysmatic vein. Proximal and distal controls are obtained with vessel loops. By tightening the distal vessel loop, the inflow in the AVF is stopped, the vein is emptied of blood by applying gentle pressure along its wall and finally, the proximal vessel loop is tightened. Each aneurysmal segment of the anterior wall of the vein is lifted up with two haemostatic forceps and the GIA 60 stapler is applied so as to allow only 1 cm of remnant vein. After stapling and excising the first part of the aneurysm by firing the GIA 60, the procedure is repeated more proximally until all the aneurysmatic segments of the vein are excised. The layer of the staple line is reinforced with one layer of 6/0 prolene continuous suture.

We have performed ten operations using this technique. The primary indication for surgery was severe steal syndrome ($N = 4$), bleeding after the removal of the dialysis needles ($N = 3$), pain over the aneurysmatic area ($N = 2$) and disability using the arm ($N = 1$). After completion of the procedure, the size of the vein was reduced by approximately 50%. There were no post-operative complications. The AVFs were successfully reused for dialysis within four weeks post-operatively. The use of the GIA 60 surgical stapler for the refashioning of an aneurysmatic AVF is a safe and effective technique which preserves the function of the AVF and its suitability for dialysis.

8.6. Post-Operative Care After AVF Construction

The ipsilateral upper extremity is kept elevated to minimise oedema and discomfort. The patient is taught how to recognise the thrill over the proximal vein and the importance of unimpeded flow through the fistula to maintain patency. The patient is advised to wear loose sleeved garments to avoid lying on the fistula-bearing extremity, and is forbidden to use the fistula limb for blood pressure measurements, blood drawing or intravenous infusions except in extreme emergencies.

Exercises, such as squeezing a squash ball, which increase flow through the fistula to accelerate maturation of the fistula are encouraged and should be instituted seven days following fistula formation.

Premature cannulation of the fistula may result in vessel damage and haematoma formation, which predisposes to early fistula failures and the formation of venous stenosis at the puncture site. Therefore, an internal AVF should not be used for six weeks after operation. In any case, the proximal veins must not be cannulated for dialysis until they have acquired the resilient consistency of a well-arterialised vein. This is discussed extensively in Chap. 16, Nursing Care.

8.7. Conclusion

The choice of access modality varies widely with a predominance of prosthetic grafts in the United States while endogenous AVF predominate in Europe. The advantages offered by endogenous AVF underlines the need to improve construction and maturation of these fistulas. As highlighted by Besarab and Escobar (16), the success of AVFs depends on adequate arterial in-flow, venous anatomy and venous out-flow. A good radial pulse with a normal Allen's test and a patent cephalic vein will certainly lead to the maturation of an AVF with adequate intra-access flow for dialysis. Where forearm vessels are not suitable for AVF construction, the surgeon should proceed to evaluate elbow AVFs with confidence knowing that the results are equally good.

References

1. Brescia, M.J. *et al.* (1966). Chronic hemodialysis using venapuncture and surgically created arteriovenous fistula. *Engl J Med*, **275**, 1089–1092.
2. Bourquelot, P. and Wolfeler, L. (1981). Microsurgery for hemodialysis distal arteriovenous fistulas in children weighing less than 10 kg. *Proc EDTA*, **18**, 537.
3. Owens, M.L. and Bower, R.W. (1980). Physiology of arteriovenous fistulas. In (eds. S.E. Wilson and M. Owens) p. 101 *Vascular Access Surgery*, Year Book Medical Publisher: Chicago.
4. Bonalumi, V. *et al.* (1982). Nine years' experience with end-to-end arteriovenous fistula at the anatomical snuffbox for maintenance haemodialysis. *Br J Surg*, **69**, 486–488.
5. Ryan, J.J. and Dennis, M.J.S. (1990). Radial-cephalic fistula in vascular access. *Br J Surg*, **77**, 1321–1323.
6. Elcheroth, J., de Parev, L. and Kinnaert, P. (1994). Elbow arteriovenous fistulas for chronic haemodialysis. *Br J Surg*, **81**, 982–984.
7. Gracz, K.C. (1977). Proximal forearm fistula for maintenance hemodialysis. *Kidney Int*, **11**, 71.
8. Dagher, F. *et al.* (1976). The use of basilic vein and brachial artery as an A-V fistula for long term haemodialysis. *J Surg Res*, **20**, 373–376.
9. Hakaim, A.G. *et al.* (1998). Superior maturation and patency of primary brachiocephalic and transposed basilic vein arteriovenous fistula in patients with diabetes. *J Vasc Surg*, **27**, 154–157.
10. Butterworth, P.C. *et al.* (1998). Arteriovenous fistula using transposed basilic vein. *Br J Surg,* **85**, 653–654.
11. Butterworth, P.C. *et al.* (1997). Postoperative regional analgesia following basilic vein transposition for vascular access. *Br J Surg*, **84**, 560–561.
12. Kirsch, W.M. (1992). A new method for microvascular anastomosis: Report of experimental and clinical research 5. *Am Surg*, **58**, 722–727.
13. Kirsch, W.M. *et al.* (1995). Comparative evaluation of sutures and non-penetrating metal clips for vascular reconstruction. *Cardiovasc Surg*, **26**, 136–139.

14. Kirsch, W.M. *et al.* (1992). Tissue reconstruction with non-penetrating arcuate-legged clips potential endoscopic applications. *J Repr Med*, **37**, 581–586.
15. Hakim, M.S. *et al.* (1997). Refashioning of an aneurysmatic AVF by using the GIA 60 surgical staples. *Int Surg*, **82**, 376–377.
16. Besarab, A. and Escobar, F. (1999). A glimmer of hope: Increasing the construction and maturation of autologous arteriovenous fistulas. *Am J Kidney Dis*, **33**, 977–979.

CHAPTER 9

USE OF AUTOGENOUS VEIN OR SYNTHETIC GRAFTS

Albert G Hakaim MD, MSc, FACS and W Andrew Oldenburg MD
Mayo Clinic, Jacksonville, Florida 32224, USA

9.1. Indications/Sites/Patency Rates

This section describes the peri-operative evaluation and sequential planning for autogenous and non-autogenous haemodialysis access. Representative patency estimates are presented in Table 9.1.

Upper Extremity Vascular Access Sites

Autogenous primary fistulae

snuffbox
radial artery to cephalic vein
brachial artery to cephalic vein
transposed forearm vein
transposed basilic vein

Autogenous secondary fistulae

greater saphenous vein grafts

Table 9.1. Representative cumulative patency estimates for autogenous and non-autogenous arteriovenous fistulae.

Fistula Type	Patency/ Interval	Author(s)
Autogenous		
Snuffbox	87%/5 yr	Horimi *et al.* 1996 (10)
Radiocephalic	67%/2 yr	Mandel *et al.* 1977 (13)
Brachiocephalic	90%/1 yr	Nazzal *et al.* 1990 (21)
	78%/1.5 yr	Hakaim *et al.* 1998 (17)
Transposed basilic vein	83%/1 yr	Davis *et al.* 1986 (7)
	79%/1.5 yr	Hakaim *et al.* 1998 (17)
	70%/8 yr	Dagher, 1986 (25)
Non-autogenous		
Polytetrafluorosthylene (PTFE)	76%/1 yr	Rizutti *et al.* 1988 (32)
	74%/1 yr	Tordior *et al.* 1988 (35)
	70%/1 yr*	Hakaim and Scott, 1997 (85)
	88%/1 yr	Windus *et al.* 1992 (36)
	50%/3 yr	Rizutti *et al.* 1988 (32)
Denatured homologous vein	57%/1 yr	Heintjes *et al.* 1995 (43)
	30%/1 yr	Bosman *et al.* 1998 (44)
Cryopreserved saphenous vein	100%/2 yr	Ahmed *et al.* 1976 (45)
Human umbilical vein	57%/2 yr	Wellington, 1981 (48)
Bovine heterografts	79%/1 yr	Brems *et al.* 1986 (55)
	69%/2 yr	Brems *et al.* 1986 (55)
Sheep collagen	71%/1 yr	Enzler *et al.* 1996 (56)
	45%/3 yr	Enzler *et al.* 1996 (56)
Silicone composite	63%/1 yr	Schanzer *et al.* 1989 (59)
Polyurethane	53%/1 yr	Nakagawa *et al.* 1995 (60)

*Cannulated at 72 hours

Non-autogenous fistulae

- polytetrafluoroethylene (PTFE)
  - expanded standard
  - expanded thin wall
- denatured homologous vein
- cryopreserved saphenous vein
- human umbilical vein
- bovine heterografts
- sheep collagen grafts
- Dacron composite
- silicone composite grafts
- Corethane grafts

There are two guiding principles for the creation of arteriovenous access for chronic haemodialysis. First, to perform the best operation first. This will result in the greatest interval of intervention-free patency. In a recent study, patients who were introduced to dialysis with inadequate function or access failure, from either an arteriovenous fistula or a PTFE graft, had increased morbidity and anxiety about dialysis treatments (1). It is also undeniable that the creation of haemodialysis access is expensive. In 1990, Medicare, (a health care administration of the government responsible for the care of renal failure patients) spent $383 million on placing vascular access grafts (2). Currently, this figure has exceeded $1 million for access placement and associated interventions (3).

Second, to begin vascular access procedures as distal in the upper extremity as possible. This preserves all available proximal sites for future access creation. This infers that autogenous fistulae are created prior to the use of non-autogenous grafts. Again, as is outlined below, primary autogenous fistulae, at all locations, are superior in durability and patency, to non-autogenous grafts.

9.1.1. *Peri-operative evaluation*

The creation of a durable arteriovenous fistula depends upon adequate arterial in-flow and out-flow. Traditionally, the assessment of in-flow has been limited to physical examination and demonstration of palmar arch supply via the Allen test. Venous out-flow has been assessed in a number of ways. Traditionally, either pre-operatively or as the initial stage of the procedure, contrast venography can be performed to delineate the diameter and course of the cephalic, anticubital and basilic veins. In general, a minimal diameter of 2 mm is required for successful primary fistula construction. Recently, pre-operative ultrasound examination of both arterial in-flow and venous out-flow has been described (4). This study concluded that a minimal vein diameter of 2.5 mm was required for a successful PTFE-vein anastomosis.

Pre-operative ultrasound of the axillo-subclavian venous system is useful in patients with a prior history of central venous dialysis catheters, Swan–Ganz catheters, or prior median sternotomy. In addition, patients who have had a percutaneously inserted central catheters (PICC) should also undergo central venous imaging. The increased morbidity of subclinical central venous stenosis ipsilateral to a newly created AVF has been well documented (5–7).

The utility of intra-operative blood flow measurement has also been reported (8). Blood flow was determined immediately following creation of the arteriovenous fistula (AVF), prior to closing the wound. Blood flow increased progressively from distal to proximal access sites. This was found to be independent of age, race, sex or the presence of diabetes. However, our prior investigations have shown a decrease in maturation of primary radiocephalic AVF in diabetic patients. This discrepancy may be explained by the fact that it is the arterial dilation and subsequent increase in flow to support an AVF which is lacking in the calcified forearm arteries of diabetic patients. Regarding the prognostic value of intra-operative fistula blood flow measurements, autogenous AVF with flow rates less than 320 ml/min and polytetrafluoroethylene (PTFE) grafts with flows less than 400 ml/min had decreased primary and secondary patency rates.

Our strategy for the sequential consideration of dialysis access sites is illustrated in Fig. 9.1. Again, based upon the guiding principle that the best operation be performed first, for diabetic patients, an autogenous forearm can be expected to function to the point of cannulation only 30% of the time. Therefore, a forearm loop PTFE graft would be recommended. This would be followed by autogenous upper arm brachiocephalic or transposed

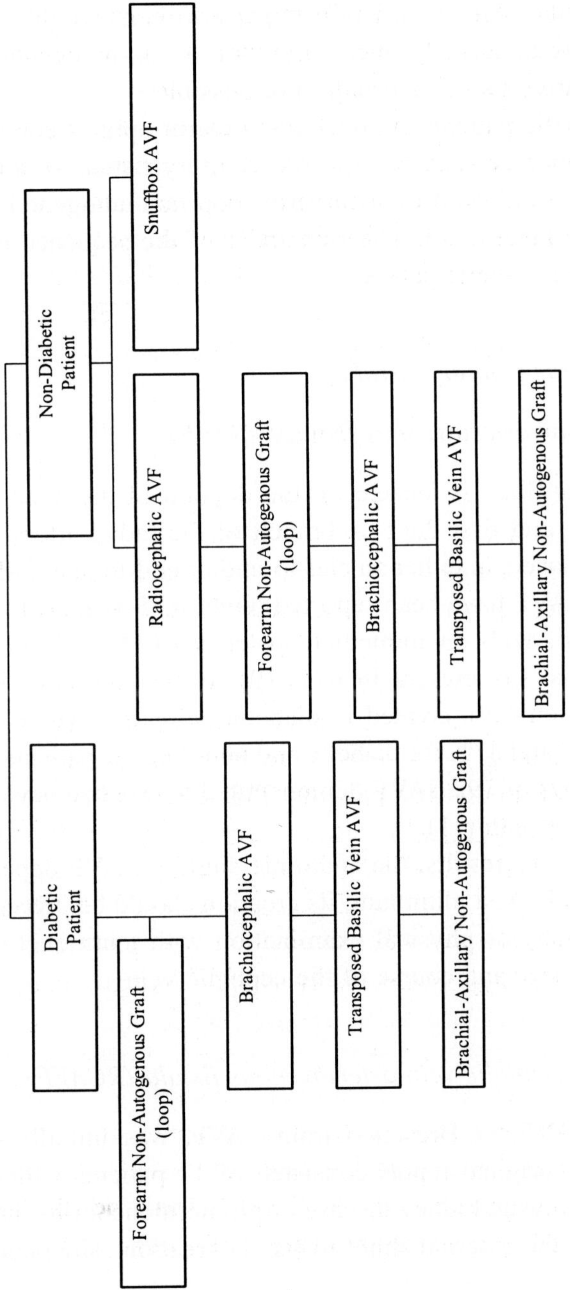

Fig. 9.1. Strategy for chronic haemodialysis access planning.

basilic vein fistulae. Finally, a PTFE upper arm graft would be considered. This sequence is important, since following a non-autogenous upper arm fistula with a native fistula is usually not possible.

For non-diabetic patients, autogenous forearm options commence with a snuffbox, or branch cephalic vein to radial artery fistula, or a radiocephalic fistula. These are followed by a forearm loop non-autogenous graft, either PTFE or bovine heterograft. The remainder of the sequence is identical to that described for diabetic patients.

9.1.2. *Autogenous primary fistula*

Anatomic snuffbox arteriovenous fistulae (SAVF)

The anatomic snuffbox is located on the dorsum of the hand, between the thenar eminance and first digit. A branch of the radial artery and cephalic vein are anastomosed, in either an end-to-end or end-to-side fashion. Several large series of SAVF have been reported. Over a 25-year period, 131 SAVF were constructed, with an immediate patency of 95%. Flow suitable for high-flux dialysis was present in 67% (9). A second series included 139 SAVF, 39 of which were created in diabetic patients. Five-year cumulative patency was 72% and 87% for diabetic and non-diabetic patients, respectively (10). A third series of 192 SAVF demonstrated a 95% patency and ability to cannulate at one month (11).

Based upon these results, the anatomic snuffbox AVF appears to be the optimal distal AVF. At a minimum, its creation should be preceded by either contrast venography or physical examination with tourniquet compression, to determine the size and course of the cephalic vein.

Radial artery to cephalic vein arteriovenous fistula (RCAVF)

The classic RCAVF, or Brescia–Cimino AVF, was initially described in 1966 (12). Their original report consisted of 13 patients with renal failure due to either polycystic kidney disease or glomerulonephritis, and initiated a trend away from the external shunt to fistula creation. The popularity of the

RCAVF is evidenced by the fact that, between 1977 and 1989, published reports have detailed a two-year patency of 67%. In addition, early thrombosis or non-maturation was documented in 11 to 27% (13–16).

However, in this collected series, only 6–13% had a history of diabetes. In our recent series, ten RCAVF were created in patients with a history of diabetes. Non-maturation to the point of cannulation occurred in 72%, while 18-month cumulative patency was only 33%.

These results were significantly worse than the 27% non-maturation for brachial artery to cephalic vein arteriovenous fistulae (BCAVF), and the 78% and 79% 18-month cumulative patency for BCAVF and transposed basilic vein arteriovenous fistulae (TBAVF) (17). Based upon these results, the radiocephalic AVF appears to be optimal for non-diabetic patients in whom a snuffbox AVF is not an option. However, either BCAVF or TBAVF appear optimal for patients with a history of diabetes.

Brachial artery to cephalic vein arteriovenous fistulae (BCAVF)

BCAVF are constructed at the antecubital fossa in an end-to-side fashion. Initially, BCAVF were successfully created in nine patients with radial artery occlusion, precluding a more distal arteriovenous fistula (18). A review of 306 BCAVF reported in the literature indicates a 1-year patency between 54 and 90% (19–21). Their utility in children has also been described (22). More recently, a series of 44 BCAVF created between 1983 and 1987 demonstrated a cumulative primary patency of 50% and 38% at one and two years, respectively (23). An additional report included 52 BCAVF with a cumulative primary patency of 74% and 61% at one and four years, respectively. As was indicated above, our recent results indicate a 78% cumulative primary patency at 18 months (17).

Transposed basilic vein arteriovenous fistulae (TBAVF)

Initially, TBAVF were considered as secondary access procedures after exhaustion of distal sites. The initial series included 23 TBAVF as the primary procedure in 21 patients, none of whom were diabetic (24). These patients were followed for eight years, at which time the functional patency was reported as 70% (25). This procedure has also been reported in a series of 66 patients, age 11 to 19 years. Cumulative primary one-year patency was 83% (7). More recently, a series of 31 TBAVF were reported, 28 of which were used for haemodialysis (26). In a prospective randomized comparison of PTFE versus TBAVF, the latter displayed one- and two-year cumulative primary patencies of 90% and 86%, respectively. Whereas, PTFE had one- and two-year cumulative primary patencies of 70% and 49%, respectively (27). Lastly, our recent report documented an 18-month cumulative primary patency of 79% for 26 TBAVF placed in patients with a history of diabetes (17). Therefore, equivalent patency for BCAVF has been described for diabetic and non-diabetic patients.

9.1.3. *Autogenous secondary fistulae*

Within this category are two small clinical series detailing the use of autogenous grafts for haemodialysis access. These include basilic and greater saphenous vein grafts.

The use of basilic vein loop grafts in 15 patients has been described. Cumulative primary patencies at one and two years were 70% and 50%, respectively (28). Greater saphenous vein has become an increasingly scarce resource, given its widespread use in peripheral vascular and coronary arterial reconstructions. A recent retrospective review of a 12-year experience with greater saphenous vein grafts in the forearm position. Grafts were placed in 29 patients. Cumulative primary patencies at one, two and three years were 89.4%, 89.4% and 71.5%, respectively (29).

9.1.4. *Non-autogenous secondary fistulae*

Polytetrafluoroethylene (PTFE) grafts

As early as 1977, PTFE grafts had been used for the creation of secondary arteriovenous fistulae (AVF). Based upon an early series of 11 PTFE grafts, 6 mm diameter PTFE grafts are preferred, compared to 8 mm grafts, due to more desirable flow velocities of the former (30). An analysis of 154 Cimino fistulae and 163 PTFE grafts has clarified early and late access failures. Failure of the Cimino fistulae usually occurred early in the post-operative period, secondary to use of inadequate veins. Failure of PTFE grafts was primarily due to venous anastomotic stenosis, correction of which was required to extend graft survival (31).

Few studies have compared the patency and durability of PTFE grafts of various configurations. One such retrospective review of 189 PTFE grafts placed over a seven year period has been reported. Overall patencies were 76%, 50% and 40% at one, three and five years, respectively (32). In addition, forearm loop grafts demonstrated greater overall patency and required fewer revisions than their straight counterparts. The latter has also been our impression. Therefore, a loop configuration is our preference. However, straight radial artery to anticubital vein PTFE grafts have been demonstrated to provide adequate flow for high-flux dialysis (33).

Non-thrombotic complications of PTFE grafts include infection, pseudoaneurysm formation, secondary to repeated cannulation, and vascular steal phenomenon. These complications have been found to be more frequent in PTFE versus autogenous access (34).

In a review of 100 PTFE grafts placed over a ten year period, the cumulative patency of these grafts was reported. At one, two and three years, cumulative patencies were 74%, 59% and 47% (35). The effect of diabetes on graft patency has also been investigated. A series of 50 non-diabetic and 51 diabetic patients were compared. By life-table analysis, one- and two-yr graft survivals of 88% and 77% were reported for non-diabetics, significantly greater than the 70% and 67% for diabetic patients (36). As has been mentioned above, the major cause of PTFE graft thrombosis is the development of stenosis at the venous anastomosis. The aetiology of this phenomenon is not completely understood. However, the effect of haemodialysis upon platelet deposition in PTFE grafts has been investigated. A series of nine patients with PTFE grafts were compared with two native fistulae. Platelets were labelled with oxine-111 indium. Compared to native fistulae, there was a marked increase in dialysis-associated platelet deposition at all sites along the prosthetic graft, compared to the native fistulae (37). Whether anti-platelet agents may be beneficial remains unknown.

Polytetrafluoroethylene (PTFE) is available in two different wall configurations; standard wall and thin wall. Few studies have been performed to determine the optimal configuration for haemodialysis access. Recently, a prospective study was done where 108 patients were randomised over a two-year period to receive either a standard wall ($n = 56$) or thin wall ($n = 52$) 6 mm stretch PTFE graft (W.L. Gore and Associates, Flagstaff, AZ, USA). Mean primary (18.2 months versus 12.1 months), secondary (20.9 months versus 13.7 months), and cumulative patency rate (22.2 months versus 15.2 months) were significantly greater for standard wall than for the thin wall configuration. In addition, there were no significant differences regarding pseudoaneurysm or infectious complications (38). Therefore, it would appear that the standard wall configuration is optimal for dialysis access.

A comparison of 6 mm standard wall PTFE haemodialysis access grafts from two different manufacturers (W.L. Gore and Associates, Flagstaff, AZ, USA and Impra, Inc., Tempe, AZ, USA) has also been reported. All grafts were placed in the radial artery to antecubital vein "straight" forearm position. 117 patients were randomized to receive 131 grafts. Life-table primary patency estimates at one (Impra 43% versus Gore 47%) and two years (Impra 30% versus Gore 26%) were not significantly different. In addition, secondary patencies at one (Impra 49% versus Gore 69%) and two years (Impra 33% versus Gore 41%) were not significantly different (39).

A second characteristic of available PTFE grafts has also been evaluated. Non-reinforced and reinforced grafts were retrospectively compared for 632 grafts placed between 1987 and 1995, in a retrospective fashion. One half of the patients were diabetic. Non-reinforced PTFE demonstrated superior primary patency, while secondary patency was not significantly different (non-reinforced 88% versus reinforced 77%). It would appear that non-reinforced PTFE is optimal for dialysis access (40).

A third manufacture modification of PTFE is the stretch configuration (WL Gore and Associates, Flagstaff, AZ, USA). Recently, stretch PTFE and standard wall PTFE have been compared in a prospective, randomised trial. Over a two-year period, 37 patients received 17 stretch and 20 standard wall PTFE grafts. Thrombotic events occurred in significantly fewer stretch grafts (12% versus 40%), while cumulative primary patency was higher for stretch versus standard wall PTFE grafts (59% versus 29%). In addition, duplex surveillance scans demonstrated a greater number of stenoses in the standard-wall PTFE grafts (41).

The majority of PTFE grafts have been placed in the upper extremity as either primary or secondary access procedures. However, for tertiary procedures, when lower extremity sites are not acceptable, other configurations have been used. In a recent single-centre retrospective review, 26 axillary artery to axillary vein, or axillary artery to jugular vein PTFE grafts have been placed for haemodialysis. At the time of graft placement, patients had undergone dialysis for a mean of 77 months (range 5 to 256 months). In addition, patients had undergone a mean of 9.4 prior access procedures. By life-table analysis, three-year patency estimate was 60% (42).

Denatured homologous vein grafts

Two large clinical studies have recently investigated the clinical utility of denatured homologous vein (DHV) grafts for haemodialysis access. Based upon a retrospective review, 125 patients received 195 grafts over a five-year period. Primary patencies at one and three years were 57% and 25%, respectively, while secondary patencies at one and three years were 76% and 52%, respectively (43). A second study was a well-designed prospective randomised multicentre trial comparing DHV and PTFE grafts. Between 1994 and 1997, 63 DHV and 68 PTFE grafts were placed in 60 males and 71 females. The one-year patency rates were 30% and 40% for DHV and PTFE, respectively. Non-thrombotic complications due to infection were more frequent for PTFE grafts while aneurysm formation requiring removal was more common in DHV (44). Therefore, while patency rates were not significantly different, no advantage was demonstrated for DHV grafts.

Cryopreserved saphenous vein grafts

Earlier studies of cryopreserved saphenous vein grafts (CSVG) utilised grafts obtained from the proximal 30 cm at the time of varicose vein stripping and ligation procedures. Grafts were removed and stored at –30°C. The CSVG recipients had major blood group antigens identical to the donors. The CSVG were placed in the brachial artery to brachial or cephalic vein position. Over a three-year period, 70 patients underwent CSVG placement. The two-year cumulative patency was 100%, and no patient was reported to experience allograft rejection (45, 46).

More recently, a detailed immunologic analysis of CSVG has been undertaken. Over a two-year period, 16 grafts were placed as a "third choice" vascular access. Blood group ABO and human lymphocyte antigen (HLA) A and B compatibilities were determined but not considered for graft placement. Recipient T-lymphocyte subsets (CD3–CD4–CD8 and CD4/8) were examined. In addition, lymphocytotoxic antibodies were determined pre-operatively, and at days 15 and 30 post-implantation. No evidence of immunologic activation was found. Also, 10 of 16 patients had a segment of CSVG implanted subcutaneously, which was subsequently examined. Both alkaline-phosphatase–anti-alkaline-phosphatase and a panel of monoclonal antibodies were used. Minimal infiltration of the outer layer of adventita was found; mainly due to monocyte macrophages (Leu M3+) with few T lymphocytes (CD3+). Therefore, rejection did not appear to be a major cause of CSVG failure (47).

Human umbilical vein grafts

Two early studies serve to illustrate the use of human umbilical vein grafts (HUVG) for haemodialysis access. In a series of 21 patients, 23 HUVG were placed, when a native arteriovenous fistula could not be constructed. Overall patency from 2 months to 26 months was 57%. Three infected grafts required removal, which was complicated by dense adhesions adjacent to the conduit. It was concluded that HUVG provided no advantage compared to commercially available teflon grafts (48).

In a second study, 33 HUVG were implanted in 24 patients. Thrombosis occurred 25 times during a mean 8.5 month follow-up period. Seven grafts were removed due to infection while aneurysm formation required removal of an additional two grafts. At the end of the observation period, 8 of the 33 grafts were used for haemodialysis (49).

Bovine heterografts

The earliest report of bovine heterografts (BHG) for haemodialysis access detailed the results of 33 BHG placed in 28 patients. Thirty-two grafts were constructed in a forearm loop configuration. Overall, 68% of grafts placed were eventually used for haemodialysis (50). A second report detailed the results of 39 grafts placed in 36 patients for haemodialysis. Follow-up ranged from 2 months to 12 months. At a mean of six months, 90% of the grafts were patent (51).

In a retrospective review of 100 BHG placed for haemodialysis, 28 patients required reoperation for graft thrombosis, stenosis, infection or haemorrhage. The chief advantage of BHG was the ability to cannulate them immediately for dialysis (52). More recently, a prospective randomised comparison of BHG and PTFE grafts resulted in no statistically significant difference in cumulative patency, complications, or secondary patency (53). A four-year retrospective review of BHG and PTFE dialysis grafts indicated an average graft use of 27 months in either group. However, the forearm loop configuration in either group had superior results (54).

A retrospective review of 385 BHG placed in 331 patients has been reported. There were 251 grafts based upon radial arterial inflow, while 127 had the proximal anastomosis at the brachial artery. Overall patencies at one, two and three years were 79%, 69%, and 63%, respectively. These patency rates were not affected by graft location or presence of diabetes (55).

Sheep collagen grafts

A recent retrospective review of 291 non-autogenous grafts for haemodialysis included 59 sheep collagen grafts (SCG). Patencies at one and three years, were 71% and 45%, respectively. This was not significantly different than other grafts or primary arteriovenous fistulae (56).

Dacron composite grafts

Plasma-TFE is a thin-walled woven dacron graft to which an ultrathin layer of tetrafluoroethylene is bonded via glow-discharge polymerisation. In a retrospective review, 19 plasma-TFE grafts were compared to 28 ePTFE grafts. Early use, defined as cannulation within seven days of construction, was not associated with graft failure. However, at one year, primary cumulative patencies for plasma-TFE and ePTFE grafts were 47% and 79%, respectively (57). A similar prospective study of 21 plasma-TFE and 19 ePTFE grafts demonstrated a one-year cumulative patency of 47% and 55%, respectively (58). It would appear that any early cannulation advantage was offset by decreased graft patency.

Silicone composite grafts

A graft composed of an inner and an outer layer of PTFE, with a silicone rubber middle layer was initially determined to be self-sealing following needle puncture, in a canine model. In a randomised prospective study, 30 patients received the composite graft while 35 PTFE grafts were placed as controls. One-year cumulative patency was 63% for the composite graft, not significantly different from the 66% for the PTFE control group. In addition, the composite grafts were cannulated a mean of 1.3 days after placement. No bleeding complications requiring surgical treatment occurred in either group (59).

Polyurethane graft

This graft is composed of polyurethane coated with gelatin and reinforced with knitted polyester fibers. In a randomised prospective trial, 39 polyurethane grafts (PUG) were compared to 18 PTFE grafts in 52 patients. Cumulative patencies at one year for PUG and PTFE grafts were 53.2% and 70.8%, respectively (60). Therefore, at the present time, polyurethane does not demonstrate a clinical advantage to available non-autogenous grafts.

Corethane/polyester graft

This experimental graft was developed in an effort to provide a self-sealing prosthesis with equivalent patency to available PTFE grafts. The graft consists of an inner blood-interfacing layer of spun Corethane fibres impregnated with a gelatin–heparin complex. The outer layer is composed of a knitted polyester sheath.

In a canine study, 26 Corethane grafts were compared to 8 PTFE grafts, in the femoral artery-to-vein loop configuration. Cumulative patencies at one year for Corethane and PTFE were 73% and 63%, respectively. Differences in both mean hemostasis time after needle removal (Corethane 3.1 min versus PTFE 21.2 min) and blood loss in achieving haemostasis (Corethane 5.7 g versus PTFE 47.9 g) were significant (61). Clinical studies remain to be performed.

Elastomeric graft

This experimental graft material is the result of electrostatic spinning technology. The graft is primarily microfibrous polydimethylsiloxane spun onto a mandrel, with a smaller diameter polyester yarn for additional strength. A helical bead provides crush and deformity resistance. In a canine model, 18 grafts were implanted and followed for 12 months. The femoral artery to vein grafts were cannulated using 16 gauge needles, weekly. Cumulative patencies at 6 and 12 months were 88% and 80%, respectively. At graft removal, histologic examination revealed tissue ingrowth in the intersticies of the graft (62). Again, clinical studies remain to be performed.

9.2. Results and Complications of Access Grafts in the Lower Extremity

Lower extremity haemodialysis access consists of both primary; autogenous, and secondary; non-autogenous, arteriovenous fistulae (AVF). In general, the lower extremity is not preferred as an initial site for AVF, due to the risk of infectious complications. These AVF require pre-operative arterial imaging, either magnetic resonance angiography (MRA) or conventional angiography, to ensure that a proximal stenosis does not exist, which could result in limb ischemia secondary to a "steal phenomenon".

9.2.1. *Autogenous lower extremity arteriovenous fistulae (AVF)*

One such configuration is a direct anastomosis between the distal superficial femoral or popliteal artery and the saphenous vein. The surgical details and results have been previously described (63, 64). Alternatively, the saphenous vein may be transposed, resulting in a more superficial conduit. Recently, the results of five such fistulas have been reported, with four remaining patent at a mean follow-up of 11 months (65). A novel approach has been the construction of a spiral vein graft from the greater saphenous vein, which was interposed between the superficial femoral artery and sapheno-femoral vein junction. This graft functioned for over five years (66). Since the anastomosis is created distal to the groin, the infectious potential of these AVF may be decreased, compared to the standard femoral incision.

9.2.2. *Non-autogenous lower extremity arteriovenous fistulae (AVF)*

The majority of such AVF have been created in a loop configuration between the common or superficial femoral artery and common femoral vein. In most cases, upper extremity sites for construction of AVF have been exhausted.

A retrospective review of 45 grafts consisting of 39 polytetrafluoroethylene (PTFE) and 6 bovine heterografts has recently been reported (67). One- and two-year cumulative patencies were 44% and 47%, respectively. In addition, 15 grafts (33%) had thrombosed at least once. Complications included graft infection, severe leg ischaemia and congestive heart failure in 18%, 16% and 4%, respectively. Similar results have been reported in a retrospective review of 49 PTFE grafts over a 12-year period (29). Thrombosis and infection occurred in 52% and 35%, respectively. An additional retrospective study was based upon 74 PTFE loop grafts placed in 61 patients between 1981 and 1995. Cumulative patency, expressed as mean loop survival time, was 1.9 years. The perceived advantage of these grafts was the ease of self-cannulation. However, the infection rate was 16% (68). It would appear that lower extremity nonautogenous grafts should only be considered when upper extremity options for AVF do not exist.

Recently, however, a small series questions this methodology. This retrospective review consists of 37 PTFE femoral-artery–femoral-loop grafts in 35 patients over a 14-year period (69). Primary cumulative patencies at one and five years were 73% and 33%, respectively. Interestingly, only two patients (5.7%) suffered an apparent graft infection. Based upon these results, the authors conclude that such grafts are comparable to upper extremity PTFE grafts and should be considered as a primary form of haemodialysis access.

Based upon the data provided by the above studies, it would seem prudent to reserve lower extremity nonautogenous grafts for those patients who no longer have upper extremity sites and cannot undergo lower extremity autogenous fistula creation.

9.3. Neointimal Hyperplasia

A complete review of the accumulated data regarding neointimal hyperplasia and peripheral arterial bypass grafting is beyond the scope of this section. However, studies limited to non-autogenous AVF form the basis for this section.

Endothelial cell transplantation as a means of providing a non-thrombogenic surface on polytetrafluoroethylene (PTFE) grafts has recently been reported. In a canine model, autogenous, fat-derived microvessel endothelial cells (MVEC) have been used to line the luminal surface of 4 mm grafts (70). After implantation, animals received aspirin and persantine. While implanted, grafts were evaluated using magnetic resonance angiography (MRA) and ultrasound. Upon removal, graft segments were analysed using scanning electron microscopy, immunohistochemistry and conventional light microscopy. A cellular lining was observed on the luminal surface of the graft. In addition, these cells reacted positively with antibodies to von Willebrand factor, supporting their endothelial character. These results indicate that this canine model is useful in determining the effect of MVEC "sodding" on the thrombogenicity and hyperplastic response of prosthetic arteriovenous grafts.

A similar study of MVEC "seeded" in patients was undertaken. In a randomised prospective clinical trial, half of the patients received seeded PTFE grafts, while an equal number received non-seeded, control, grafts (71). Patients were to undergo a one-year post-implantation biopsy. A total of nine patients were enrolled in the study.

Unfortunately, only two patients agreed to the biopsy; one seeded and one control graft. The seeded graft showed significant concentric intimal thickening at 20 months and had thrombosed by 24 months. The non-seeded graft at 16 months had irregular areas of intimal thickening. Interestingly, the intima of the seeded graft was twice the thickness of the non-seeded graft. Based upon this very small series, the clinical effect of MVEC seeding on PTFE graft patency may be detrimental.

In a clinical pilot study, autologous saphenous vein endothelial cells were harvested and cultured on PTFE grafts in seven patients requiring haemodialysis (72). The patients were followed by duplex ultrasound and clinical evaluation. One graft was excised at five weeks post-implant, due to infection. Microscopic examination of the explant revealed 85% of the luminal surface was covered by endothelial cells.

The advantage of using fat-derived MVEC is the ability to obtain these cells from the subcutaneous tissue at the time of arteriovenous fistula construction. This would omit the havesting, culturing and potential contamination of autologous endothelial cells. As further clinical experience with this methodology is obtained, the efficacy of placing endothelial cells on the luminal surface of PTFE grafts may be elucidated.

9.4. Preventing Stenosis at the Graft–Vein Anastomosis

Compared to pharmacological or genetic manipulation to decrease the occurrence of intimal hyperplasia, alteration in the geometry and fluid mechanics of the venous anastomosis has received little attention. However, several important concepts are evolving and will be discussed below.

Flow diffusers are common in commercial oil and gas pipelines to accommodate the mismatch between two dissimilar areas. The flow diffuser is an enlargement in circumferential diameter of about 16 degrees, at the point of insertion. This results in a decrease in velocity with a simultaneous increase in pressure. These forces prevent boundary layer separation. The end result being decreased turbulance, which decreases the energy requirement in commercial applications. This concept was first introduced in haemodialysis access surgery as an anecdotal finding. It was observed that when femoral artery to femoral vein fistulas were constructed with such a diffuser, one year patency increased from 53% to 88% (73).

Until recently, little data was available regarding the effect of the angle of intersection of the graft with the native vessel and the incidence of intimal hyperplasia. This angle is inversely proportional to the length of the venotomy; as the angle is decreased, the length of the venotomy is increased. Experimental studies of the effect of anastomotic angle manipulation on turbulence and boundary layer separation have been conducted *in vitro* (74). Using planar illumination in a suspended tracer particle model, anastomotic angles of 15, 30 and 45 degrees were analysed. Flow separation was minimal with the 15-degree anastomotic angle. In an additional study, using an *in vitro* PTFE model, anastomotic angles of 15, 30, 45, 60, 90 and 120 degrees were compared. The calculated impedance index decreased as the anastomotic angle was reduced (75).

Both canine and pig models have also been employed to determine the anastomotic angle which results in the least turbulence, boundary layer separation and recirculation, all of which may be prerequisites for intimal hyperplasia. Using a porcine aortic model, an 8 mm diameter polyurathane interposition was placed in an end-to-side configuration. Distal anastomoses were created at 15-, 45- or 90-degree angles. The 45- and 90-degree anastomoses displayed zones of recirculation, which were absent at the 15-degree anastomosis (76).

Few clinical studies have been conducted to determine the effects of the anastomotic angle and flow diffusers, on the development of intimal hyperplasia. We have recently reported the results of our initial experience with PTFE — axillary-vein anastomoses incorporating a flow diffuser and

15-degree anastomotic angle (77). Five patients underwent construction of brachial artery to axillary vein 6 mm stretch-PTFE grafts. The graft alterations required to obtain a 15-degree anastomotic angle and 16-degree enlargement of the area of the anastomosis was determined *in vitro*. Several different combinations of anastomotic length and angle were compared. Digital images (Sony MVC-FD7, Japan) of the open end of the graft and anastomoses completed to a second 6 mm PTFE graft (to simulate the vein) were obtained. The exact anastomotic angle and increase in area at the anastomosis were computed, using a computer-based tool (University of Texas Health Sciences Campus at San Antonio, Texas, USA). A reproducible 15-degree anastomotic angle and 16-degree area enlargement resulted when the 6 mm graft was transected at an angle based at 2 cm proximal to the open end of the graft. The graft was then incised for 1 cm at the middle of the anterior aspect. The venotomy consisted of a 2 cm longitudinal incision and an additional 1 cm extension at 45 degrees to the midline. The anastomosis was completed with continuous 7-0 polypropylene suture (Ethicon, Johnson and Johnson, New Brunswick, NJ, USA).

The degree of stenosis at the venous anastomosis was determined at six-month intervals (ATL, Ultramark 9, Bothell, WA, USA) and contrast fistulography. These patients were compared to a historic group of 20 brachial-axillary 6 mm stretch-PTFE AVF. Primary cumulative patency at 18 months for the experimental group ($n = 5$) was 100%, compared to the historic controls ($n = 20$) of 32% ($p < 0.05$).

In addition, four of five experimental anastomoses had minimal stenosis, mean 30% area reduction (range 20–45%) at 18 months (77). Therefore, in this small series, it appeared that incorporation of a flow diffuser and 15-degree anastomotic angle delayed anastomotic stenosis. A randomised prospective trial of this type of anastomosis and the conventional end-to-side anastomosis are currently underway.

9.5. Requirements for a Successful Graft

The ideal conduit for haemodialysis access is the autogenous arteriovenous fistula. The obvious advantage being the lack of a venous anastomosis and associated intimal hyperplasia-induced stenosis. Whether the vein is left undisturbed in its native position or transposed, the patency achieved is superior to currently available non-autogenous grafts.

Although polytetrafluoroethylene (PTFE) grafts may have a higher blood flow rate compared to native fistulae in similar anatomic locations, the haemodialysis dose has been shown to be independent of the type of fistula used (78).

Recently, a study of 430 randomly selected haemodialysis patients for vascular access and non-access related diagnoses has been reported. This group required 508 hospitalisations in one year; 322 (63%) for at least one week duration. Independent risk factors for access-related hospitalisation were female sex and white race. In all, 48% of hospitalisations were for access-related diagnoses (79). In a review of 92 consecutive patients for determinants of the type of initial vascular access, women were more likely to receive a PTFE graft at the initial procedure (80).

In a national cross-sectional study from the United States Renal Data System Special Studies section, a comparison was made between patients starting haemodialysis from 1986–1987 ($n = 2741$) and 1990 ($n = 1409$). Graft use had increased from 51 to 65%. This trend away from fistulas could not be explained by co-morbid patient factors. It is imperative that this trend be reversed to provide the best access procedure first.

In this regard, the effect of a concerted effort to increase autogenous AVF use as the initial access has recently been described. This is as part of a combined effort between nephrology and surgery, with the primary objective to increase the total number of native AVF used, compared to non-autogenous grafts (81). As a result of this initiative, the number of autogenous, AVF increased from 28% in 1990–92 to 44% in 1993–95.

Patient factors can also determine the patency of AVF. Diabetes mellitus is now the most common cause of end-stage renal disease (ESRD) in the United States (82). Suprisingly, few data exist regarding the patency of autogenous AVF in these patients. The original report describing the construction and patency of autogenous radiocephalic AVF was published in 1966 (12). All patients suffered end-stage renal failure from either chronic glomerulonephritis or polycystic kidney disease. We have recently reported our investigation of patency and maturation of either radiocephalic, brachiocephalic or transposed basilic vein autogenous AVF (17). Based upon a prospective analysis of 58 AVF, we concluded that in patients with a history of diabetes, the majority of radiocephalic AVF functioned. However, only 30% matured to the point of useful cannulation for haemodialysis. In addition, the cumulative patency for these fistulas was 33% at 18 months. However, both brachiocephalic and transposed basilic vein autogenous AVF had a non-maturation rate of 27% and zero, respectively. Lastly, cumulative primary 18-month patency was 78% for brachiocephalic and 79% for transposed basilic vein AVF. These findings have led us to consider brachiocephalic and transposed basilic vein AVF to be the optimal autogenous accesses for diabetic patients.

Although speculative, the lack of maturation of radiocephalic AVF may be due to the decreased ability of calcified forearm arteries to dilate, as is required to increase flow and maintain flow in the fistula. In addition, there may be a defect in the forearm veins of patients with diabetes. A recent study may provide circumstantial support for the latter. Using an isolated muscle bath technique, the reactivity of vein segments from patients undergoing AVF construction with and without diabetes was determined. The vasoactivity of the vein segments was measured as isometric tension development. The response of vein segments from diabetics to noradrenaline and acetylcholine was significantly less than that of vein from non-diabetic patients (83).

Recently, whether diabetes has an adverse effect on non-autogenous graft patency has been addressed. A retrospective review of 51 non-diabetic and 50 diabetic patients receiving PTFE grafts was undertaken (36). Cumulative graft patencies at one year for non-diabetics and diabetics were 88% and 70%, respectively. At two years, cumulative patencies for non-diabetics and diabetics were 77% and 67%, respectively.

As with all peripheral vascular reconstructions, cigarette smoking negatively impacts access patency. Although this has not been extensively studied, a significant increase in early and late autogenous AVF thrombosis in active cigarette smokers has been established (84). Our recent investigation of a series of 79 PTFE dialysis grafts concluded that a history of peripheral vascular disease, due presumably in part to cigarette smoking, decreased cumulative primary patency from 74 to 60% (85).

The traditional practice has been to construct arteriovenous fistulae in the non-dominant upper extremity. Theoretically, this would allow for the unobstructed use of the dominant hand during the dialysis session. In a recent review of factors that affect fistula patency, hand dominance and access site were not related to a survival difference for autogenous fistulae. However, sleep preference on the side opposite the vascular graft trended towards longer access survival time (86).

The effect of recombinant human erythropoietin (EPO) on thrombosis of vascular access has recently been addressed. A prospective cohort study compared EPO-treated haemodialysis patients with a group matched for type of vascular access, clinical centre, and age. In total, a series of 64 matched fistula pairs and 38 matched graft pairs were included. For the fistula group, in regard to time to first thrombosis, there was no difference between EPO and matched controls. However, EPO treatment increased the probability of graft thrombosis (87). Therefore, EPO therapy may increase the likelihood of non-autogenous graft thrombosis, compared to autogenous fistulae.

Whether early cannulation of non-autogenous dialysis grafts negatively impacts the short and long-term successes of these accesses has been investigated. The primary advantage being the omission of temporary central venous catheters and the associated risk of central vein stenosis (5).

One of the initial reports of early dialysis access cannulation included a variety of autogenous (saphenous vein) and non-autogenous (bovine heterograft, polytetrafluoroethylene) grafts. In addition, early cannulation was defined as cannulation within 24 hours of fistula creation (88). It was concluded that the principle requirement for early cannulation appeared to be a tunnel diameter of 7 mm; allowing little space between the conduit and the subdermal tissue.

Other reports have attempted to determine the relative benefits of plasma-polytetrafluoroethylene (plasma-TFE), or expanded-polytetrafluoroethylene (ePTFE) for early cannulation. Plasma-TFE is a thin-walled woven Dacron graft to which an ultrathin layer of tetrafluoroethylene is bonded via glow-discharge polymerisation. In a retrospective review, 19 plasma-TFE grafts were compared to 28 ePTFE grafts. Early use, defined as cannulation within seven days of construction, was not associated with graft failure. However, at one year, primary cumulative patencies for plasma-TFE and ePTFE grafts were 47% and 79%, respectively (57). However, a similar prospective study of 21 plasma-TFE and 19 ePTFE grafts demonstrated a one year cumulative patency of 47% and 55%, respectively (58). It would appear that any early cannulation advantage was offset by decreased graft patency.

A novel approach has been to incorporate the perceived early cannulation benefit of plasma-TFE and durability of ePTFE. Grafts were constructed to consist of 10 to 12 cm of plasma-TFE with the remainder being ePTFE. Post-operatively, the plasma-TFE segment was cannulated within 72 hours. These grafts were compared to a historic series of conventional ePTFE grafts . There were no significant differences in complications or graft patency. Again, the value of plasma-TFE remains uncertain.

Recently, we have reported a non-randomised prospective trial of early, defined as cannulation after 72 hours of implantation, and late, defined as cannulation after 14 days. All grafts were placed in the brachial artery to axillary vein position. Overall, 80% of patients had a prior distal arteriovenous fistula in the ipsilateral extremity. All grafts were 6 mm stretch PTFE (W.L. Gore and Associates, Flagstaff, AZ, USA) and were placed within a subdermal tunnel created with a 6 mm tunneling device (Impra, Inc., Tempe, AZ, USA). Forty-eight grafts were cannulated early while 31 grafts were accessed after 14 days. There were no episodes of cannulation hemorrhage or wound infection in either group. Cumulative primary patency estimates at one year were not significantly different; 70% for early and 74% for late cannulation.

In addition, central venous catheters were omitted in 47 of 48 patients in the early cannulation group (85). Based upon these results, we have incorporated early cannulation of PTFE grafts as a means of omitting central venous catheters and subsequent central venous stenosis.

References

1. Bay, W.H., Van Cleef, S. and Owens, M. (1998). The haemodialysis access: Preferences and concerns of patients, dialysis nurses, and technicians, and physicians. *Am J Nephrol*, **18**, 379–383.

2. Neumann, M.E. (1995). RAND report: Medicare spent $383 million on placing vascular access grafts in 1990. *Nephrol News Issues*, **9**, 38–39.
3. Hakim, R. and Himmelfarb, J. (1998). Haemodialysis access failure: A call to action. *Kidney Int*, **54**, 1029–1040.
4. Silva, M.B. *et al.* (1998). A strategy for increasing use of autogenous hemodilaysis access preocedures: Impact of perioperative noninvasive evaluation. *J Vasc Surg*, **27**, 302–307.
5. Cimochowski, G.E. *et al.* (1990). Superiority of the internal jugular over subclavian access for temporary haemodialysis. *Nephron*, **54**, 154–156.
6. Glaze, R.C., MacDougall, M.L. and Weigmann, T.B. Thrombotic arm edema as a complication of subclavian vein catheterisation and arteriovenous fistula formation for haemodialysis. *Am J Kidney Dis*, **7**, 439–441.
7. Davis, J.B., Howell, C.G. and Humphries, A.L. Jr. (1986). Haemodialysis access: Elevated basilic vein arteriovenous fistula. *J Pediatr Surg*, **21**, 1182–1183.
8. Johnson, C.P. *et al.* (1998). Prognostic value of introperative blood flow measurements in vascular access surgery. *Surgery*, **124**, 729–737.
9. Boccardo, G. *et al.* (1998). High-flux arteriovenous fistula at the anatomic snuffbox. *Minerva Urol Nefrol*, **50**, 39–43.
10. Horimi, H. *et al.* (1996). Clinical experience with an anatomic snuff box arteriovenois fistula in dialysis patients. *ASAIO J*, **42**, 177–180.
11. Sekar, N. (1993). Snuff-box arteriovenous fistulas. *Int Surg*, **78**, 250–251.
12. Brescia, S.B., Cimino, J.E. and Appel, K. (1966). Chronic haemodialysis using venipuncture and a surgically created arteriovenous fistula. *N Eng J Med*, **275**, 1089–1092.
13. Mandel, S.R. *et al.* (1977). Vascular access in a university transplant and dialysis program. *Arch Surg*, **112**, 1375–1380.
14. Winsett, O.E. and Wolma, F.J. (1985). Complications of vascular access for hemodialysis. *South Med J*, **78**, 513–517.
15. Zibari, G.B, Rohr, M.S. and Landrenau, M.D. (1988). Complications from permanent haemodialysis vascular access. *Surgery*, **104**, 681–686.
16. Wehrli, H., Chenevard, R. and Zaruba, K. (1989). Surgical experiences with the arteriovenous haemodialysis shunt (1970–1988). *Helv Chir Acta*, **56**, 621–627.
17. Hakaim, A.G., Nalbandian, M. and Scott, T. (1998). Superior maturation and patency of primary brachiocephalic and transposed basilic vein arteriovenous fistulae in patients with diabetes. *J Vasc Surg*, **27**, 154–157.
18. Cascardo, S., Acchiardo, S. and Beven, E.G. (1970). Proximal arteriovenous fistulae for haemodialysis when radial arteries are unavailable. *Proc Eur Dial Transplant Assoc*, **7**, 42–46.

19. Bender, M.H.M., Bruyninckx, C.M.A. and Gerlag, P.G.G. (1994). The brachiocephalic elbow fistula; a useful alternative angioaccess for permanent haemodialysis. *J Vasc Surg*, **20**, 808–813.
20. Tautenhahn, J., Heinrich, P. and Meyer, F. (1994). Arteriovenous fistulas for haemodialysis-patency rates and complications — A retrospective study. *Zentralbl Chir*, **119**, 506–510.
21. Nazzal, N.M. *et al.* (1990). The brachiocephalic fistula: A successful secondary vascular access procedure. *Vasa*, **19**, 326–329.
22. Kinnaert, P., Janssen, F. and Hall, M. (1983). Elbow arteriovenous fistula (EAVF) for chronic haemodialysis in small children. *J Pediatr Surg*, **18**, 116–119.
23. Gade, J., Aabech, J. and Hansen, R.I. (1995). The upper arm arterio-venous fistula — An alternative for vascular access in hemodialysis. *Scand J Urol Nephrol*, **29**, 121–124.
24. Dagher, F. *et al.* (1976). The use of basilic vein and brachial artery as an AV fistula for long-term haemodialysis. *J Surg Res*, **20**, 373–376.
25. Dagher, F.J. (1986). The upper arm AV haemoaccess: Long-term follow-up. *J Cardiovasc Surg*, **27**, 447–449.
26. Butterworth, T.M. *et al.* (1998). Arteriovenous fistula using transposed basilic vein. *Br J Surg*, **85**, 653–654.
27. Coburn, M.C. and Carney, W.I. Jr. (1994). Comparison of basilic vein and polytetrafluoroethylene for brachial arteriovenous fistula. *J Vasc Surg*, **20**, 896–902
28. Hibberd, A.D. (1991). Barchiobasilic fistula with autogenous basilic vein: Surgical technique and pilot study. *Aust N Z J Surg*, **61**, 631–635.
29. Bhandari, S., Wilkinson, A. and Sellars, L. (1995). Saphenous vein forearm grafts and goretex thigh grafts as alternative forms of vascular access. *Clin Nephrol*, **44**, 325–328.
30. Elliott, M.P. *et al.* (1977). Use of expanded polytetrafluoroethylene grafts for vascular access in haemodialysis: Laboratory and clinical evaluation. *Am Surg*, **43**, 455–459.
31. Palder, S.B. *et al.* (1985). Vascular access for hemodialysis. Patency rates and results of revision. *Ann Surg*, **202**, 235–239.
32. Rizzuti, R.P., Hale, J.C. and Burkart, T.E. (1988). Extended patency of expanded polytetrafluoroethylene grafts for vascular access using optimal configuration and revisions. *Surg Gynecol Obstet*, **166**, 23–27.
33. Pontari, M.A. and McMillen, M.A. (1991). The straight radial-anticubital PTFE angio-access graft in an era of high-flux dialysis. *Am J Surg*, **161**, 450–453.

34. Kherlakian, G.M. *et al.* (1986). Comparison of autogenous fistula versus expanded polytetrafluoroethylene graft fistula for angioaccess in haemodialysis. *Am J Surg*, **152**, 238–243.
35. Tordoir, J.H. *et al.* (1988). Long-term follow-up of the polytetrafluoroethylene (PTFE) prosthesis as an arteriovenous fistula for haemodialysis. *Eur J Vasc Surg*, **2**, 3–7.
36. Windus, D.W., Jendrisak, M.D. and Delmez, J.A. (1992). Prosthetic fistula survival and complications in haemodialysis patients: Effects of diabetes and age. *Am J Kid Dis*, **19**, 448–452.
37. Windus, D.W., Santoro, S. and Royal, H.D. (1995). The effects of hemodialysis on platelet deposition in prosthetic graft fistulas. *Am J Kid Dis*, **26**, 614–621.
38. Lenz, B.J. *et al.* (1998). A three-year follow-up on standard versus thin wall ePTFE grafts for haemodialysis. *J Vasc Surg*, **28**, 464–470.
39. Kaufman, J.L. *et al.* (1997). A prospective comparison of two expanded polytetrafluoroethylene grafts for linear forearm haemodialysis access: Does the manufacturer matter? *J Am Coll Surg*, **185**, 74–79.
40. Schuman, E.S. *et al.* (1997). Reinforced versus nonreinforced polytetrafluoroethylene grafts for haemodialysis access. *Am J Surg*, **173**, 407–410.
41. Tordoir, J.H. *et al.* (1995). Early experience with stretch polytetrafluoroethylene grafts for haemodialysis access surgery: Results of a prospective randomised study. *Eur J Vasc Endovasc Surg*, **9**, 305–309.
42. McCann, R.L. (1996). Axillary grafts for difficult haemodialysis. *J Vasc Surg*, **24**, 457–461.
43. Heintjes, R.J. *et al.* (1995). The results of denatured homologous vein grafts as conduits for secondary haemodialysis access surgery. *Eur J Vasc Endovasc Sug*, **9**, 58–63.
44. Bosman, P.J. *et al.* (1998). A comparison between PTFE and denatured homologous vein grafts for haemodialysis access: A prospective randomised multicentre trial. The SMASH Study Group. Study of graft materials in access for hemodialysis. *Eur J Vasc Endovasc Surg*, **16**, 126–132.
45. Ahmed, N. *et al.* (1976). Brachial artery to brachial vein preserved vein allograft fistulas for haemodialysis. *J Cardiovasc Surg*, **17**, 483–488.
46. Piccone, V.A. Jr. *et al.* (1978). Preserved saphenous vein allografts for vascular access. *Surg Gynecol Obstet*, **147**, 385–390.
47. Baraldi, A. *et al.* (1989). Absence of rejection in cryopreserved saphenous vein allografts for haemodialysis. *ASAIO Trans*, **35**, 196–199.

48. Wellington, J.L. (1981). Umbilical vein grafts for vascular access in patients on long-term dialysis. *Can J Surg*, **24**, 608–609.
49. Jorgensen, L. *et al.* (1985). Human umbilical vein for vascular access in chronic haemodialysis. *Scand J Urol Nephrol*, **19**, 49–53.
50. Biggers, J.A. (1975). Bovine graft fistulas in patients with vascular access problems receiving haemodialysis. *Surg Gynecol Obstet*, **140**, 690–692.
51. Sterling, W.A., Taylor, H.L. and Diethelm, A.G. (1975). Vascular access for haemodialysis by bovine graft arteriovenous fistulas. *Surg Gynecol Obstet*, **141**, 69–72.
52. Johnson, J.M. *et al.* (1976). The modified bovine heterograft in vascular access for chronic haemodialysis. *Ann Surg*, **183**, 62–66.
53. Hurt, A.V. *et al.* (1983). Bovine carotid artery heterografts versus polytetrafluoroethylene grafts. A prospective, randomized study. *Am J Surg*, **146**, 844–847.
54. Doyle, D.L. and Fry, P.D. Polytetrafluoroethylene and bovine grafts for vascular access in patients on long-term haemodialysis. *Can J Surg*, **25**, 379–382.
55. Brems, J., Castaneda, M. and Garvin, P.J. (1986). A five-year experience with the bovine heterograft for vascular access. *Arch Surg*, **121**, 941–944.
56. Enzler, M.A. *et al.* (1996). Long-term function of vascular access for haemodialysis. *Clin Transplant*, **10**, 511–515.
57. Farmer, D.L. *et al.* (1993). Failure of glow-discharge polymerisation onto woven Dacron to improve performance of haemodialysis grafts. *J Vasc Surg*, **18**, 570–575.
58. Helling, T.S., Nelson, P.W. and Shelton, L. (1992). A prospective evaluation of plasma-TFE and expanded PTFE grafts for routine and early use as vascular access during haemodialysis. *Ann Surg*, **216**, 596–599.
59. Schanzer, H.M. *et al.* (1989). Clinical trials of a new polytetrafluoroethylene-silocone graft. *Am J Surg*, **158**, 117–120.
60. Nakagawa, Y. *et al.* (1995). Clinical trial of new polytetrafluoroethylene vascular grafts for haemodialysis: Compared with expanded polytetrafluoroethylene grafts. *Artif Organs*, **19**, 1227–1232.
61. Wilson, G.J. *et al.* (1995). A Corethane/polyester composite vascular prosthesis for vascular access. Comparison with expanded polytetrafluoroethylene grafts in a canine model. *ASAIO J*, **41**, M728–M734.
62. Drasler, W.J. *et al.* (1990). Venturi grafts for haemodialysis access. *ASAIO Trans*, **36**, 753–757.

63. DeRosa, P. *et al.* (1986). A simple vascular access in "desperate" cases. *Ital J Surg Sci*, **16**, 209–210.
64. Mandel, S.R. and McDougal, E.G. (1985). Popliteal artery to saphenous vein vascular access. *Surg Gynecol Obstet*, **160**, 358–359.
65. Gorski, T.F. *et al.* (1998). Lower-extremity saphenous vein transposition arteriovenous fistula: An alternative for haemodialysis access in AIDS patients. *Am Surg*, **64**, 338–340.
66. Cimochowski, G.E. *et al.* (1991). Use of the spiral vein graft as an arterial substitute for secondary access. *Am J Nephrol*, **11**, 64–66.
67. Taylor, S.M. *et al.* (1996). Results and complications of arteriovenous access dialysis grafts in the lower extremity: A five year review. *Am Surg*, **62**, 188–191.
68. Khadra, M.H., Dwyer, A.J. and Thompson, J.F. (1997). Advantages of polytetrafluoroethylene loops in the thigh for haemodialysis access. *Am J Surg*, **173**, 280–283.
69. Korzets, A. *et al.* (1998). The femoral artery–femoral vein polytetraflurorethylene graft: A 14 year experience. *Nephrol Dial Transplant*, **13**, 1215–1220.
70. Williams, S.K., Jarrell, B.E. and Kleinert, L.B. (1994). Endothelial cell transplantation onto polymeric arteriovenous grafts evaluated using a canine model. *J Invest Surg*, **7**, 503–517.
71. Schmidt, S.P. *et al.* (1998). Evaluation of expanded polytetrafluoroethylene arteriovenous access grafts onto which microvessel-dervied cells were transplanted to "improve" graft performance: Preliminary results. *Ann Vasc Surg*, **12**, 405–411.
72. Swedenborg, J. *et al.* (1997). *In vitro* endothelialization of arteriovenous loop grafts of haemodialysis. *Eur J Vasc Endovasc Surg*, **13**, 272–277.
73. Bell, D.B. and Rosenthal, J.J. (1988). Arteriovenous graft life in chronic hemodialysis. *Arch Surg*, **123**, 1169–1172.
74. Hughes, P.E. and How, T.V. (1996). Effects of geometry and flow division on flow structures in models of the distal end-to-side anastomosis. *J Biomech*, **29**, 855–872.
75. Wijesinghe, L.D., Smye, S.W. and Scott, J.A. (1998). Impedance index measurements of *in vitro* PTFE end-to-side anastmoses: Effect of angle and Miller cuff. *Eur J Vasc Endovasc Surg*, **16**, 65–79.
76. Staalsen, N.H. *et al.* (1995). The anastomosis angle does change the flow fields at vascular end-to-side anastomoses *in vivo*. *J Vasc Surg*, **21**, 460–471.

77. Hakaim, A.G. (1997). Superior patency of arteriovenous anastomoses employing the diffuser principle. *Microcirculation*, **4**, 163.
78. Ifudu, O. *et al.* (1998). Haemodialysis dose is independent of type of surgically crated vascular access. *Nephrol Dial Transplant*, **13**, 2311–2316.
79. Ifudu, O. *et al.* (1996). Correlates of vascular access and nonvascular access-related hospitalisations in haemodialysis patients. *Am J Nephrol*, **16**, 118–123.
80. Ifudu, O. *et al.* (1997). Determinants of type of initial haemodialysis access. *Am J Nephrol*, **17**, 425–427.
81. Sands, J. and Miranda, C.L. (1997). Increasing numbers of AV fistulas for haemodialysis access. *Clin Nephrol*, **48**, 114–117.
82. Leehey, D.J. (1994). Haemodialysis in the diabetic patient with end-stage renal disease. *Ren Fail*, **16**, 547–553.
83. Nishi, K. *et al.* (1985). Vascular response to vasoactive agents in dialysed patients with chronic renal failure with and without diabetes mellitus. *Nephrol Dial Transplant*, **9**, 10–15.
84. Wetzig, G.A., Gough, I.R. and Furnival, C.M. (1985). One hundred cases of arteriovenous fistula for haemodialysis access: The effect of cigarette smoking on patency. *Aust NZ J Surg*, **55**, 551–554.
85. Hakaim, A.G. and Scott, T.E. (1997). Durability of early prosthetic dialysis graft cannulation: Results of a prospective, nonrandomized clinical trial. *J Vasc Surg*, **25**, 1002–1006.
86. Diskin, C.J., Stokes, T.J. and Panus, L.W. (1996). Effects of hand dominance and sleep side preference on haemodialysis vascular access survival. *Nephron*, **74**, 724–728.
87. Churchill, D.N. *et al.* (1994). Probability of therombosis of vascular access among haemodialysis patients treated with recombinant human erythropoietin. *J Am Soc Nephrol*, **4**, 1809–1813.
88. Taucher, L.A. (1985). Immediate, safe haemodialysis in arteriovenous fistulas created with a new tunneler. *Am J Surg*, **150**, 212–215.

CHAPTER 10

COMPLICATIONS OF VASCULAR ACCESS

Keith M Rigg MD, FRCS
Nottingham City Hospital
Nottingham NG5 1PB, UK

10.1. Introduction

Vascular access is the Achilles heel for all involved in the care of haemodialysis patients. The initial planning, timing, operative technique and long-term care of vascular access are all important factors in the outcome. As the dialysis population becomes more elderly with more co-morbidity, there will be a smaller proportion of patients suitable for transplantation. Consequently, an increasing proportion of patients will be on haemodialysis for the rest of their lives. It becomes more important that these patients have good vascular access that is complication free and has a high long-term success rate. There is a subgroup of patients who have a high risk of complications and failure with vascular access. This includes patients with diabetes, arteriosclerosis, previous problems with vascular access, those who are elderly, and patients from the Black and Asian populations (1–3). A multidisciplinary approach to planning dialysis access and subsequent care for these patients is essential.

Complications of vascular access are inevitable, but surgeons, physicians, dialysis nurses and patients have a responsibility to minimise those complications that are potentially avoidable. Surgeons should have a commitment to vascular access, be involved in the planning of access, possess good technical skills, and be able to recognise and promptly treat any complications. The surgeon has been shown to be an independent factor for vascular access success (4). Nephrologists should refer patients early for access assessment when possible; avoid temporary subclavian lines, instead favouring internal jugular or femoral lines; and recognise problems early with subsequent treatment or referral. Dialysis nurses should develop needling techniques that reduce the risk of infection, haemorrhage and aneurysm formation, and recognise and report dysfunctional fistulas and grafts at an early stage. Patients, prior to dialysis, should be taught not to allow venepuncture and blood pressure recording on their non-dominant arm, keep well hydrated, and again to report any changes in their vascular access. A

Table 10.1. Complications of vascular access.

Wound complications	Delayed wound healing	
	Wound breakdown	
	Wound infection	
	Haemorrhage	
	Lymphocoele	
Complications of the access conduit	Early thrombosis	
	Late thrombosis	
	Stenosis	
	Aneurysm	
	Infection	
	Perigraft seroma	
Haemodynamic complications	Venous hypertension	
	Ischaemia	Arterial steal
		Ischaemic monomyelic neuropathy*
		Carpal tunnel syndrome*
		Reflex sympathetic dystrophy*
	Cardiac failure	
Functioning but unusable access	Failure to mature	
	Vein/graft too deep	
	Vein/graft too superficial	
	Painful cannulation	
	Recurrent clotting	

*Although these are neurological complications, ischaemia is thought to be an aetiological factor.

number of centres have introduced graft surveillance programmes using techniques such as ultrasound flow dilution and Doppler ultrasound (5–7). This allows dysfunctional vascular access to be identified at an early stage when surgical or radiological intervention can successfully treat stenoses before graft failure intervenes (8). Although this approach requires extra resources and further validation, it is likely to be cost effective if longer-term patency can be achieved with a concurrent reduction in the use of temporary venous catheters for access.

The complications of vascular access that will be discussed in this chapter are listed in Table 10.1.

10.2. Patency Results

Before considering the complications of vascular access in greater detail, it is important to consider patency rates. This is the most significant clinical outcome measure for vascular access. If a fistula or graft fails, it can be because of any of the complications detailed in the rest of this chapter. It is clear that patency rates are better with arteriovenous fistulae when autologous vein is used, compared to grafts, when synthetic material is used. Primary patency is when the access is still functioning without any intervention, secondary patency when the access has required some intervention to remain patent. Secondary patency rates reflect the workload generated by complications. Secondary patency rates from a number of centres for arteriovenous fistulae and grafts are summarised in Tables 10.2. and 10.3. It is clear that patency rates are greater for brachial than radial arteriovenous fistulas due to the size of the vessels involved. The relatively low one-year patency rates reflects the 20–40% of early failures but once the fistula is working, it can be used for many years. Most of the long-term results for grafts have been obtained using polytetrafluorethylene (PTFE) grafts that have been consistently used. Other graft materials such as bovine heterografts and denatured autologous vein have been used less frequently. With the development of new types of PTFE grafts, it remains to be seen if patency rates can be improved and become equivalent to fistulae in the long term.

Table 10.2. Secondary patency rates of arteriovenous fistulae.

Type of Fistula and Number	One Year (%)	Two Years (%)	Three Years (%)	References
Brachial (73)	84		78	(9)
Brachial (59)	90	86		(10)
Brachial (81)	70	57	50	(11)
Radial & brachial (429)	74		64	(12)
Radial (154)	64	53	45	(13)

Table 10.3. Secondary patency rates of arteriovenous grafts.

Graft Material and Number	One Year (%)	Two Years (%)	Three Years (%)	Reference
PTFE (100)	74		59	(14)
PTFE (189)	76		50	(15)
Homologous vein (195)	76		52	(16)
PTFE (69)	58		40	(12)
Homologous vein (63)	63			(17)

10.3. Wound Complications

Wound complications tend to occur early following access formation but, if not dealt with promptly and effectively, can lead to access failure.

Many patients with end-stage renal failure have tissues that bruise and traumatise easily. Wound healing may be delayed as a result of malnutrition and previous use of steroids and as a result be more prone to infection. A number of steps should be taken to minimise delayed wound healing, wound breakdown and wound infection.

Incisions should be placed in the direction of Langers lines where possible. They should be closed with good tissue approximation without tension. A single-layer skin closure is perfectly satisfactory for an arteriovenous fistula but when a synthetic graft is used, it should be closed in several layers. Wound breakdown in the presence of a graft can be extremely difficult to deal with. An absorbable subcuticular suture is a well-tried and tested method of wound closure and further support and approximation can be provided with steristrips. For those patients with a particularly friable skin, the wound can be closed with tissue glue.

Haemorrhage can be minor and simply result in bruising or a small haematoma, although pressure should only be applied judiciously or it will result in access failure. In the first 24–48 hours, haemorrhage can be more catastrophic if there is a breakdown of the anastomosis or a ligature comes off a venous or arterial branch. If the wound remains intact, bleeding can stop by tamponade but this may result in compression and subsequent thrombosis of the fistula or graft. In this situation or where there is significant bleeding from the wound, the patient should be returned to the theatre for re-exploration.

Lymphocoeles occur at the anastomotic site and arise from disruption of local lymphatics during mobilisation of the vessels. The common sites tend to be the antecubital fossa and the groin. They present soon after surgery as a tense swelling under the wound and have a transmitted pulsation from underlying vessels. It can sometimes be difficult to confidently exclude a false aneurysm and a Doppler ultrasound should resolve the diagnostic dilemma. Treatment depends on the size and local pressure effect. Small lymphocoeles should resolve spontaneously but larger ones may require drainage. The author prefers an operative approach to this, leaving a small suction drain *in situ* rather than repeated aspiration that may encourage infection.

The incidence of wound infection following access surgery should be low as the operation can be defined as clean surgery, although when the groin is used, the procedure may be classified as clean-contaminated surgery. When infection does occur, the offending organism is likely to be *Staphylococcus aureus*. The best method of preventing wound infection is the use of strict aseptic technique but prophylactic antibiotics should be considered in addition. Studies looking at the use of prophylactic antibiotics in reducing the risk of infection have been contradictory; some showing an advantage and others showing no effect. This may be more to do with the site and type of access in that wound infections are more common in the groin where synthetic graft is used than in the arm where the autologous vein is commonly used (18). However, the use of a single-shot prophylactic antibiotic to cover *S. aureus* is prudent to minimise the incidence of infection and this is particularly the case if prosthetic material is being used. When a graft is being placed in the groin, antibiotic prophylaxis should cover *S. aureus* and gram-negative organisms.

10.4. Complications of the Access Conduit

10.4.1. *Early thrombosis*

Thrombosis may occur in the first 24 hours post-operative. It occurs more commonly in fistulas based on the radial artery rather than fistulas or grafts based on larger arteries with usually larger run-off veins (19). It can be due to many factors that result in poor flow through the fistula or graft. This can be summarised as either inadequate inflow or inadequate venous run-off. Inadequate inflow can be due to atherosclerotic and calcified arteries, hypotension or hypovolaemia, or technical problems in creating the anastomosis. Inadequate venous run-off can be due to the run-off vein being kinked or under tension, in spasm, scarred from previous venepuncture, or of inadequate size. When a graft is placed, kinking should be avoided by ensuring it does not cross a joint and if placed in loop configuration, by ensuring the apex of the loop is not too tight. The use of compression dressings post-operatively should be avoided. In theory, many of the factors identified above can be minimised by careful assessment, good planning, technical expertise and vigilant post-operative care, but in practice, early thrombosis still occurs.

The diagnosis is usually clear in that a palpable thrill is absent although this may be more difficult to determine when a graft has been placed. If there is doubt, then examination with a handheld Doppler or a formal Doppler ultrasound can help to clarify the situation. If the fistula is only weak, the patient can be reviewed in clinic after two weeks and if there is no thrill palpable, further surgery can be planned. The results of re-exploration for early thrombosis of fistulae are dependent on the reasons for failure. If the fistula was initially working well after clamp release, then re-exploration is warranted. Gentle thrombectomy of the vein and, on occasions, the artery, followed by re-anastomosis with good hydration will often restore flow. If the vessels were of poor quality and there was poor initial function, re-exploration is unlikely to salvage the fistula and is probably not warranted. All early thromboses of grafts should be re-explored unless there were clear reasons at the initial operation why this would not be effective. Either or both of the anastomoses should be taken down, the graft milked free of thrombus, flushed with heparinised saline and the anastomosis redone.

10.4.2. *Late thrombosis*

Thrombosis can occur at any time after formation of vascular access, but from the comparative patency rates of fistulae and grafts, it is more likely to occur earlier in the patient with a graft. Although thrombosis is the acute event, aetiological factors may have been present for some time and are not always the same for fistulae and grafts.

Similar factors for both types of access include progressive arteriosclerosis of the inflow artery with reduced flow in the access conduit, needle trauma resulting in fibrosis and consequent stenosis and extraluminal haematoma causing compression. Patients can become hypovolaemic through concurrent illness such as diarrhoea or through having a target weight set too low, or hypotensive; both of which may result in thrombosis. Other patients have hypercoaguable states that may be primary or secondary. Secondary states may be induced by surgery, infection and the use of erythropoeitin. Patients with a fistula can develop stenosis of the run-off vein, which can be caused by repeated needling and trauma at one site. In comparison, patients with synthetic grafts are prone to a stenosis at the venous anastomosis secondary to neointimal hyperplasia. Stenosis in either setting can progress and lead ultimately to thrombosis and this is discussed further below.

The diagnosis of late thrombosis is usually very clear; the patient has noticed that fistula thrill is absent or the dialysis nurse is unable to cannulate. Treatment options depend on whether the access is a fistula or graft, how long it has been thrombosed for, and whether there had been progressive dysfunction. There are essentially three treatment options: do nothing and establish further access, surgical revision or radiological intervention. Once a fistula has thrombosed, there is usually little to be gained from re-exploration and thrombectomy unless this is done very early in conjunction with dealing with the underlying aetiology. However, the use of bridge grafts or a more proximal anastomosis is appropriate if there is a proximal vein that has not thrombosed. There is however some evidence to show that pharmacological thrombolysis associated with angioplasty of stenoses can restore patency of fistulas. Initial success and one-year patency rates of 59% were seen after thrombolysis, percutaneous angioplasty and thromboaspiration in one study of 64 thrombosed fistulas (20). Local thrombolysis within 48 hours of thrombosis using recombinant tissue plasminogen activator (rTPA) followed by immediate use of the fistula for haemodialysis was successful in a group of 11/16 patients with a median functioning time of five months (21). This approach has not been reproduced in other studies although the importance of early use of the fistula for dialysis and subsequent investigation and treatment of stenoses was emphasised.

The techniques of pharmacothrombolysis are more applicable to synthetic grafts than fistulae and therefore most of the published work and experience are on these. When a graft thromboses, surgical thrombectomy has been the mainstay of treatment until recently. When performed, it should be done as soon as possible although satisfactory results can be obtained up to one week post-thrombosis. It is usually necessary to combine this with dealing with the underlying stenosis at the venous anastomosis. This can be done using a patch angioplasty to widen the outflow vein or by using a jump graft to bypass the stenotic area. This technique is however being superseded as primary treatment by pharmacological thrombolysis with or without mechanical thrombectomy. Although the result of pharmacological thrombolysis is equivalent to surgical thrombectomy in some centres, the former has the advantage that overnight hospitalisation is not usually required and the graft can always be used straightaway (22, 23). Pharmacological thrombolysis is usually performed using urokinase or tissue plasminogen activator (TPA), and increasingly using a crossed-catheter pulse spray technique. This allows maximum exposure of the thrombus to the thrombolytic agent and can be combined with angioplasty of any associated stenotic lesion. Mechanical thrombectomy can be performed using a number of different devices that macerate and aspirate the thrombus. These include the embolectomy catheter (24); the hydrodynamic catheter (25) which utilises the Venturi effect; and oscillating or rotational devices (26). Details and results of the techniques will be described elsewhere.

10.4.3. *Stenosis*

It is obviously preferable to identify stenosis within fistulae or grafts before thrombosis occurs and when therapeutic options are more likely to be successful. Although this can be recognised by the dialysis nurse when dialysis becomes suboptimal or by the clinician on careful clinical examination, these methods are not very sensitive. An increasing number of units now have in place graft surveillance programmes, using techniques such as ultrasound flow dilution and Doppler ultrasound. This enables stenoses to be identified at an earlier stage when angioplasty or surgical revision can be employed with good effect. Although these techniques may suggest stenosis, a fistulogram or angiogram should clearly identify the stenosis and the associated venous anatomy.

Within an arteriovenous fistula, stenoses can be singular or multiple and are usually caused by fibrosis resulting from repeated cannulation at the same site or extravasation of blood from needling. Stenosis within a synthetic graft is usually at the venous anastomosis but can occur anywhere along its length. Stenosis at the venous anastomosis is secondary to neointimal hyperplasia. This is a multifactorial response to endothelial injury, the shear stress of blood flow and a change in vessel wall compliance as pulsatile arterial blood enters the vein. It is also influenced by the angle of the anastomosis and probably by the luminal size of the graft and run-off vein. Stenosis within the lumen of the graft usually arises from repeated needling at a site where thrombus forms leaving an intraluminal thrombogenic surface. Regenerative fibrosis takes place resulting ultimately in stenosis.

The current treatment of choice for a stenosis in a fistula or graft is percutaneous balloon angioplasty. Immediate success rates in excess of 90% can be obtained, with six-month and one-year patency rates of 67% and 44% for anastomotic stenoses and 56% and 18% for midgraft stenoses (27). When restenosis occurs, further angioplasty can be performed. The use of stents after balloon angioplasty is still being evaluated. Some centres report improved patency after stent insertion (28, 29) whereas others have not shown any benefit over balloon angioplasty alone (30, 31). Stent insertion can be of benefit in venous rupture following balloon angioplasty where one-year secondary patency rates of 56% have been reported (32). Atherectomy of stenoses is also being evaluated and early results are mixed (33, 34).

10.4.4. *Aneurysm*

Aneurysm formation can complicate an arteriovenous fistula or graft. Two types of aneurysm can be encountered:

(i) True aneurysmal dilatation, usually of a fistula, but rarely can occur in a longstanding graft, when the graft material has disintegrated. This should now be an infrequent finding with modern graft materials.
(ii) False aneurysm arising from the access conduit due to recurrent needling trauma or at the anastomosis due to partial anastomotic breakdown. This can subsequently become infected resulting in a mycotic aneurysm.

Aneurysms are more commonly encountered in the AV fistula where they may be true or false. Distinguishing between the two is rarely of any consequence. It can be difficult to say whether veins are aneurysmal or merely dilated (Fig. 10.1). As veins become more dilated, they become more tortuous and can become more difficult to needle. Flow disturbance occurs within the vein, resulting in further weakening of the vein wall and thrombus may also be laid down around the periphery of the lumen. This type of dilatation rarely causes symptoms to the patient apart from the cosmetic appearance. Localised aneurysmal areas often arise from repeated needling trauma at the same site and result from weakening of the vessel wall. These may require surgery if they continue to enlarge or cause fistula dysfunction.

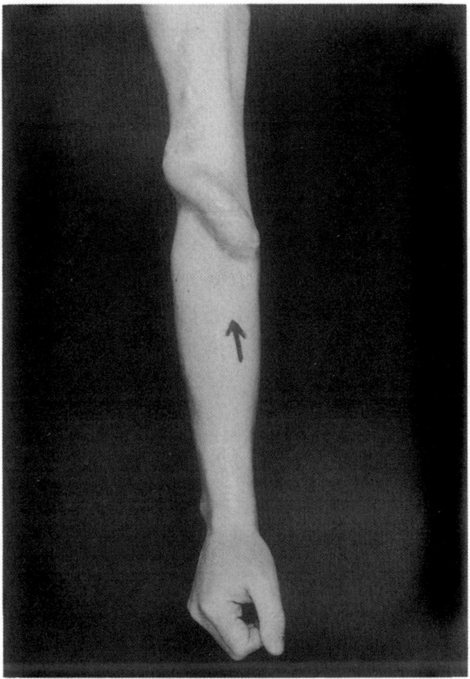

Fig. 10.1. Aneurysmal dilatation of vein in a brachiocephalic arteriovenous fistula.

Various surgical procedures have been described which include ligation of the aneurysm with associated bypass, aneurysmorrhapy and more recently using vascular staples. In addition, some success has been reported using endoluminal stents (35).

True aneurysmal dilatation of grafts is now not seen with modern grafts although it was seen with the early thin-walled synthetic grafts. False aneurysms of grafts are of more concern and nearly always require surgical intervention as the risk of infection is high. These are found at either a needling site on the graft or at an anastomosis. A misplaced dialysis needle can lacerate grafts and this leads to extravasation of blood, the "blow". Unless this is adequately compressed, a false aneurysm subsequently develops. The expansile nature of the swelling can usually distinguish it clearly from a haematoma but, if there is doubt, Doppler ultrasound can confirm the diagnosis. If surgical intervention takes place before infection of the aneurysm supervenes, it can be possible to primarily close the defect in the graft. Once infection has supervened, the infected portion of the graft with the aneurysm needs to be resected and a new piece of graft inserted as a jump graft to bypass the infected area. False aneurysms occurring at the anastomosis result from a partial anastomotic breakdown which may occur early due to a technical problem with the anastomosis or later due to infection. Infected aneurysms in this site are a surgical challenge. Graft reconstruction and salvage is not always possible and the inflow artery may require ligation.

Aneurysmal dilatation of the inflow artery can also be seen when the fistula has been present for some time. This does not usually manifest itself until the fistula is ligated and the vessel is seen directly.

10.4.5. *Infection*

Infection of the access conduit is seen much more commonly with synthetic grafts than with fistulas, and is also more common in patients with diabetes. The site of access formation also influences the incidence of infection, with groin grafts having a higher incidence. The infection can be localised, systemic or metastatic. Infection rates for autologous vein fistulas should be 2–3%, but may be 11–35% when synthetic grafts are used (36).

Localised infection may manifest itself as an area of cellulitis over the access conduit or, as already mentioned, with a false aneurysm. Localised cellulitis should be treated promptly to reduce the risk of the access conduit becoming colonised, with resultant systemic or metastatic infection. Treatment comprises antibiotic therapy and avoiding the site of infection for needling. Depending on the extent and severity of infection, the patient may require intravenous antibiotics and avoidance of using the fistula or graft. The causative organism in fistulas is *S. aureus* in over 90% of cases (19), but this should be confirmed by culture where possible. Other studies have demonstrated a wide range of responsible organisms including streptococcal and gram-negative organisms. Flucloxacillin is a good first-line antibiotic that can be supplemented with rifampicin if clinical response is slow. A two-week course of antibiotics is usually required but may need to be longer if clinical circumstances dictate. The treatment of mycotic aneurysms has already been detailed above.

When grafts or, uncommonly fistulas, become colonised, then systemic infection may ensue. This should always be suspected when haemodialysis patients present with systemic signs of infection. There is usually erythema and/or tenderness over a portion of the access conduit. In addition, there may be abscess formation, a rapidly expanding false aneurysm and, on occasions, rupture with catastrophic haemorrhage. Once graft colonisation has occurred, it can be extremely difficult to eradicate the infection even with prolonged courses of antibiotics. Patients require admission and the use of intravenous antibiotics. Unless the responsible organism has been identified, treatment should be initiated with broad-spectrum antibiotics covering the likely pathogens. In the likely event of it being a staphylococcus infection, the preferred antibiotic is now flucloxacillin. Vancomycin used to be the drug of choice, but with the development of vancomycin-resistant enterococci, it should be used more sparingly. In dialysis patients, it needs to be given in a reduced dose which usually coincides with a dialysis day. If infection cannot be eradicated, then surgical revision is necessary. Graft salvage is attempted where possible, with excision of the infected segment and a new segment of graft inserted, circumventing the infected area. Depending on the site and severity of infection, it may be necessary to remove the graft. Infections at the anastomoses can rarely be salvaged and require excision of the graft with either a repair of the artery or ligation of the artery. Although repair of the artery either directly or using an onlay patch is the ideal, there is a risk of re-infection with subsequent anastomotic breakdown and haemorrhage. Ligation of the artery is the safest means of controlling the situation and results in surprisingly few problems with ischaemia. In the arm, the brachial artery can safely be ligated but the groin can present more problems. The superficial femoral artery should be used where possible for grafts in the thigh, since this can be safely ligated if necessary, allowing the profunda femoris to supply the lower limb. Despite that, it is possible to ligate the common femoral artery without resultant problems, provided there were no pre-existent ischaemic problems.

Metastatic infections, such as metastatic abscesses or endocarditis, can also arise as a result of an infected access conduit with bacteraemia or septicaemia. The treatment of these needs to be directed at both the metastatic and primary sites of infection.

10.4.6. *Perigraft seroma*

Repeated needling of grafts is essential for haemodialysis and it is preferable for needling sites to be rotated to avoid excess damage at one site. Implantation of a synthetic graft induces an inflammatory response in the perigraft tissues. This results in the formation of a perigraft fibrous capsule, and in porous grafts, tissue ingrowth allowing the development of a pseudointima. After needling, the hole in the graft is plugged by fibrous tissue ingrowth. Inevitably, needling of the graft results in some extravasation of blood and can lead to a significant perigraft haematoma and seroma. Not only can this result in infection with subsequent false aneurysm formation but also in the perigraft fibrous cuff becoming separated from the graft. This may subsequently lead to the pseudointimal layer sloughing and predisposing to thrombus formation (37).

10.5. Haemodynamic Complications

10.5.1. *Venous hypertension*

When the entire limb becomes swollen after creation of the vascular access, it indicates a significant stenosis or occlusion of the proximal vein. This usually occurs within days if there is a pre-existing lesion, but will occur later if there is an evolving stenosis or new thrombosis.

As most access is created in the upper limb, this usually indicates subclavian vein stenosis or occlusion. The commonest aetiology for this is previous insertion of temporary subclavian venous catheters. There are a number of important aetiological factors and these include catheters that have been in place for a prolonged period of time, repeated attempts at cannulation, those catheters that have been infected and the physical properties of the catheter itself. Since this complication of subclavian vein catheterisation has been recognised, there has been a move to abandon this approach in favour of an internal jugular approach. It is hoped that the incidence of subclavian vein stenosis will decrease as a result. In those patients who have had previous subclavian cannulation and in whom an upper limb fistula or graft is planned, imaging of the subclavian veins is recommended to avoid the potential sequelae. Swelling of the lower limb occurs uncommonly after insertion of a leg graft, where there is an occlusion of the common femoral or iliac vein.

A rapid swelling of the whole limb post-access formation is virtually pathognomonic of a proximal venous outflow problem. It needs to be distinguished from the temporary swelling of the limb that can occur after insertion of a graft, particularly where a deep vein is used as a run-off vein. When the swelling develops over a longer period of time, it is accompanied by venous collaterals over the upper arm and chest. Doppler ultrasound and venography can confirm the clinical diagnosis.

There are a number of treatment options available. Up until recently, the only guaranteed way to reverse the effect and restore the arm to normal size was to ligate the fistula. This rules out that limb from availability for future access but does not deal with the underlying problem. In those patients where access is problematical, where the swelling is not extreme, and where the measures outlined below are unsuccessful, it may be possible to continue dialysis for some time after due discussion with the patient. However, with the development of interventional radiology, balloon angioplasty of stenoses in the subclavian vein and stent insertion have been increasingly used, although with varying success. As the aetiology of subclavian vein stenosis is related to catheterisation, the stenosis is typically a short segment. It is possible to pass a guidewire across the stenosis allowing it to be dilated and an expandable metallic stent to be placed. When there is occlusion rather than stenosis, it is sometimes possible to pass the guidewire through it, but not always. The guidewire can either be introduced from the same arm or proximally via the femoral vein. There is a good initial response rate after angioplasty but a high rate of restenosis. An early study showed an initial success in 76% of stenoses, but with a one-year patency rate of 34% and at two years, 6% (38). Other studies have shown one-year patency rates of 12–36% (39–41). Central venous stenoses tend to be fibrotic and have elastic recoil properties and therefore, angioplasty has an initial failure rate of 30% (42). As angioplasty alone does not have good *long-term* results, there is an increasing use of balloon angioplasty associated with insertion of an expandable metal stent. Long-term results after the insertion of Gianturco stents are no better than angioplasty alone (40). However, there are better results being reported for the self-expandable Wallstent™ (39, 43) and the balloon expandable Palmaz stent (44) with one- and two-year patency rates of 60–70% and 50–60%, respectively. When restenosis occurs, despite the presence of a stent, it is possible to insert an overlapping stent. Cumulative secondary one-year (two-year) patency rates of 100% (85%) have been reported (39). If the stenosis or occlusion is long, balloon angioplasty and insertion of stent are not successful. In this situation, veno-venous bypass surgery should be considered if function of the fistula or graft is to be maintained. However, this is a major and demanding surgery that is not without significant morbidity and mortality, and in these situations, ligation of the fistula may be more prudent.

Although proximal venous occlusion or stenosis as described above causes venous hypertension, the associated chronic changes are not usually seen unless no treatment is offered. Venous hypertension is also seen in a different setting where it involves the forearm and hand. As with other chronic venous disease, it may lead to oedema, skin changes and eventual venous ulceration. This occurs rarely in less than 1% of all fistulae and in those where there is a side-to-side anastomosis. It is caused by retrograde venous flow in arterialised veins where the valves are incompetent. When a side-to-side radial or Brescia–Cimino fistula has been fashioned, venous hypertension may involve the whole hand (45, 46) or selective digits (47). The hand and forearm may develop changes of venous hypertension when a side-to-side brachial fistula has been created. In mild cases, no treatment may be necessary, but in more severe cases with pain and ulceration, the distal limb of the fistula should be ligated.

10.5.2. *Arterial insufficiency*

Arterial insufficiency usually manifests itself as a steal syndrome. Less frequently, neurological complications occur which may have a component of vascular insufficiency. These include carpal tunnel syndrome in dialysis patients, ischaemic monomyelic neuropathy and reflex sympathetic dystrophy. Although the diagnosis of these conditions can sometimes be difficult, the central role of ischaemia should be recognised, which usually requires timely surgical intervention.

Haemodynamic studies have demonstrated that in the majority of radiocephalic fistulas, there is a 40% reduction of the thumb blood pressure, which can be restored by occluding the fistula. Occlusion of the radial artery distal to the fistula resulted in a significant increase in thumb blood pressure, indicating radial steal (48). Thus, although a functional steal may often occur after access surgery, this only rarely results in symptoms. The steal syndrome may occur early or late following access formation with signs and symptoms of ischaemia, which may be worse during dialysis. In milder forms, patients may complain of cold and pale fingers during dialysis. In its more severe forms, patients may have the classical features of pain, paralysis, paraesthesia, pallor and coldness. In the larger series, steal syndrome occurs after 1.8–9% of vascular access operations (49–51). In the upper limb, it occurs least frequently in the wrist fistulas and more often in the elbow fistula or graft. It can also occur in the lower limb after arteriovenous graft insertion (Fig. 10.2). It is more common in patients who have diabetes or peripheral vascular disease (50). As the dialysis population becomes older and with more associated co-morbidity, it is likely that the incidence of this complication will increase. A number of steps can be taken to reduce the risk of vascular steal in high-risk patients. These include making the fistula as distal in the limb as possible, e.g. preferring wrist to elbow, performing end-to-side rather than side-to-side anastomoses, and ensuring the arteriotomy is no more than 5–7 mm in length.

It is important to make the diagnosis early and decide on a treatment option before irrecoverable damage occurs to the limb. Diagnosis can be difficult particularly if pain is the only symptom. Vascular steal, carpal tunnel syndrome and venous hypertension can all cause pain. A thorough clinical history and examination, knowledge of the anatomy of the access, and use of investigations such as Doppler, monitoring digital pressures and plethysmography can help to clarify the cause of the symptoms (52). Mild

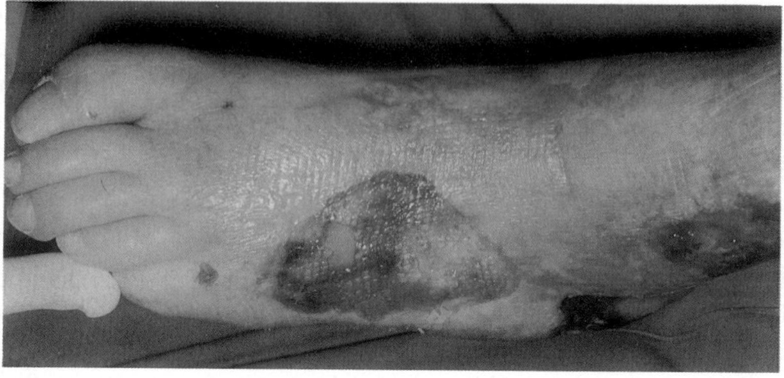

Fig. 10.2. Ulceration of the lower limb three weeks after femoro-femoral arteriovenous loop thigh graft in a patient with diabetes.

steal symptoms that are reversible after dialysis can usually be managed conservatively, provided the patient is kept under regular review. Moderate to severe steal may occur within the first 24 hours but more often at a later time, and should be treated. When severe ischaemia occurs, it is usually seen within 12–24 hours of access construction. For this reason, it is important that the patient is kept under surgical review for this period. If not treated promptly, compartment compression syndrome and irreversible ischaemia can develop, causing permanent functional impairment and possible amputation.

There are a number of surgical treatment options available.

- *Ligation of the fistula*

Ligation of the fistula should be performed for any patient manifesting signs and symptoms of severe ischaemia. This is the only guaranteed way to restore blood flow to the extremity and must be done without delay to prevent functional impairment. Ligation of the fistula should also be considered in less severe ischaemia if the access-preserving measures described below are ineffective. At surgery, the vein can be ligated and divided at the anastomosis although this does pose a theoretical risk of aneurysmal dilatation of the vein stump. However, this is a simple, quick, safe and effective method of reversing the steal. Alternatively, the anastomosis can be taken down and the artery reconstructed, usually necessitating the use of a vein patch to prevent any narrowing of the artery.

- *Ligation of the distal venous limb if there is side-to-side anastomosis*

For vascular steal that is not limb threatening, ligation of the distal venous limb can be effective in sufficiently improving blood supply. This converts the fistula from a side-to-side anastomosis to an end-to-side anastomosis.

- *Plication or banding to reduce flow through the conduit*

A number of techniques have been described to reduce the blood flow through the conduit and therefore to reduce symptoms of steal. Banding or plication of the outflow tract (50, 53, 54) has been performed with varying degrees of success. The problem with these techniques has been in determining how much the outflow tract should be narrowed in order to abolish the signs of ischaemia but at the same time not causing thrombosis. The use of intra-operative pulse volume recording has been advocated as a scientific way to determine how much banding/plication should be performed (54).

- *Use of stepped interposition grafts*

Although this technique is now not routinely used, case reports have demonstrated efficacy of the technique in flow reduction. A stepped interposition graft is inserted at the arterial end of the access and the desired flow reduction can resolve the symptoms of steal (55).

- *Distal arterial ligation and bypass*

This technique has recently been described and combines correction of the vascular steal with continuing use of the access site (53, 56). The technique involves ligating the artery just distal to the anastomosis and placing a bypass graft of vein or synthetic graft from above the access anastomosis to just below the site of ligation. Results show a cumulative access patency of 73% at one year and 45.5% at two years from one centre (53) and 94% at 18 months in another (56). In addition, resolution of ischaemic symptoms was seen in nearly all patients. This method deserves further evaluation and may become the treatment of choice for most patients with vascular steal.

Carpal tunnel syndrome occurs infrequently in patients on haemodialysis and the aetiology is multifactorial. One aetiological factor is relative ischaemia of the median nerve caused by vascular steal, while compression of the nerve in the carpal tunnel by venous hypertension and fluid overload is the other. When suspected, nerve conduction studies should be performed. If confirmed, good results are obtained from surgical decompression of the carpal tunnel.

Ischaemic monomyelic neuropathy is a rare complication (< 0.5%) which occurs in diabetic patients with pre-existing neuropathy and atherosclerotic vascular disease after creation of access using a major proximal limb artery, e.g. brachial artery (57, 58). It is diagnosed by the onset of severe, acute, painful weakness of the distal extremity with wrist drop, minimal wrist flexion and lack of movement of the intrinsic muscles (49). It often occurs immediately after access formation and the diagnosis can be difficult. The pathophysiology is probably peripheral nerve ischaemia, which may be irreversible at the time of diagnosis. Early diagnosis and ligation of the fistula or graft are essential if there is going to be any chance of preserving neurological function. Because diagnosis is often delayed, subsequent ligation of the access often does not reverse the neurological deficit (57).

Reflex sympathetic dystrophy (RSD) following vascular access surgery has been reported in the literature in four patients (59, 60). It is a syndrome of burning pain, hyperalgesia, swelling, hyperhydrosis and trophic changes in the skin and bone of the affected extremity. It is usually seen after bony trauma and surgery and although the mechanism is not clearly known, nerve injury is the initiating factor. Three of the described cases had ischaemia secondary to placement of a graft. In these, RSD is likely to have developed secondary to nerve damage caused by ischaemia rather than direct nerve damage. There is no successful treatment but sympathetic blockade can help the symptoms. This again emphasises the need for early treatment when ischaemia is recognised.

10.5.3. *Cardiac failure*

Congestive cardiac failure is again a rare complication following formation of a graft or fistula. When it does occur, it is more likely when

- A proximal artery, such as the brachial or femoral artery, has been used rather than the radial artery
- A side-to-side anastomosis has been performed
- There is a wide arteriotomy
- There is pre-operative echocardiographic evidence of impaired left ventricular function or hypertrophy

Haemodynamic studies show that flow rates in radiocephalic fistulas are in the range of 200–400 ml/min when cardiac failure is unlikely. It has been estimated that cardiac failure only occurs when 20–50% of the cardiac output goes through the fistula. However, flow rates in fistulas and grafts based on the brachial or femoral artery are twice as great as those based on radial arteries. As the outflow vein in a fistula dilates over time, the flow rate through that will increase, and at some stage, may lead to high output cardiac failure (61). It has been shown that two weeks after construction of an arteriovenous fistula or graft that there is an echocardiographic increase in shortening fraction, stroke volume, ejection fraction and cardiac output, and a decrease in systemic vascular resistance (62). In the short term at least, there is a state of mild volume overload that is offset by the unloading effect of the decreased peripheral vascular resistance.

It should be remembered that cardiac failure secondary to a fistula or graft can occur both in patients on haemodialysis and in patients with functioning transplants who still have working access. It can easily be forgotten that a patient's fistula or graft may be responsible for or contribute to his/her cardiac failure, although intrinsic cardiac disease, volume overload and anaemia are more likely causes (63). There are a number of non-invasive tests that can be performed to confirm the diagnosis. Temporary digital occlusion of the fistula may result in bradycardia, the Nicoladoni–Branham sign. The use of echocardiography and duplex flow measurement before and after temporary occlusion of the fistula can help to confirm the diagnosis. Although the majority of patients who develop high output cardiac failure ideally need to keep their access for haemodialysis, patients with functioning transplants can have their fistula or graft ligated (64). In the dialysis patient, treatment should be directed, where possible, at reducing the flow through the access conduit and therefore maintaining its use for access. The anastomotic opening can be reduced in size by direct suture or the outflow vein narrowed by banding. The extent of these techniques can be measured using intra-operative electromagnetic flow studies. If the flow is reduced to below 300 ml/min, there is an increased risk of thrombosis (61). There will be occasions when these techniques are not appropriate and the fistula or graft requires ligation to treat the cardiac failure.

10.6. Functioning but Unusable Access

From the surgeon's perspective, access may be technically successful with none of the above complications, but from the perspective of the patient and dialysis nurse, it may be a failure because the access is unusable. There are a number of patient- and access-related factors as to why this situation may arise.

The fistula may fail to mature adequately and although there may be a palpable thrill, it cannot be needled or inadequate flows are obtained. Up to 24% of arteriovenous fistulas never mature for some of the following reasons (13, 49). These include poor arterial inflow and small or damaged outflow veins. In this situation, it can be difficult knowing when to wait for further maturation or to abandon the fistula and construct a new one. Alternatively, veins may mature in less accessible sites such as the extensor surface of the forearm where needling is difficult or where there is only a short length of suitable vein available.

The elderly or obese patient also poses particular problems. In the elderly, the skin is often thin, the veins mobile and friable, and fibrous growth around the fistula/graft deficient. These factors can lead to difficulty in needling and an increased risk of extravasation of blood at post-needling with consequent haematoma formation. If this occurs repeatedly, the access may become unusable and consideration needs to be given to insertion of a permanent venous catheter or peritoneal dialysis, as a further fistula or graft is likely to lead to a similar problem. In the obese, the fistula vein or graft can be too deep to needle without recurrent trauma. Access needs to be planned very carefully in the obese where the procedure of choice is either a radiocephalic fistula or a forearm graft. There is an optimal depth for the graft so that needling is easy, but not so superficial that a fibrous cuff does not develop.

Needling of the fistula or graft can be very painful in some patients and it can be very difficult to elucidate a precise cause for this, but a vascular cause needs to be excluded. This may be caused by damage or close proximity of cutaneous nerves to the access site. If this is the case, altering the needling site may be sufficient to relieve symptoms.

Other accesses can be problematical due to recurrent clots. This may be indicative off an underlying stenosis or impending thrombosis, but may simply reflect difficult and traumatic cannulation. The use of aspirin or low-dose warfarin may help but the access should be imaged for an underlying problem. It may also indicate that the patient's target weight has been set too low or there is an increased thrombogenic potential, such as the use of erythropoeitin.

10.7. Summary

As the dialysis population becomes more elderly and with more co-morbidity, the provision of a successful access will entail an ever increasing workload and it is inevitable that more complications will be seen. One centre has shown that 25% of access procedures performed were thrombectomies and revisions (49). However, with the increasing use of thrombolysis, angioplasty and stent insertion, it is likely that treatment of complications will become an increasing workload for the radiologist rather than surgeon. It has been calculated that 20% of all admissions in haemodialysis patients are related to problems with vascular access (61). It is important that all access procedures are well planned, well performed and well monitored for complications. When complications do occur, they need to be recognised and treated early, to preserve access if possible and to avoid irrecoverable damage to the limb.

References

1. Leapman, S.B. *et al.* (1996). The arteriovenous fistula for hemodialysis access: Gold standard or archaic relic? *Am Surg*, **62**, 652–656; discussion 656–657.
2. Windus, D.W. (1994). The effect of comorbid conditions on hemodialysis access patency. *Adv Ren Replace Ther*, **1**, 148–154.
3. Woods, J.D. *et al.* (1997). Vascular access survival among incident hemodialysis patients in the United States. *Am J Kidney Dis*, **30**, 50–57.
4. Prischl, F.C. *et al.* (1995). Parameters of prognostic relevance to the patency of vascular access in hemodialysis patients. *J Am Soc Nephrol*, **6**, 1613–1618.
5. Bay, W.H. *et al.* (1998). Predicting hemodialysis access failure with color flow Doppler ultrasound. *Am J Nephrol*, **18**, 296–304.
6. May, R.E. *et al.* (1997). Predictive measures of vascular access thrombosis: A prospective study. *Kidney Int*, **52**, 1656–1662.
7. Safa, A.A. *et al.* (1996). Detection and treatment of dysfunctional hemodialysis access grafts: Effect of a surveillance program on graft patency and the incidence of thrombosis. *Radiology*, **199**, 653–657.
8. Roberts, A.B. *et al.* (1996). Graft surveillance and angioplasty prolongs dialysis graft patency. *J Am Coll Surg*, **183**, 486–492.
9. Bender, M.H., Bruyninckx, C.M. and Gerlag, P.G. (1995). The Gracz arteriovenous fistula evaluated. Results of the brachiocephalic elbow fistula in haemodialysis angio-access. *Eur J Vasc Endovasc Surg*, **10**, 294–297.
10. Coburn, M.C. and Carney, W.I., Jr. (1994). Comparison of basilic vein and polytetrafluoroethylene for brachial arteriovenous fistula. *J Vasc Surg*, **20**, 896–902; discussion 903–904.
11. Dunlop, M.G., Mackinlay, J.Y. and Jenkins, A.M. (1986). Vascular access: Experience with the brachiocephalic fistula. *Ann R Coll Surg Engl*, **68**, 203–206.
12. Enzler, M.A. *et al.* (1996). Long-term function of vascular access for hemodialysis. *Clin Transplant*, **10**, 511–515.

13. Palder, S.B. *et al.* (1985). Vascular access for hemodialysis. Patency rates and results of revision. *Ann Surg*, **202**, 235–239.
14. Tordoir, J.H. *et al.* (1988). Long-term follow-up of the polytetrafluoroethylene (PTFE) prosthesis as an arteriovenous fistula for haemodialysis. *Eur J Vasc Surg*, **2**, 3–7.
15. Rizzuti, R.P., Hale, J.C. and Burkart, T.E. (1988). Extended patency of expanded polytetrafluoroethylene grafts for vascular access using optimal configuration and revisions. *Surg Gynecol Obstet*, **166**, 23–27.
16. Heintjes, R. J. *et al.* (1995). The results of denatured homologous vein grafts as conduits for secondary haemodialysis access surgery. *Eur J Vasc Endovasc Surg*, **9**, 58–63.
17. Bosman, P.J. *et al.* (1998). A comparison between PTFE and denatured homologous vein grafts for haemodialysis access: A prospective randomised multicentre trial. The SMASH Study Group. Study of graft materials in access for haemodialysis. *Eur J Vasc Endovasc Surg*, **16**, 126–132.
18. Bennion, R.S. *et al.* (1985). A randomized, prospective study of perioperative antimicrobial prophylaxis for vascular access surgery. *J Cardiovasc Surg* (*Torino*), **26**, 270–274.
19. Kherlakian, G.M. *et al.* (1986). Comparison of autogenous fistula versus expanded polytetrafluoroethylene graft fistula for angioaccess in hemodialysis. *Am J Surg*, **152**, 238–243.
20. Poulain, F. *et al.* (1991). Local thrombolysis and thromboaspiration in the treatment of acutely thrombosed arteriovenous hemodialysis fistulas. *Cardiovasc Intervent Radiol*, **14**, 98–101.
21. Vlassopoulos, D. *et al.* (1996). Local thrombolysis with recombinant tissue plasminogen activator for thrombosed vascular access in hemodialysis patients [letter]. *Clin Nephrol*, **46**, 77–78.
22. Beathard, G.A. (1995). Thrombolysis versus surgery for the treatment of thrombosed dialysis access grafts. *J Am Soc Nephrol*, **6**, 1619–1624.
23. Polak, J.F. *et al.* (1998). Comparative efficacy of pulse-spray thrombolysis and angioplasty versus surgical salvage procedures for treatment of recurrent occlusion of PTFE dialysis access grafts. *Cardiovasc Intervent Radiol*, **21**, 314–318.
24. Beathard, G.A., Welch, B.R. and Maidment, H.J. (1996). Mechanical thrombolysis for the treatment of thrombosed hemodialysis access grafts. *Radiology*, **200**, 711–716.
25. Wada, H. *et al.* (1996). Immediate postoperative complications following hemodialysis access procedures. *Int Surg*, **81**, 99–101.

26. Trerotola, S.O. *et al.* (1998). Treatment of thrombosed hemodialysis access grafts: Arrow–Trerotola percutaneous thrombolytic device versus pulse-spray thrombolysis. Arrow–Trerotola Percutaneous Thrombolytic Device Clinical Trial. *Radiology*, **206**, 403–414.
27. Beathard, G.A. (1992). Percutaneous transvenous angioplasty in the treatment of vascular access stenosis. *Kidney Int*, **42**, 1390–1397.
28. Gray, R.J. *et al.* (1995). Use of Wallstents for hemodialysis access-related venous stenoses and occlusions untreatable with balloon angioplasty. *Radiology,* **195,** 479–484.
29. Turmel-Rodrigues, L.A. *et al.* (1997). Wallstents and Craggstents in hemodialysis grafts and fistulas: Results for selective indications. *J Vasc Interv Radiol*, **8**, 975–982.
30. Beathard, G.A. (1993). Gianturco self-expanding stent in the treatment of stenosis in dialysis access grafts. *Kidney Int*, **43**, 872–877.
31. Hoffer, E.K. *et al.* (1997). Prospective randomized trial of a metallic intravascular stent in hemodialysis graft maintenance. *J Vasc Interv Radiol,* **8**, 965–973.
32. Funaki, B. *et al.* (1997). Wallstent deployment to salvage dialysis graft thrombolysis complicated by venous rupture: Early and intermediate results. *AJR Am J Roentgenol*, **169**, 1435–1437.
33. Gray, R.J. *et al.* (1994). Phase I results of pullback atherectomy for hemodialysis access. *J Vasc Interv Radiol*, **5**, 581–586.
34. Mizumoto, D. *et al.* (1996). The treatment of chronic hemodialysis vascular access by directional atherectomy. *Nephron*, **74**, 45–52.
35. Rabindranauth, P. and Shindelman, L. (1998). Transluminal stent-graft repair for pseudoaneurysm of PTFE hemodialysis grafts. *J Endovasc Surg*, **5**, 138–141.
36. Ready, A.R. and Buckels, J.A.C. (1996). Management of infection: Vascular access surgery in hemodialysis. In *Vascular Access: Principles and Practice*, (ed. S.E. Wilson), pp. 198–211, Mosby: St Louis.
37. Back, M.R. and White, R.A. (1996). The biologic response of prosthetic dialysis grafts. In *Vascular Access: Principles and Practice*, (ed. S.E. Wilson), pp. 137–149. Mosby, St Louis
38. Glanz, S. *et al.* (1988). Axillary and subclavian vein stenosis: Percutaneous angioplasty. *Radiology*, **168**, 371–373.
39. Mickley, V. *et al.* (1997). Stenting of central venous stenoses in haemodialysis patients: Long-term results. *Kidney Int*, **51**, 277–280.

40. Quinn, S.F. *et al.* (1995). Percutaneous transluminal angioplasty versus endovascular stent placement in the treatment of venous stenoses in patients undergoing hemodialysis: Intermediate results. *J Vasc Interv Radiol*, **6**, 851–855.
41. Wisselink, W. *et al.* (1993). Comparison of operative reconstruction and percutaneous balloon dilatation for central venous obstruction. *Am J Surg*, **166**, 200–205.
42. Kovalik, E.C. *et al.* (1994). Correction of central venous stenoses: Use of angioplasty and vascular Wallstents. *Kidney Int*, **45**, 1177–1181.
43. Vorwerk, D. *et al.* (1995). Venous stenosis and occlusion in hemodialysis shunts: Follow-up results of stent placement in 65 patients. *Radiology*, **195**, 140–146.
44. Shoenfeld, R. *et al.* (1994). Stenting of proximal venous obstructions to maintain hemodialysis access. *J Vasc Surg*, **19**, 532–538; discussion 538–539.
45. Irvine, C. and Holt, P. (1989). Hand venous hypertension complicating arterio-venous fistula construction for haemodialysis [see comments]. *Clin Exp Dermatol*, **14**, 289–290
46. Wood, M.L., Reilly, G.D. and Smith, G.T. (1983). Ulceration of the hand secondary to a radial arteriovenous fistula: A model for varicose ulceration. *Br Med J*, **287**, 1167–1168.
47. Swayne, L.C. *et al.* (1983). Selective digital venous hypertension: A rare complication of hemodialysis arteriovenous fistula. *Cardiovasc Intervent Radiol*, **6**, 61–62.
48. Duncan, H., Ferguson, L. and Faris, I. (1986). Incidence of the radial steal syndrome in patients with Brescia fistula for hemodialysis: Its clinical significance. *J Vasc Surg*, **4**, 144–147.
49. Ballard, J.L., Bunt, T.J. and Malone, J.M. (1992). Major complications of angioaccess surgery. *Am J Surg*, **164**, 229–232.
50. Morsy, A.H. *et al.* (1998). Incidence and characteristics of patients with hand ischemia after a hemodialysis access procedure. *J Surg Res*, **74**, 8–10.
51. Zibari, G.B. *et al.* (1988). Complications from permanent hemodialysis vascular access. *Surgery*, **104**, 681–686.
52. Rutherford, R.B. (1997). The value of noninvasive testing before and after hemodialysis access in the prevention and management of complications. *Semin Vasc Surg*, **10**, 157–161.
53. Haimov, M., Schanzer, H. and Skladani, M. (1996). Pathogenesis and management of upper-extremity ischemia following angioaccess surgery. *Blood Purif*, **14**, 350–354.

54. Rivers, S.P., Scher, L.A. and Veith, F.J. (1992). Correction of steal syndrome secondary to hemodialysis access fistulas: A simplified quantitative technique. *Surgery*, **112**, 593–597.
55. Kirkman, R.L. (1991). Technique for flow reduction in dialysis access fistulas. *Surg, Gynae & Obstet*, **172**, 231–233.
56. Berman, S.S. *et al.* (1997). Distal revascularization-interval ligation for limb salvage and maintenance of dialysis access in ischemic steal syndrome. *J Vasc Surg*, **26**, 393–402; discussion 402–404.
57. Hye, R.J. and Wolf, Y. G. (1994). Ischemic monomelic neuropathy: An under-recognized complication of hemodialysis access. *Ann Vasc Surg*, **8**, 578–582.
58. Riggs, J.E. *et al.* (1989). Upper extremity ischemic monomelic neuropathy: A complication of vascular access procedures in uremic diabetic patients. *Neurology*, **39**, 997–988.
59. Kemler, M. A., & Tordoir, J. H. (1998). Reflex sympathetic dystrophy following vascular access surgery for haemodialysis: influence of peripheral ischaemia? *Nephrol Dial Transplant,* **13**, 784–786.
60. Weisse, W.J. and Bernard, D.B. (1991). Reflex sympathetic dystrophy syndrome of the hand after placement of an arteriovenous graft for haemodialysis. *Am J Kidney Dis*, **18**, 406–408.
61. Wilson, S.E. (1996). Complications of vascular access procedures: Thrombosis, venous hypertension, arterial steal, and neuropathy. In *Vascular Access: Principles and Practice*, (ed. S.E. Wilson), pp. 212–224, Mosby, St Louis.
62. Ori, Y. *et al.* (1996). Haemodialysis arteriovenous access — A prospective haemodynamic evaluation. *Nephrol Dial Transplant*, **11**, 94–97.
63. Anderson, C.B. *et al.* (1976). Cardiac failure and upper extremity arteriovenous dialysis fistulas. Case reports and a review of the literature. *Arch Intern Med*, **136**, 292–297.
64. Young, P.R., Jr., Rohr, M.S. and Marterre, W.F., Jr. (1998). High-output cardiac failure secondary to a brachiocephalic arteriovenous hemodialysis fistula: Two cases. *Am Surg*, **64**, 239–241.

CHAPTER 11

REVISION ACCESS SURGERY

Paul A Lear FRCS
Renal Transplant Unit
Southmead Hospital
Westbury-on-Trym
Bristol BS10 5NB, UK

The arteriovenous fistula of the wrist remains the cornerstone of vascular access for haemodialysis (1). This form of conduit is very durable and has a half-life of ten years, but many different clinical scenarios may herald failure or at least a need for surgical revision or radiological intervention.

It is my intention to cover the chronic and acute-on-chronic problems which occur in association with vascular access and not to consider the early failure of unestablished fistulae which merely require re-operation at another more favourable site. The problems associated with arteriovenous fistulae are frequently site dependent and it is therefore easiest to cover them from this anatomical perspective, and to consider the many options open to practitioners interested in access salvage. The provision of renal replacement therapy to an increasing elderly population has put fresh demands on surgeons to provide quality access, and because of the better survival rates on dialysis, the requirement for revision access surgery has become significant in recent years.

11.1. Clinical Problems Within the Haemodialysis Unit

A greater awareness of potential complications by dialysis nursing staff has streamlined the referral for access revision. The identification of high venous pressures and "recirculation" leading to poor quality and inefficient dialysis results in referral for surgical assessment. The use of high-flow dialysis (using 350–450 ml/min as opposed to standard flows of 200–250 ml/min) has meant earlier detection of anatomical stenoses, both anastomotic and within the fistula vein itself.

Elevated venous pressure is a reliable method of detecting fistula stenosis and the elective treatment of these stenoses significantly decreases the rate of fistula thrombosis and fistula loss (2).

All patients with deteriorating dialysis figures from their vascular access are evaluated clinically and then their fistula visualised in the vascular laboratory by a duplex scanner. Great experience is needed in interpreting the findings.

11.2. Duplex Scanning

Duplex scanning provides an accurate assessment of haemodialysis fistulae, certainly sufficient to plan revision surgery or interventional radiology. Ultrasound (without Doppler) determines the anatomy of the access and identifies undilated venous segments or major branches diverting flow. Using a high resolution linear probe in conjunction with a stand-off where necessary, venous diameters can be measured at points of interest, particularly at needle sites and strictures, and the depth of the vein determined; an upper arm fistula may be functioning well but unusable due to a deep-lying cephalic vein. Once the anatomy has been mapped, careful doppler measurements identify sites of stenosis. Most commonly, stenoses occur adjacent to the anastomosis in the first four centimetres of cephalic vein where it has been mobilised during surgery. By their very nature, all fistula anastomoses are stenotic, so exceptionally turbulent flow is encountered around that site. Measurements of peak systolic velocities are made along the entire fistula including the anastomosis; assessment of the arterial inflow may also be needed in the presence of peripheral vascular disease.

For patients with previous subclavian or jugular line placements, the patency of the axillary and subclavian veins is also determined. A rough estimate of volume flow is useful, and is taken both from the supplying artery and from a straight section of vein; where the flow rate is below 300 ml per min, there is an increased risk of thrombosis and a need for urgent revision (3). Volume flow measurements taken from the artery distal to the anastomosis give an indication of the significance of steal for radial fistulae.

Duplex examination of access grafts enables identification of false aneurysms and sites of stenoses, the latter occurring primarily around the venous anastomosis and in the efferent vein (Figs. 11.1–11.4). Peak systolic velocities are measured throughout the graft and both anastomoses, and also in the draining vein. Investigation of the axillary and subclavian veins is merited where there are symptoms of arm swelling or the patient has had previous line placement. Volume flow is of particular interest and must be calculated from a straight, undamaged section of the graft. Flow rates below 450 ml per min indicate a graft in jeopardy which therefore requires urgent revision (4, 5).

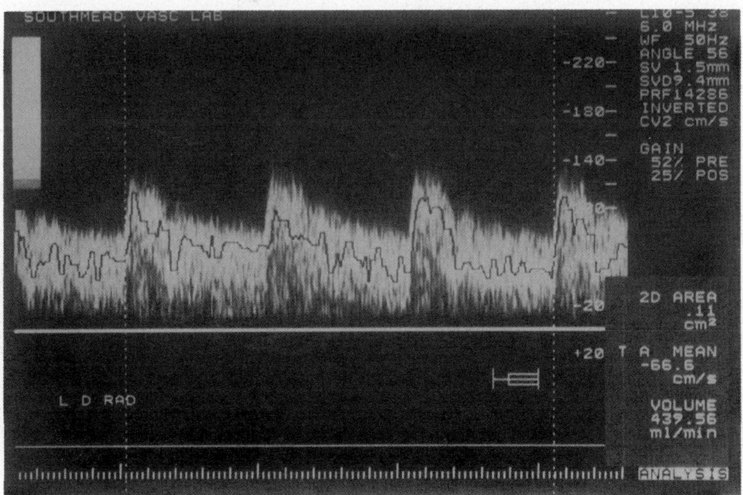

Fig. 11.1. Spectral trace from duplex scan showing fistularised radial flow and approximate volume measurement.

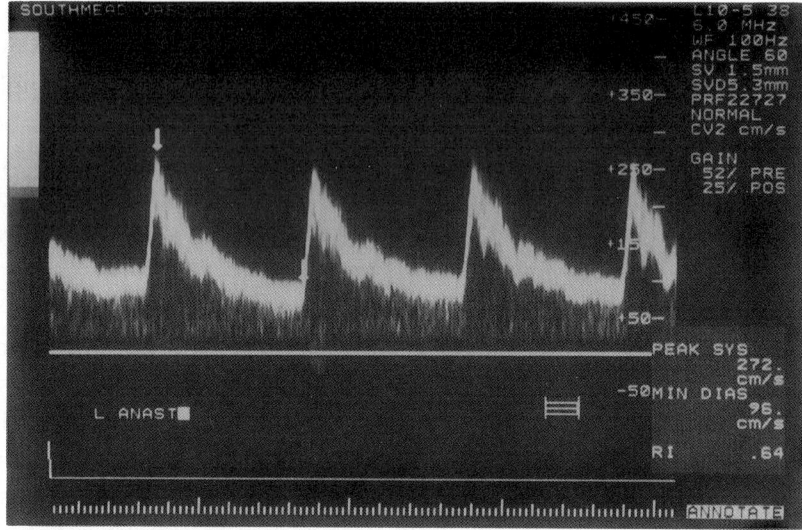

Fig. 11.2. Peak systolic and end diastolic velocities at fistula anastomosis.

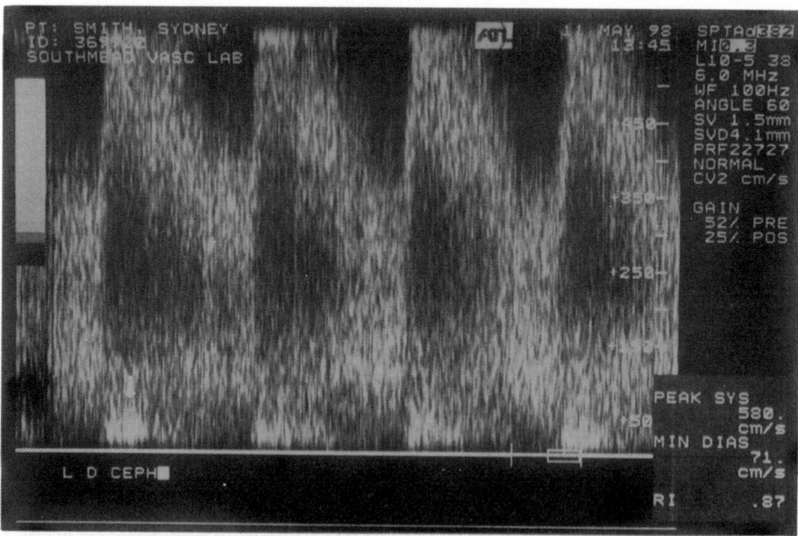

Fig. 11.3. Raised peak systolic velocity in distal cephalic vein at site of stenosis.

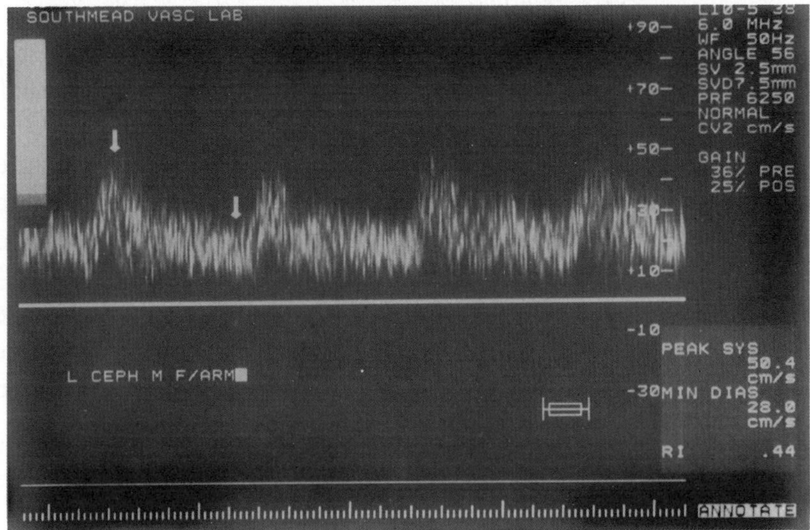

Fig. 11.4. Low cephalic vein flow velocities mid-forearm beyond tight stenosis.

11.3. The Brescia–Cimino (Wrist Fistula)

The complications of this fistula are somewhat determined by the nature of the surgical anastomosis. The now more popular end-cephalic vein to side-radial artery anastomosis has largely seen the end of venous hypertension in the hand, although this is occasionally encountered. Complications occur due to stenosis in the different afferent arterial or venous components of the fistula.

Stenosis of the anastomosis itself will usually result in deteriorating flow and the failure to successfully dialyse. Cannulation of the fistula vein is not uncommonly complicated by encountering *in situ* thrombosis. The diagnosis of anastomotic stricture is easily made on clinical examination by emptying the fistula vein while the anastomosis is compressed and subsequent release and observation of the refilling time. Duplex scanning will confirm the diagnosis. Our experience in Bristol suggests that the best treatment is more proximal surgical revision of the fistula: under these circumstances, the already mature vessel is needleable straightaway. When planning surgery, it is important to avoid incising the skin at sites of frequent needling. The fistula vein is very adherent to the skin at these points and, therefore, the tissues are friable and do not take sutures well. The skin is liable to breakdown as it is largely devitalised during surgery.

Arterial stenosis proximal to the anastomosis results in an entirely different clinical pattern. There is either poor flow or the supply is principally via the ulnar artery and the patent palmar arch; this results in adequate flow through the fistula but symptoms of "steal" develop particularly during dialysis. This can be very severe in the elderly and diabetics and may advance to ischaemic tissue damage at the finger tips (Fig. 11.5). Clinically, the diagnosis is reached by manual compression of the ulnar artery, emptying the fistula vein by

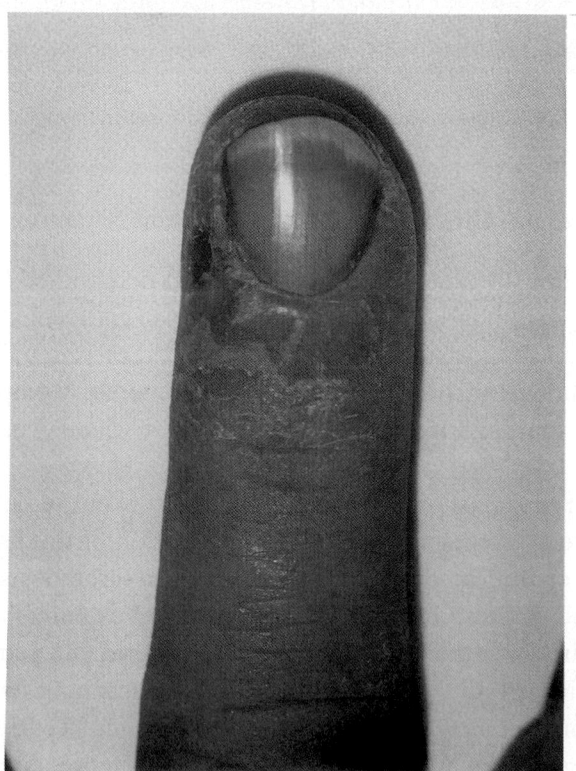

Fig. 11.5. Digital ischaemia in a diabetic patient with a radial fistula.

elevating the arm and identifying poor refilling of the fistula vein. Surgical correction is by creating a new anastomosis at a more proximal site above the stenosis. Ligation of the distal radial artery will prevent the recurrence of steal, although this is rarely a problem. If the arterial stenosis is extensive, a PTFE graft may be interposed from the brachial artery down to a suitable segment of the forearm fistula vein, preserving as much needleable length as possible.

Venous stenosis above the anastomosis results in poor fistula blood flow beyond it but a high pulse pressure between the anastomosis and the stricture. When this is left over a considerable period, aneurysmal dilation will occur. This will then either proceed to thrombosis or require surgical correction due to unsightliness or pain some point along the way. Aneurysmal dilatation is frequently seen in those patients with successful renal transplants where the flow along the fistula is no longer of clinical relevance. Prior to aneurysm formation, however, duplex scanning can confirm the extent of a venous stenosis and a decision can be made between percutaneous, intraluminal balloon angioplasty and surgical revision. Stenosis at the angle between the mobilised cephalic vein (at the time of fistula surgery) and the remainder of the vein is common [Fig. 11.6(a)]. Balloon angioplasty is performed to good effect from the arterial or venous end [Fig. 11.6(b)]. It does however require a distance of 2–3 cm beyond the anastomosis to allow the safe use of the Grundzig balloon away from the puncture hole. On occasions when the stricture is close to the arterial anastomosis, an open exploration and vein patch plasty are performed. Longer lesions require an interposition autogenous vein or PTFE graft.

The needling of arteriovenous fistula over many months and years results in dense adherence of the fistula vein to the surrounding subcutaneous tissue and the skin. Together with aneurysmal change, the depth of tissue between the skin and the fistula lumen can be reduced to little more than 2 mm. Bypass of strictured sites must obviously avoid the aneurysmal segments when considering re-anastomosis. Indeed, it is important to maximise graft cover to facilitate suturing in the first instance in order to get subsequent sound healing.

Side-to-side arteriovenous anastomoses at the wrist can result in significant venous hypertension of the hand. This may be most evident in the thenar

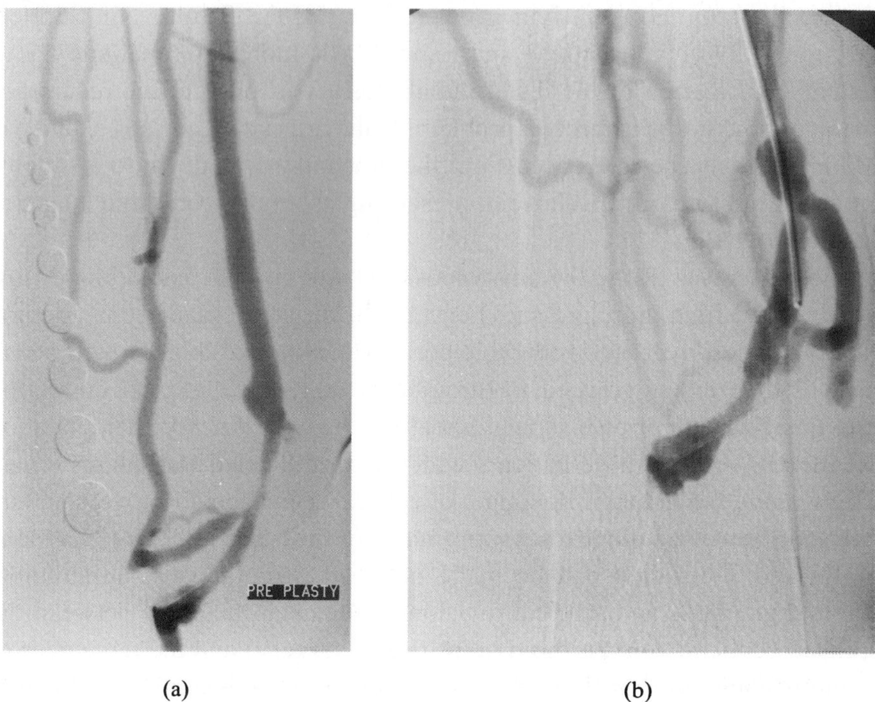

(a) (b)

Fig. 11.6. (a) Fistulogram showing stenosis within the cephalic vein just distal to the anastomosis with collateral drainage of the flow into the deep forearm veins. (b) Post balloon fistulaplasty radiograph showing principal fistula drainage along the cephalic vein and immediate reduction in collateral drainage.

eminence. The clinical appearance is one of obvious hand swelling and an inability to form a fist. Ligation of the distal venous limb of the fistula results in rapid settling of the symptoms and the hand oedema will resolve literally overnight.

Local venous hypertension is now more commonly seen due to an occlusion of the fistula vein at some point beyond the anastomosis. The high-pressured vein back-fills some of the tributaries to provide collateral drainage. Contrast fistulography is, in our experience, frequently needed to clearly identify the anatomical situation and facilitate the placement of an interposition graft to restore fistula flow and reduce the venous back pressure.

It has already been suggested that severe ischaemia can develop in elderly patients with peripheral vascular disease and in diabetics. I will deal with this particular area later in the chapter.

11.4. Antecubital Fossa Fistula

This is a very common site for secondary access. The configuration of available veins is highly variable and the complications may be to some extent anticipated. Too high a flow within these fistulae can result in steal syndromes and, very occasionally, high output heart failure in the elderly. We can measure quite accurately the flow within the brachial fistula vein via duplex scanning. High output flow causing cardiac problems can be alleviated by compressive occlusion of the anastomosis. Flow reduction can only be achieved currently via open surgery and some form of "plastic" procedure. I prefer to use a 1–1.5 cm segment of a 4-mm PTFE interposition graft running from the immediate post-anastomotic venous segment. Others have found that simple plication ligature close to the anastomosis will suffice; an operative doppler assessment is important to quantify an adequate reduction in flow.

Arm oedema is associated with these high flow fistulae. It is particularly encountered in those patients with previous subclavian catheters for temporary dialysis. The stenosis within the subclavian vein rarely reveals itself prior to fistula formation because of the development of collateral vessels which are quite capable of providing adequate arm drainage under normal circumstances (Figs. 11.7 and 11.8). When the arm swelling becomes severe, there is skin breakdown at the site of frequent fistula needling and eventually a large necrotic ulcer will develop (Fig. 11.9). The only treatment may be closure of the fistula and recourse to a further site for vascular access, but assessment of the feasibility for subclavian venoplasty is important (6).

In attempts to maximise the development of a brachial fistula, multiple drainage veins are often left intact. With retrograde arterialisation of the forearm veins and dissipation of blood flow into more than one upper arm vein, there can be poor fistula development. Revision surgery should aim to focus flow along one vein, preferably the cephalic vein. Using an operative Doppler probe, improved flow in the vein can be measured and surgery therefore tailored to meet the patient's requirements.

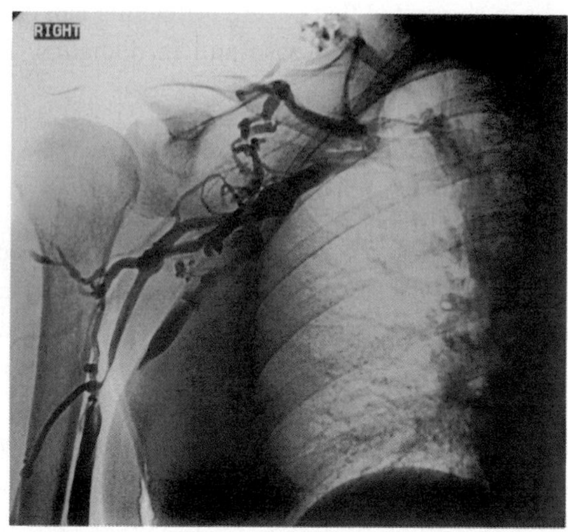

Fig. 11.7. Occlusion of the subclavian vein secondary to a previous haemodialysis catheter. Well-developed collaterals are seen radiologically.

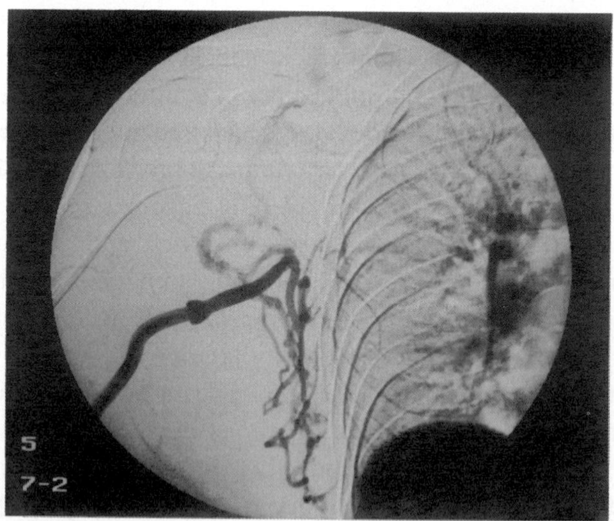

Fig. 11.8. Axillary vein occlusion with early collateral formation on the lateral chest wall.

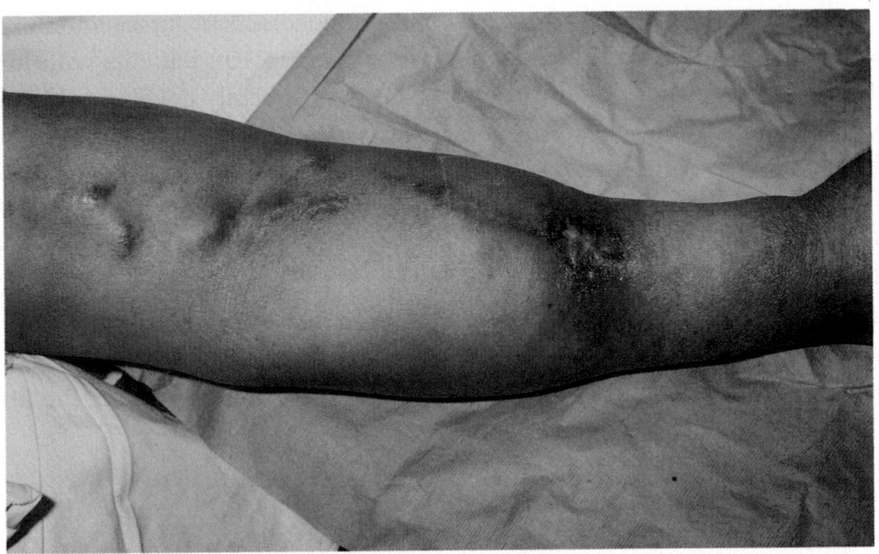

Fig. 11.9. Venous hypertension with arm swelling, resulting in ulceration at needle puncture sites.

11.5. Angioplasty

Percutaneous transluminal angioplasty is used wherever safe and practical; surgery inevitably results in the loss of some needleable fistula length. The combination of interventional angioplasty and earlier clinical referral can reduce the thrombosis rate by 60% from 0.6 to 0.2% patient years, and the need for access replacement similarly from 0.26 to 0.07% patient years. Angioplasty will correct the majority of strictures at the anastomosis or sites close to it (85%), and provided that more central veins are still patent, it will improve 95% of stenoses at these sites. Re-stenosis rates are high (20–33%) at or close to the anastomosis, and nearly 50% in the central vein. However, the majority are suitable for further dilatation (7).

11.6. Aneurysm Formation

High pressure fistula and a weakened wall result in aneurysm formation. Patients with extensive aneurysmal changes throughout a long fistula are usually reviewed on a regular basis, and provided access for haemodialysis is satisfactory, we leave the situation alone (Fig. 11.10). Patients with long-standing fistulae and a successful graft do not infrequently develop aneurysms or pseudoaneurysms near the anastomosis (Fig. 11.11). We usually treat these by simple vein ligation and local arterial repair or arterial excision (at the wrist), or excision and brachial artery repair at the elbow. Repair is not particularly straightforward because of the arterial calcification and the vessel wall kinking that takes place over time, and we now replace the brachial artery segment with a short vein graft. Repair of the radial artery should be attempted in patients with peripheral vascular disease and diabetes to avoid possible ischaemia, and in any patient to allow the option of using the ulnar vessel in future access surgery.

Aneurysms can become troublesome because of *in situ* thrombosis. This usually develops at a time of relative hypotension at the end of dialysis. Initial treatment is with systemic tissue plasminogen activator (TPA). Occasionally, local treatment is feasible by catheter delivery. Ten milligrams of recombinant TPA (rTPA) is given as a ten-minute infusion, and repeated as necessary at two and four hours with repeated angiography to assess progress (8). The use of aggressive thrombolysis in our hands results in haemorrhage from multiple recent dialysis needle puncture sites. Overall, I prefer an operative approach to remove the thrombus and to modify the aneurysm simultaneously. This may even require a small "inlay" PTFE graft between two reasonable quality bits of vein at either end. Completion angiography does not infrequently reveal a fistula stenosis above the segment and an operative balloon angioplasty is advised to prevent further thrombosis.

Extensive vascular modification to preserve useful dialysis vein is very much at the operator's discretion. The use of long PTFE conduits from brachial artery to mature wrist cephalic fistula vein provides salvage and more needleable access. This is appropriate for patients with proximal arterial disease within the radial artery which is not amenable to angioplasty. Along similar lines, long segments of stenosed, or stenosed and aneurysmal fistula can be replaced and bypassed using PTFE between the arterialised nubbin of fistula vein at the wrist and the antecubital fossa.

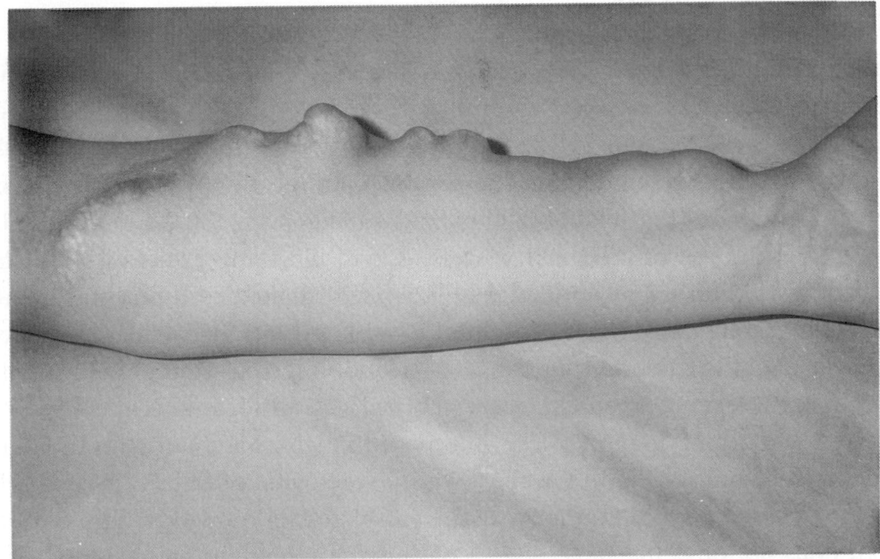

Fig. 11.10. Multiple aneurysmal dilatations in a long standing forearm fistula.

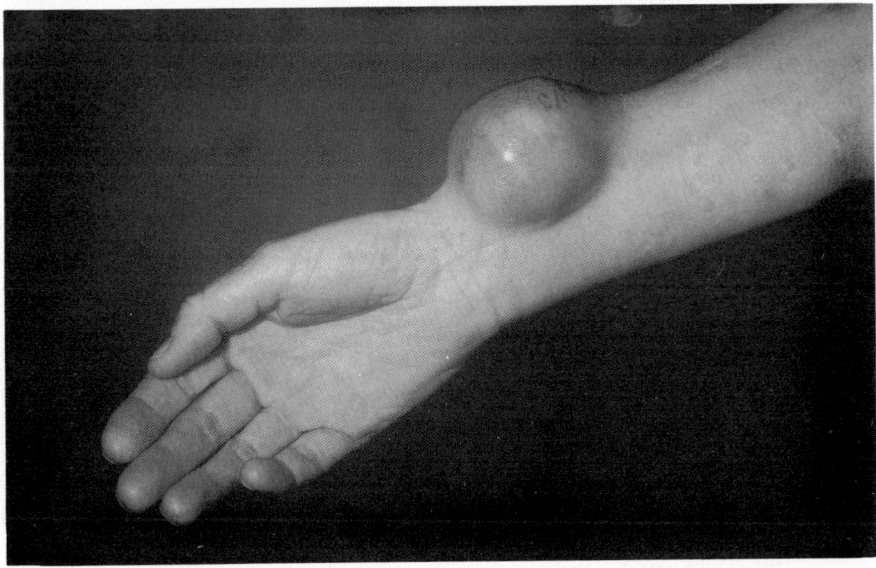

Fig. 11.11. Large aneurysm of radial fistula in a long-term successful transplant recipient.

11.7. Diabetes and Microvascular Disease

Patients with end-stage diabetic renal disease, who are not suitable for peritoneal dialysis, must be carefully worked up for vascular access for haemodialysis. The same is true for patients suffering from connective tissue disorders such as scleroderma. The serious complication of digital necrosis is not uncommon (Fig. 11.12). The radial fistula appears to be a more serious threat than the use of proximal vessels, due to the sump effect of the distal anastomosis. Closure of a radial fistula with formation of a brachial fistula has been accompanied by resolution of on-going tissue ischaemia. However, in my view, it is important to consider arterial inter-position grafts in these high-risk patients. The subclavian artery acts as an ideal site; the artery is divided and a long loop of PTFE interposed much as a shunt with the graft looping subcutaneously on the chest wall. In the event of failure, particularly due to sepsis, this needs removal and vessel repair. However, this type of graft does not raise the same degree of concern with hand ischaemia.

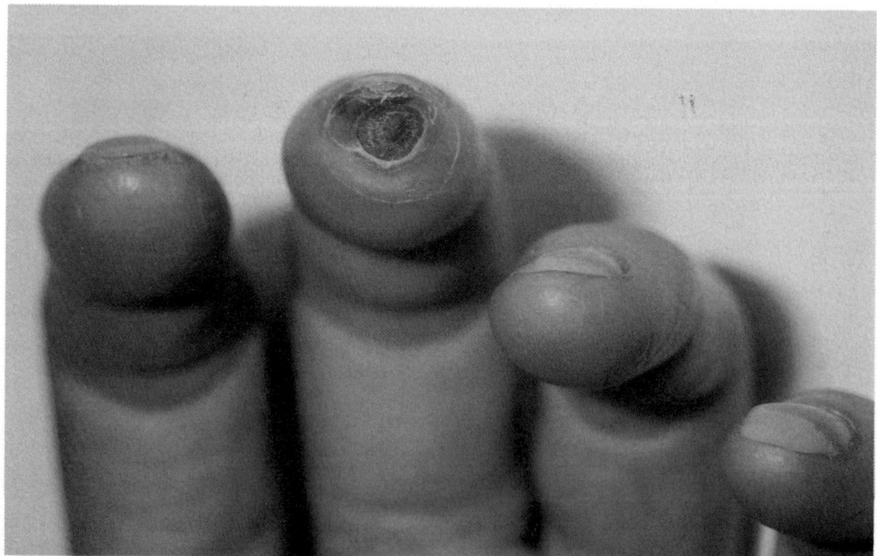

Fig. 11.12. Digital ulceration in a patient with microvascular disease.

11.8. Revision of Synthetic Vascular Access

PTFE grafts fail principally due to thrombosis. This may be sudden, usually post-dialysis during a period of hypotension or preceded by increasing problems with raised venous pressure while on dialysis, and poor flow. Intimal hyperplasia of the distal anastomosis is common due to the significant turbulence at the end of the graft [Figs. 11.13(a) and 11.13(b)]. This is occasionally amenable to angioplasty, particularly when at the groin site, but will often require the use of some form of stent to maintain a good lumen.

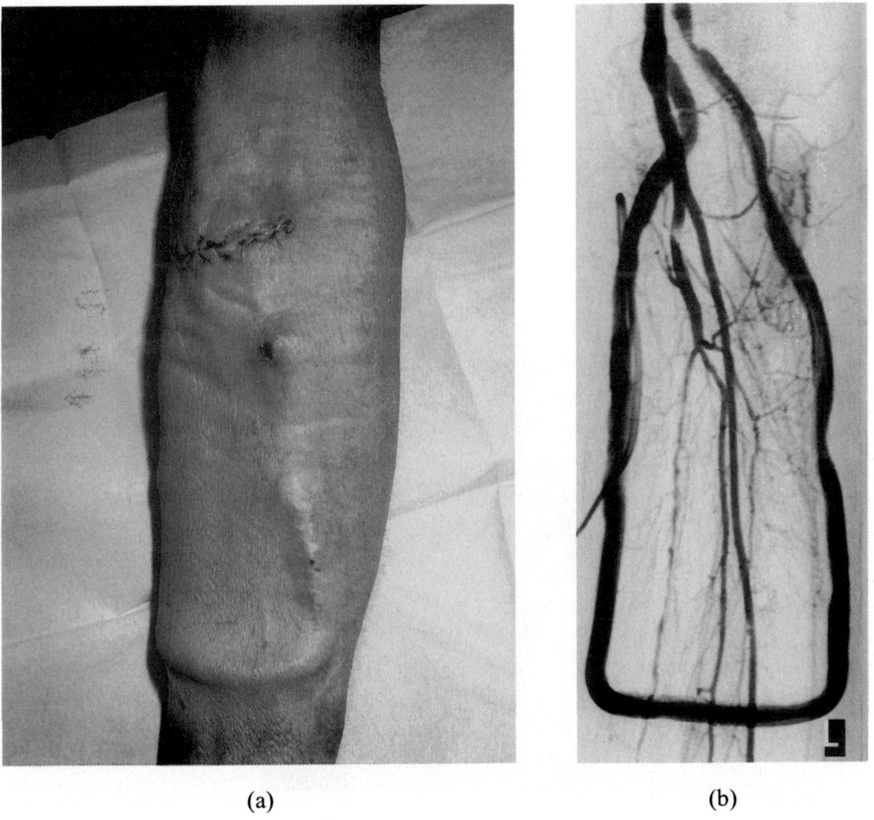

(a) (b)

Fig. 11.13. (a) Forearm loop access graft with local "blow". (b) Post-operative radiograph identifying a stenosis at the venous end of the loop graft suitable for angioplasty.

The diagnosis of intimal hyperplasia is most frequently made when the thrombosed graft is explored. Exploration of these grafts should be made at a site remote from needling because, frequently, the graft wall is deficient here and it is not possible to effectively repair it. A special stripper is now commercially available which facilitates the complete clearance of all thrombotic material within the graft lumen. The graftotomy is then merely closed. Recurrent problems after exploration are not uncommon and, therefore, Warfarin therapy is usually initiated in our department. Grafts can be explored 2–3 weeks after thrombosis and cleared using these special strippers. It may not be possible to salvage the graft, however, and this is not uncommon

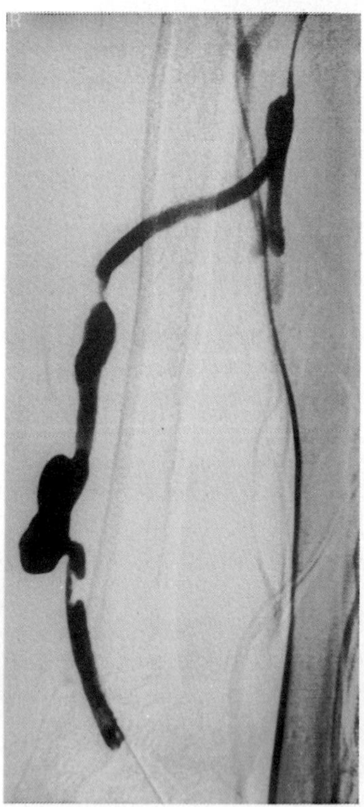

Fig. 11.14. Fistulogram of a patient with a long-standing upper arm access graft showing sites of stenoses and pseudo-aneurysm formation.

particularly in the upper arm where a skip graft to more normal vein can be employed from the original graft, thus keeping most of it intact.

Large holes in the superficial wall of the graft are also not uncommon problems, and they give rise to pseudo-aneurysms (Fig. 11.14). These are caused by recurrent needling at the same site and are associated with low-grade infection. In theory, repair is possible but, in practice, because of the inevitable local sepsis mobilisation of the graft at either end, the placement of a further interposition graft passing through clean territory is the best way forward. If it is straightforward, then the old graft which has been isolated should be removed. However, these grafts can be exceedingly adherent to the skin and surrounding tissues, and excision may require a full-length graft exposure to clear the foreign material.

11.9. Summary

The number of patients on chronic haemodialysis increases annually in every unit throughout the United Kingdom. Older patients are being accepted onto these programmes with significant co-morbid vascular disease. Inevitably, revision of vascular access creates demands on both surgical and radiological departments. It is important to maximise the life span of any single access procedure because the number of potential sites is limited yet the long-term survival of the patients is substantially better than it was 20 years ago. There is a need in every renal department for a surgeon who has interest in preserving vascular access for haemodialysis.

References

1. Bescia, M.J. *et al.* (1966). Chronic haemodialysis using venepuncture and a surgically created arteriovenous fistula. *N Engl J Med*, **275**, 1089–1092.
2. Schwab, S.J. *et al.* (1989). Prevention of haemodialysis fistula thrombosis: Early detection of venous stenosis. *Kidney Int*, **36**, 707–711.
3. Culp, K. *et al.* (1995) Vascular access thrombosis in new haemodialysis patients. *Am J Kidney Dis*, **26**, 341–346.
4. Koksoy, C. *et al.* (1995) Predictive value of colour Doppler ultrasonography in detecting failure of vascular access grafts. *Br J Surg*, **82**, 50–52.

5. Shackleton, C.R. *et al.* (1987). Predicting failure in polytetrafluoroethylene vascular access grafts for haemodialysis: A pilot study. *Can J Surg*, **30**, 442–444.
6. Newman, G. *et al.* (1991). Total central vein obstruction: Resolution with angioplasty and fibrinolysis. *Kidney Int*, **39**, 761–764.
7. Newman G. *et al.* (1991). Functional restenosis rate after hemodialysis graft angioplasty, abstracted. *J Am Soc Nephrol*, **2**, 341.
8. Abdurhman, A., Shapiro, W. and Porush, M. (1993). The use of tissue plasminogen activator to declot arteriovenous accesses in hemodialysis patients. *Am J Kidney Dis*, **21**, 38–43.

CHAPTER 12

CENTRAL VENOUS CATHETERS

Jacob A Akoh FRCSEd (Gen)
Plymouth Hospitals NHS Trust
Derriford Hospital
Plymouth PL6 8DH, UK

12.1. Introduction

As renal services expand, more elderly patients with peripheral vascular disease have been recruited onto dialysis and the proportion of patients with exhausted vascular accesses has increased. Provision of an effective and durable vascular access for patients with end-stage renal disease (ESRD) therefore continues to be a challenge. Central venous catheters (CVC) provide a means of achieving vascular access in patients considered unsuitable for conventional arteriovenous fistula, bypass graft or chronic ambulatory peritoneal dialysis. Blood flow rates of 200 to 300 ml/min are usually achieved with CVC (1–3). Higher flow rates can be achieved in some patients but inconsistently.

Permanent CVC have a number of advantages over arteriovenous access. These include ease of insertion, removal or replacement in experienced hands, immediacy of use and ability to insert into multiple sites. The absence of haemodynamic stress, steal syndrome, significant recirculation and avoidance of venipuncture make CVC an attractive option. Currently 10–22% of long-term haemodialysis patients use CVC as their permanent vascular access (4–8). Long-term haemodialysis catheters are used in 91% of renal units in the UK where 60% of haemodialysis patients would have had catheterisation of their central veins for dialysis during their treatment (9).

12.2. Types of CVC

Temporary catheters are designed for short-term haemodialysis. Such catheters posses no cuff and do not require subcutaneous tunnelling. They can therefore be easily inserted at the bedside by the Seldinger technique. In contrast, permanent catheters possess a dacron cuff which is usually located in a subcutaneous tunnel away from the venous insertion site. The cuff and the tunnel provide physical barriers to bacterial invasion, thus allowing use of catheter for a longer period without the risk of infection.

12.2.1. *Multiple-lumen tunnelled catheters*

These radiopaque catheters are made of silicone rubber or polyurethane and include: PermCath (Quinton Instruments Comp, Seattle WA); Access Cath (MEDCOMP, Harleysville PA); Soft Cell PC (Vas-Cath Incorporated, Mississauga, Ontario, Canada); HY-MED (Neostar Medical Technologies Inc, Manchester GA). The cross-sectional geometry of single- or multiple-lumen catheters is shown in Fig. 12.1 (10). Catheters with side-by-side cross-sectional configuration perform better than co-axial ones (11). Similarly, catheters with two distinct lumina rather than a midline septum provide higher and more reliable blood flows (12). To minimise recirculation or mixing of blood at the tip of the double-lumen catheters, the arterial and

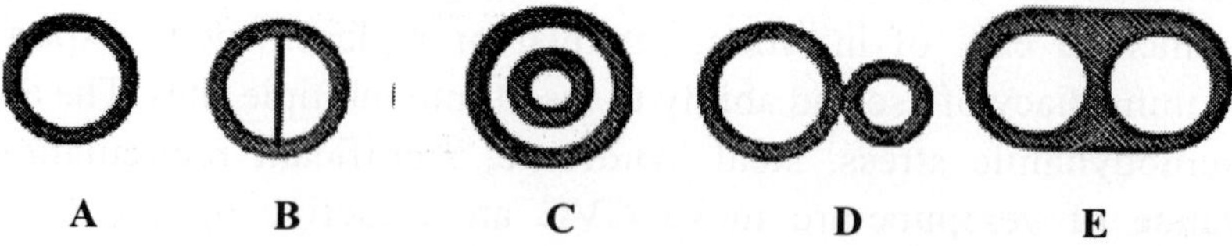

Fig. 12.1. Catheter geometry: (A) single lumen; (B) circular catheter with two D-shaped lumens; (C) co-axial catheter with inner and outer circular lumens; (D and E) double-lumen catheters.

venous holes are separated. The catheters are available in a range of lengths and sizes to permit their use in a wide variety of patients.

12.2.2. *Single-lumen tunnelled catheters*

These include the single-cuff straight Tenckhoff catheter used by Francis *et al.* (13), the reversible transparent Demers catheter (14) and the Hickman nutritional catheter (15). Single-lumen catheters are particularly very valuable in children. They provide flows in excess of 200 ml/min and require connection to a pressure–pressure single-cannula dialysis system.

Twin catheters

Tesio Twin-Cath (Medcomp, Harleysville PA) consist of two independent single-lumen catheters made of silicone rubber designed to overcome some of the problems with the larger, stiffer, less biocompatible double-lumen haemodialysis catheters (16). Blood flow rates in excess of 300 ml/min are regularly achieved with this catheter. The main distinguishing feature of the twin catheter is in the design. In contrast to double-lumen catheters with inflow holes facing one direction, twin catheters have inflow holes around the distal 4–5 cm of the tip. Whereas vigorous aspiration of the double lumen catheter may suck the unidirectional arterial lumen against the vessel wall, this is unlikely to happen with twin catheters. It is believed that the separate "arterial" and "venous" catheters are less likely to kink. Furthermore, the position of the tips can be varied to minimise recirculation (17).

Many units are reluctant to try this new system because of the theoretically increased risk involved with two venotomies and two exit sites. Many expect that switching to this system will double complication rates. This fear seems unjustified (12, 16, 18–20).

12.3. Insertion Sites

Insertion sites for CVC include right and left femoral, subclavian and internal jugular veins. If these sites are not available, the inferior vena cava can be used.

12.3.1. *Femoral vein*

The femoral route was used for temporary dialysis long before the subclavian route became popular (21). Femoral vein cannulation hinders easy movement and may be complicated by retroperitoneal haematoma, iliofemoral thrombosis, infection and traumatic arteriovenous fistula. Nowadays, the femoral route is used in patients with pulmonary oedema where the Trenderlenburg position is not tolerated. Otherwise, it is reserved for use when other sites are not available (22) or while waiting for an alternative access to be created.

12.3.2. *Subclavian vein*

Erben *et al.* (23) first described the routine use of the subclavian vein (SCV) for placement of haemodialysis catheters in 1969. The subclavian route became popular in the 1970s and 1980s due to ease of placement of the catheter, patient acceptance and low rate of acute complications (24, 25). Apart from being a simple procedure, the SCV route avoids the need for a neck wound. Once considered to be the answer to the nephrologist's prayer, SCV cannulation is now falling out of favour (26). Experience with SCV catheterisation with stiff double-lumen polytetrafluoroethylene catheters for temporary dialysis led to the conclusion that the SCV route is sub-optimal for long-term access. Clinically, silent strictures of the SCV can be present even after short-term catheters have been removed (27). Although the earlier catheters were more thrombogenic than the softer ones being used now, use of the subclavian vein is not advisable for prolonged dialysis or if ipsilateral arteriovenous fistula is planned for the future.

12.3.3. *Internal jugular vein*

Many authors (28–31) recommend the right internal jugular vein (IJV) as the preferred site for inserting permanent catheters. The popularity of this route owes a lot to the problems encountered with the SCV route. Acute insertion complications are not only rarer but potentially less dangerous with the IJV compared to the SCV route. Similarly, this route is associated with lower rates of infectious complications than the SCV route (26). There is no risk of catheter compression between rigid structures as may happen to a SCV catheter between the first rib and the clavicle (Fig. 12.2). Unlike the SCV route, IJV stenosis would not interfere with future access procedures in the upper extremities. The drawbacks of this site include neck stiffness and flow interruption with change in neck position (32).

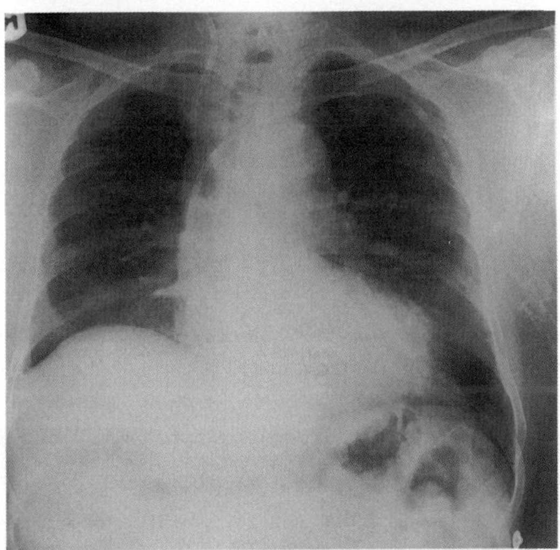

Fig. 12.2. Compression of CVC between the clavicle and first rib.

The right side is favoured because the right IJV is generally larger than the left and offers a more direct course into the right atrium (Fig.12.3). There is no risk of damage to the thoracic duct and the lower right pleural dome diminishes the risk of pneumothorax (33). Also, some patients with long-term left-sided jugular catheters have developed left brachiocephalic vein stenosis, possibly due to increased friction on the vein wall as a result of the multiple changes in catheter direction (Fig. 12.4). These complications occur only rarely and left-sided catheters, though not the first choice, are considered acceptable.

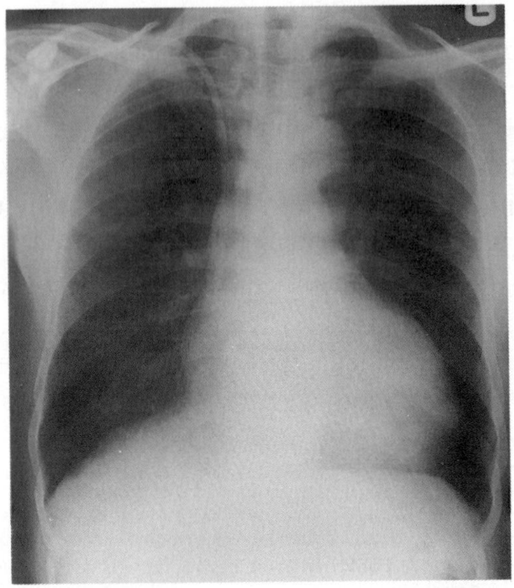

Fig. 12.3. The right internal jugular vein provides a direct route into the right atrium.

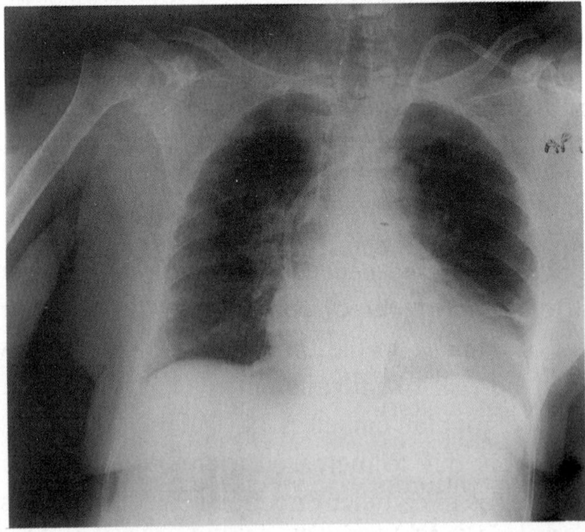

Fig. 12.4. The course of a catheter inserted via the left internal jugular vein.

12.3.4. *External jugular vein*

This approach is simple yet without the inherent dangers associated with cannulation of the central veins (13). Unfortunately, the external jugular vein (EJV) is usually not of sufficient size to accommodate a large catheter without tearing. This route can only be used by surgeons and fluoroscopy is mandatory to ensure correct placement of the catheter. Since the EJV is ligated following the cutdown procedure, this approach may only be used once. As a result, a malpositioned catheter may not be changed over a guidewire. However, for patients needing multiple catheters, this approach will not increase the likelihood of complications with other insertions.

12.3.5. *Inferior vena cava*

The use of this route is restricted to situations when vascular access is virtually impossible or too dangerous by the usual routes. The inferior vena cava (IVC) is approached via the translumbar (34) or transhepatic routes and the catheter tip is placed at the IVC/right atrial junction. Reports of the use of the IVC for vascular access are either anecdotal or of small series, making it difficult to estimate the risk of IVC thrombosis or Budd–Chiari syndrome. However, such small series (34, 35) demonstrate that large catheters can be inserted or removed safely via a percutaneous translumbar route. Extra-length catheters (35–45 cm from tip to cuff) are required for this route.

12.4. Technique of Insertion

12.4.1. *Pre-operative considerations*

Whether the procedure is carried out in the operating room, radiological suite or the bedside, a sterile field must be maintained throughout the procedure. Gowns, masks, gloves, sterile drapes and equipment should be used. Appropriate positioning of the patient is required (Trendelenburg position for SCV, IJV or EJV insertion) and standard surgical preparation of the insertion site using suitable antiseptic must be performed. When applicable, a local anaesthetic should be administered to the insertion site and the underlying tissue, including the path for the subcutaneous tunnel. A catheter of appropriate length is chosen based on the patient's size and side on which it is to be inserted — a longer catheter, e.g. 40 cm PermCath or 23 cm Vas-Cath, is used for left-sided insertion. Each lumen of the catheter is filled with heparinised saline (1 unit/ml) and the extension tubes are clamped.

Fluoroscopy is a major consideration in the decision to perform the procedure either in the operating room or radiological suite. It is absolutely essential that patients with pacemaker wires or inferior vena cava umbrella be cannulated under fluoroscopy in order to avoid inadvertent dislodgement during guidewire and catheter manipulation. Fluoroscopy ensures trouble-free insertion in other situations. Patients with previous CVC may have developed subclinical venous stenosis which may make catheter advancement difficult. Fluoroscopy helps in navigating these obstacles without injuring the vessel. Occasionally, the wire goes up the contralateral internal jugular or subclavian vein making catheter insertion difficult. Both the cutdown and percutaneous placement procedures require confirmation of placement by fluoroscopy or chest X-ray. Fluoroscopy is superior to chest X-ray in ensuring appropriate catheter positioning in that it offers instantaneous information about catheter position and allows necessary manipulation under vision. Fluoroscopic screening ensures precise positioning of the catheter tip in the right atrium (28).

The reader is advised to consult the catheter instruction leaflet for a detailed account of the steps to be taken during the insertion procedure. What is presented here is a brief account with highlights of common pitfalls.

12.4.2. *Percutaneous subclavian cannulation*

With the patient in the Trendelenburg position, the SCV is punctured with an 18-gauge needle. The needle is inserted just lateral to the mid-part of the clavicle in the infraclavicular fossa and directed forward by a finger tip that is pressed firmly into the suprasternal notch. The vein is entered when blood can be freely aspirated into the syringe. The syringe is removed and a flexible guidewire inserted through it and advanced into the right atrium.

When possible, the guidewire position is monitored by fluoroscopy. A 1-cm incision is made at the insertion site. Another incision is made at the exit site, which is placed inferior to the insertion site. The exit site should allow a gentle curve in the catheter as it lies in the subcutaneous tunnel leaving the dacron cuff to lie about 2 cm from it. The catheter is tunnelled by pulling the venous end from the exit site and out at the insertion site. After tunnelling, the catheter is flushed with heparinised saline and its external surface is cleaned. The dilator and "peel-away" sheath are then passed over the guidewire into the vein. Advancing these dilators requires a moderate amount of pressure, which may cause substantial damage if misapplied. It is crucial that free motion of the wire in the dilator be constantly verified during dilator advancement. Otherwise, the stiff dilator might follow its own straight course and puncture the innominate vein wall, rather than following the curve of the wire towards the right atrium. The guidewire and dilator are then removed and the catheter inserted through the lumen of the "peel-away" sheath while the sheath is gradually pulled apart and removed.

12.4.3. *IJV catheter insertion*

With regards to the internal jugular venous approach, three techniques may be used: percutaneous IJV cannulation, IJV cutdown, and EJV cutdown. Although a cutdown with exposure and controlled cannulation of the vein is safer than blind percutaneous insertion, it involves a venotomy closed by a purse-string suture, as well as considerable manipulation of the vein to achieve control of blood flow with a loop around the vein above and below the venotomy. This may result in scarring at the insertion site, which may make subsequent cannulation either difficult or impossible. IJV stenosis in a patient relying on CVC for long-term dialysis is a major drawback. The percutaneous approach is therefore preferred to the cutdown technique. Multiple needle passes should be avoided due to the higher risk of haematoma in uraemic patients. Haematoma will increase the risk of infection and catheter dysfunction. Early conversion to a cutdown is indicated if the percutaneous approach is difficult. A cutdown is the method of choice in patients with coagulopathy, to avoid bleeding complications that may result from blind needle passes.

12.4.4. *Percutaneous internal jugular cannulation*

This is now the most common approach because it is simple and, in contrast to a cutdown, can be used repeatedly at one site (36, 37). The technique of insertion is similar to the percutaneous subclavian cannulation described above. With the head turned 30° to the left for a right-sided placement, the sternal notch, cricoid cartilage, carotid pulse at the level of the cricoid cartilage and the sternocleidomastoid triangle are identified. The skin is punctured just lateral to the carotid pulsation at the level of the cricoid and the needle is directed caudad at 30° angle towards a point just lateral to the ipsilateral sternoclavicular joint. Carotid pulsations are a key landmark in this approach because the internal jugular vein always travels with the artery in the carotid sheath. Other approaches to the IJV have been described (33) with the central and posterior approaches being most reliable. With the central approach, the needle is inserted in the centre of the sternocleidomastoid triangle and advanced parallel to the sagittal plane and 30° to the coronal plane. In the posterior approach, the needle is introduced at the junction of the middle and lower thirds of the IJV behind the clavicular head of the sternocleidomastoid muscle. From this point, the needle is directed towards the suprasternal notch (31). From an exit site located either anteriorly below the clavicle or posteriorly, the catheter is tunnelled and brought out at the insertion site. The anteriorly placed exit site is less cumbersome for many patients. The vein is first entered with an 18-gauge needle, over which a guidewire, dilator and "peel-away" sheath are passed. The catheter is then inserted in the vein through the "peel-away" sheath.

The standard route, based on visual and anatomic landmarks, into the IJV in the mid-neck region often leads to catheter obstruction by angulation in the tissues of the neck as the catheter exits the IJV (Fig. 12.5). Furthermore, the catheter may be torqued by contraction of the sternocleidomastiod muscle or mandibular movement and the catheter tip may be displaced by neck movements (38). In contrast, the base of the neck is relatively fixed during head rotation and serves as a good exit site for the central venous catheters. The use of the low approach results in a gently curved catheter course from the IJV to through the subcutaneous tunnel that extends across the clavicle and anterior chest wall (Fig. 12.3). This approach has now been shown by

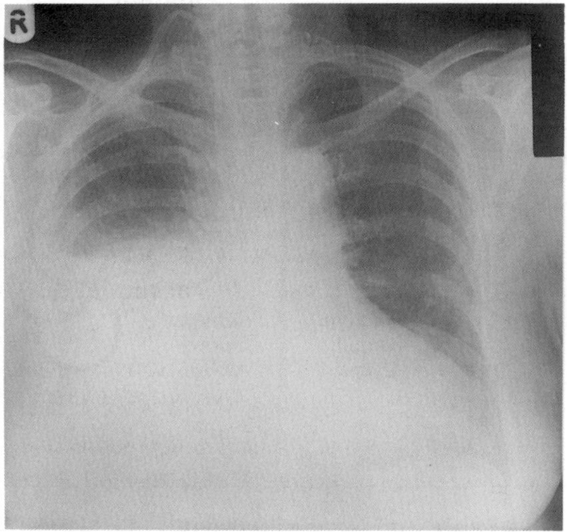

Fig. 12.5. Double trouble: haemothorax following attempted right subclavian cannulation and catheter obstruction due to sharp angulation in the neck.

many authors to be successful with a low complication rate (36, 38–41). Insertion should therefore be at the base of the neck, lower than most central venous lines, to allow tunnelling anteriorly without kinking. At this level, the IJV lies anterior and lateral to the carotid artery in 94% of cases (38) making blind landmark-guided cannulation easier than at a higher level.

Even though the artery and the vein always travel together in the carotid sheath, their relative position is not completely predictable (Fig. 12.6). This is easily understood by considering that the vein starts posterior to the internal carotid at the base of cranium and ends anterior to the common carotid in the chest. Thus, the vein spirals around the artery with different relative locations at different levels. Multiple passages of the needle may therefore be required to find the vein, increasing the risk of carotid artery puncture or pneumothorax. For this reason, Dickson and co-workers have suggested that the IJV should be cannulated first by a 22-gauge "finder" needle so that if the carotid artery is punctured, there is little risk of significant haemorrhage.

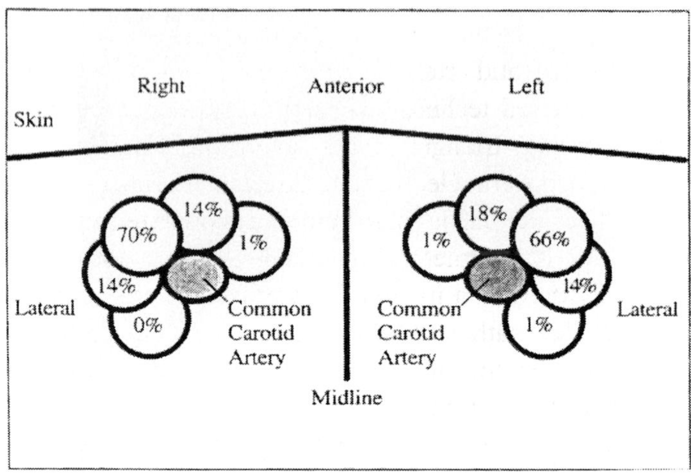

Fig. 12.6. Relationship of the internal jugular vein to the common carotid artery at the level of the cricoid cartilage in 110 patients undergoing carotid duplex ultrasonography [culled from Dickson *et al.* (31)].

The use of ultrasound-guided cannulation for catheter insertion will avoid pneumothorax, carotid puncture and multiple needle passes due to anatomic variability. It also decreases the procedure time and increases its success rate (38, 42, 43). The ultrasound device (Site-Rite, Dymax Corp., Pittsburgh PA) equipped with a 7.5 MHz transducer, covered with a sterile plastic sheath, and a sterile needle guide which snaps onto the side of the probe is used. The IJV and carotid are first demonstrated on the monitor as two adjacent round structures. They are easily distinguished from each other by the carotid pulsation and the compressibility of the vein. The needle is then advanced and can then be seen on the monitor as it penetrates the vein. This method is so reliable and simple that it can be used in patients with coagulopathy and in situations where haematoma has resulted from multiple unsuccessful needling. This technique, though very useful and much better than the blind, external landmark-guided technique, requires additional equipment and personnel and should probably be reserved for more challenging cases.

12.4.5. *IJV cutdown*

With this approach, carotid artery puncture and pneumothorax are extremely unlikely. In the standard technique, a transverse incision is made over the lower anterior cervical triangle, 2 cm above the clavicle. After incising the platysma, the two sternocleidomastoid heads are separated to expose the carotid sheath. The internal jugular vein is encountered upon entering the carotid sheath and is completely encircled. A purse-string suture of 5/0 polypropylene is stitched on its anterior surface. After obtaining distal and proximal control, the catheter, already tunnelled and flushed, is inserted through a venotomy in the middle of the purse-string. After verifying its position, the purse-string is tightened.

To reduce the degree of venous scarring, an alternative technique may be used. After entering the carotid sheath, only the anterior venous surface is exposed, without completely encircling the vein. A purse-string is sutured on this surface. As in the Seldinger technique, a needle, guidewire dilator and "peel-away" sheath are inserted into the vein. However, in contrast to the blind percutaneous technique, the insertion is under direct vision into the middle of the purse-string. The advantages of direct visualisation are preserved but without the extensive dissection of the standard cutdown. By minimising scarring and stenosis of the jugular vein, this technique may increase the likelihood of preserving it for subsequent cannulation. For both techniques, the IJV can be approached at a slightly higher level from the postero-lateral aspect of the clavicular head of the sternocleidomastoid muscle. Because the catheter exits the vein behind the muscle, it is less liable to kink.

12.4.6. *EJV cutdown*

Through a transverse incision in the neck, the EJV is exposed and loops passed around it both above and below the site of venotomy. The procedure is similar to IJV cutdown.

12.4.7. *Twin-catheter insertion*

This can be done either by using the pacemaker technique (which allows two catheters to be inserted into one venipuncture site) or by the two-puncture method. Access to the IJV is achieved by percutaneous cannulation according to the Seldinger technique. The first guidewire is inserted through the needle and the needle withdrawn, as in the standard technique. A 6-French (6F) minisheath-introducer from the Twin-Cath kit (Fig. 12.7) is then advanced over this wire. Through this, a second guidewire is inserted and the sheath is removed. Thus, two wires end up in the vein with one puncture. In a slight modification of the technique, Canaud *et al.* (16) described the insertion of two guidewires through a 16-gauge needle. Dilators with "peel-away" sheaths are then advanced over each wire, as in the standard method. An advantage of the twin catheters is that the dilators are much smaller and more flexible and, therefore, less fearsome to advance over the wire. Although this technique saves one puncture, it may result in significant

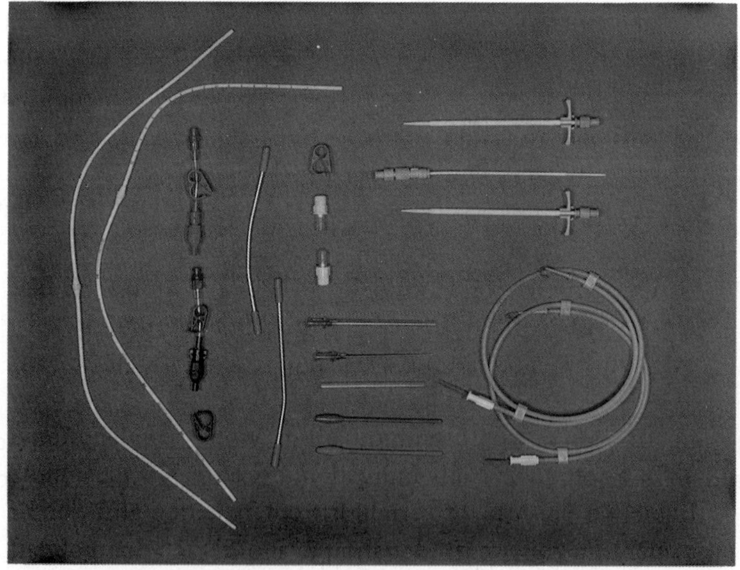

Fig 12.7. Twin-Cath kit.

bleeding, hence, the reason for the two-puncture technique. Once the location of the internal jugular vein is identified with the first puncture, it is a simple matter to puncture it a second time and bleeding around each catheter will be minimal.

Once the two catheters are inserted centrally, tips are positioned by fluoroscopy. The tip of the arterial aspiration catheter (red hub) is positioned 4 cm proximal to that of the venous catheter (blue hub) to minimise recirculation (Fig. 12.8) (19). Each catheter is back-tunnelled over the chest wall. Note that back-tunnelling is not possible with the standard dual-lumen catheter because the external portion of the dual-lumen catheter is bulky and cannot be pulled back through a subcutaneous tunnel. Instead, the venous tip of the dual-lumen catheter must be pulled through the tunnel first before inserting the catheter in the vein. Back-tunnelling is possible because both the proximal and distal ends of the Tesio catheters are of equal calibre. This is performed with a special curved introducer needle with the two exit sites 1 cm apart. The cuff is placed close to the skin exit to eliminate the portion of the subcutaneous tunnel which is prone to tunnel infections.

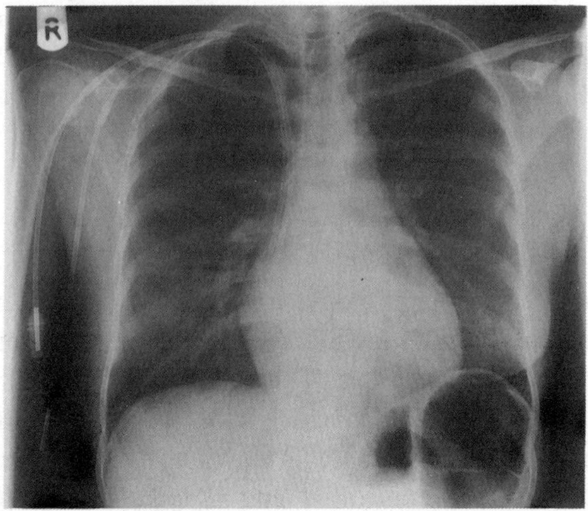

Fig. 12.8. Satisfactory position of Tesio twin catheters inserted via the right internal jugular vein.

12.4.8. *Postinsertion considerations*

For dual-lumen catheters, placing the venous lumen lateral to the arterial ensures correct orientation of the tip in the atrium. This prevents the arterial lumen from occluding against the atrial wall during dialysis. Catheter patency is checked by using a 20-ml syringe with saline to ascertain flow through each lumen. Once flow is satisfactory, both lumens are flushed with heparinised saline and clamped. At the conclusion of the procedure, heparin lock is applied. Each lumen of the catheter is filled with 1.5 ml (or amount required to fill lumen) of a saline solution containing 1000 units of heparin per millilitre. The concentration used varies from unit to unit. The heparin is withdrawn before dialysis and a fresh solution instilled after every dialysis session.

The entry site is suture-closed. With the cuffed dual-lumen catheters, the exit site is always larger than the catheter, since the larger cuff has to be pulled through it, and therefore needs to be closed with sutures. This should be done with suture material of low tissue reactivity and the sutures should be removed after one week to limit irritation at the exit site. Without careful attention to this area, some dual-lumen catheter exit sites never completely heal and become chronically irritated, setting the stage for tunnel infections. With Tesio catheters, the exit site fits snugly and neatly around the catheter, which is completely immobilised without the need for an irritating stitch.

12.5. Who Should Insert Them?

It is common practice for nephrology junior staff to insert temporary CVC and for surgeons to insert tunnelled haemodialysis catheters by either a percutaneous or cutdown technique. The status quo has been challenged by interventional radiologists and controversy now rages about who should insert permanent CVC. While there is no doubt that surgical insertion is carried out under a more controlled and aseptic setting, physicians and interventional radiologists claim equally good results as surgeons (36, 44, 45). The results presented by Lund and co-workers (45) show that catheters placed by interventional radiologists do not have a higher complication or failure rate than those inserted by surgeons (37, 46–48). Results can be improved by the use of imaging during manipulation of the guidewire, dilator or catheter. Ultrasound-guided needle insertion into the subclavian vein will eliminate pneumothorax. Trerotola *et al.* (36) reported excellent results with real-time ultrasound-guided puncture and fluoroscopic insertion of catheters via the internal jugular vein. They achieved technical success in 100% of cases with no instances of pneumothorax, haemothorax or substantial bleeding. As a result of their findings, they advocated that interventional radiologic placement should be the method of choice. The lower cost of a radiological suite with readily available imaging facilities, compared to the operating room, favours radiological insertion. However, patients who require general anaesthesia for medical or other reasons are especially suited for surgical insertion.

For those nephrologists seeking encouragement for percutaneous placement of long-term dialysis catheters in the dialysis unit, the report by Swartz (44) provides the answer. Over a 3.5-year period, 93 (79%) catheters were placed percutaneously by nephrologists while the remaining 25 (21%) were inserted surgically. This report, which covers a large number of devices over an extended period of time at a large hospital, shows that tunnelled indwelling CVC can be placed safely by the use of a bedside percutaneous technique in the majority of cases. Careful selection of cases for catheter insertion outwith the radiological suite or theatre is required as insertion technique was neither randomised nor controlled in their study. Moreover, in 6% of bedside catheter insertions, transportation of the patient to the radiology department is required for completion of the procedure (48).

Inexperienced physicians have about twice as many complications and unsuccessful cannulations as their more experienced colleagues (49). The physician, surgeon and interventional radiologist may all share the burden of catheter placement in any given centre. What is important is that whoever is responsible for inserting the catheters must be well motivated and pay attention to details of technique. Close liaison between them will lead to better patient selection and improvement of results.

12.6. Role

Inherent dangers in cannulating central veins have led to a cautious approach to the insertion of CVC for long-term dialysis. In a few instances, a permanent dual-lumen catheter may provide the only practical access for long-term haemodialysis. Frail elderly patients with multiple medical problems and a limited life expectancy fall into this group (50). The major indication for CVC is when primary and secondary vascular accesses have been exhausted in a patient deemed unsuitable for CAPD.

The majority of ESRD patients are initiated to dialysis by means of a temporary CVC while a more permanent access is created. Recent data from the United States Renal Data System (USRDS) indicate that among patients starting ESRD with haemodialysis during 1993, 48% had no access placed or attempted before the onset of ESRD (51). Delayed referral to the nephrologist is a widespread problem with the majority requiring dialysis within 24 hours (52). Such patients inevitably require placement of a temporary dialysis catheter. Tunnelled CVC provide the method of choice for temporary access of longer than three weeks duration and are very useful during the period of maturation of arteriovenous fistula.

12.7. Complications

CVC have a down side. Serious complications may occur during their insertion, some life threatening, others a mere irritation. Delayed complications may lead to loss of the catheter or patient. In the decision to use CVC, the risks must be weighed against those of other techniques and the necessity for adequate access to the circulation. Because of the attendant increase in access complications, the National Kidney Foundation expert panel recommended that less than 10% of patients should be maintained on chronic haemodialysis by means of CVC (53).

12.7.1. *Mechanical complications*

Whether the CVC is inserted percutaneously or via a cutdown onto the vein, damage to adjacent structures (arteries, nerves, other soft tissue) can result.

Failure to cannulate the subclavian vein, reported to occur in approximately 5% of attempts (54), varies widely and depends on the experience of the performing physician or surgeon. Fatal complications have been reported with SCV cannulation (46, 54, 55). In a review of 11 publications involving 1542 catheterisations and a questionnaire survey of 16 dialysis centres (4000 catheterisations), the mortality associated with catheter dialysis was found to be between 0 and 1.25 per 1000 catheterisations (46). It is important that no matter what the level of experience, the performing physician or surgeon must know when to stop. Other complications associated with SCV cannulation include: malposition of the catheter, pneumothorax, haemothorax (Fig. 12.5), perforation of the superior vena cava or atrium, venous air embolism and cardiac arrythmias (56). Arrhythmias provoked by guidewire insertion usually resolve spontaneously soon after guidewire withdrawal (57).

Following IJV cannulation, complications include pneumothorax, deep-vein thrombosis, C5 paralysis, Horner's syndrome, injury to the subclavian or vertebral artery, injury to the vagus nerve or stellate ganglion, vocal cord paralysis and mediastinal bleeding (31, 58). The most dangerous risk associated with IJV cannulation, carotid artery injury, occurs in 4% of attempts due to its close proximity to the vein (31, 59). Haematoma resulting from arterial damage can result in airway compromise. The subclavian artery is punctured in 1.4% of SCV cannulations and bleeding from it may lead to hypotension and/or haemothorax.

Other insertion complications reported include: perforation of central veins with lodgement of the catheter in the thoracic cavity (60) or the pericardial sac (61); and atrial perforation with fatal consequences (62, 63). Such catheter protrusions through the vein wall may not produce any significant symptoms until dialysis is started. Blood then starts to be pumped out of the superior vena cava or atrium via the arterial port and returned via the venous port into the extravascular space. If there is any sudden onset of chest pain, hypotension and/or breathlessness, the dialysis nurse should be alerted to this unusual problem. Dialysis must be stopped immediately, followed by appropriate action.

Avoidance of complications requires a thorough understanding of the relevant clinical anatomy and experience of inserting these catheters. Repeated attempts at vein puncture must be avoided and recourse to ultrasound-guided approach should be made in difficult cases. In a review of catheter placement in the IJV, Dickson *et al.* (31) recommend the use of hand-held Doppler ultrasound in obese patients and those with short necks. The use of soft-tipped catheters and J-tipped guidewires should reduce the incidence of SVC or atrial perforation. Under no circumstances should the catheter be forced if resistance to insertion is experienced either over the guidewire or in the "peel-away" sheath. Air embolism can be prevented by adopting the head-down position during catheter placement or removal. Although cardiac arrythymias are mainly innocuous, they could become significant in patients with electrolyte abnormalities or when a deeply positioned catheter tip irritates the tricuspid valve. Routine chest X-rays after insertion of radio-opaque catheters and repositioning, if necessary, will avoid this problem.

12.7.2. *Catheter dysfunction*

Catheter dysfunction is manifested by failure to aspirate blood from the lumen(s), inadequate blood flow or high resistance pressures during haemodialysis (64). It may also present as high rate of recirculation of dialysed blood and thus could be missed if dialysis adequacy is not monitored regularly. The recirculation rate within most central venous dialysis catheters is between 3% and 7% in most clinical settings. Values above 10–15% are thought to indicate dysfunction.

Episodes of inadequate blood flow represent the most important obstacle to efficient haemodialysis delivery. The incidence of poor or inadequate flow varies from 7 to 46% (1, 2) and depends on the definition applied. At the Duke Medical Centre — where catheter malfunction is defined as failure to achieve blood flow rates of at least 300 ml/min on two consecutive occasions or failure to achieve rates of 200 ml/min on a single occasion — malfunction was experienced in 87% of catheters (65). Most centres will intervene if the blood flow rates fell below 200 ml/min (45, 66). Poor blood flow rate is a more common problem with dual-lumen than single lumen catheters. Causal factors include malposition of catheter tip, kinking (Fig. 12.5) or knotting of the catheter and inadvertent withdrawal (Fig. 12.9).

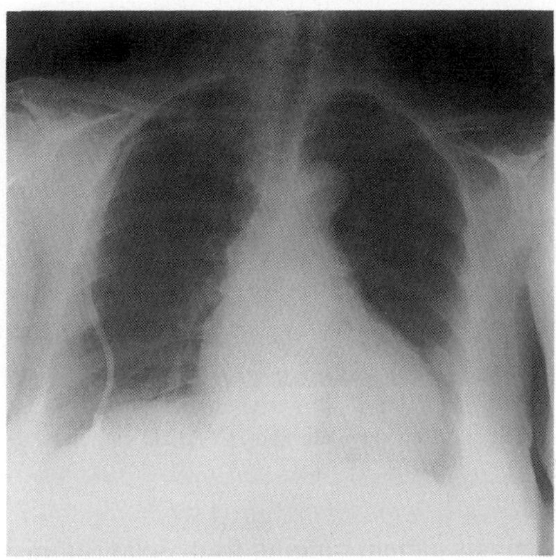

Fig 12.9. Catheter migration with its tip lying in the subclavian vein.

Function is less satisfactory when the catheter tip is positioned in the superior or inferior vena cava as compared to the right atrium or right ventricle (5). Ladocsi *et al.* (67) addressed the problem of catheter kinking by creating a subcutaneous tunnel with a laterally situated exit site inferior to the lateral third of the clavicle when the catheter is placed in the internal jugular vein. Catheter withdrawal or migration occurs in about 1–4% of patients (46), and can be prevented by fixing the catheter at the exit site. Adequate fixation of the catheter in the subcutaneous tunnel by dacron or polyester cuff interaction with tissue is not achieved until 1–2 weeks after insertion. Therefore, suture fixation at the exit site is required to obviate catheter pull-out. Some catheters like the Vas-Cath (Vas-Cath Inc., Mississauga, Ontario, Canada) are provided with holes for this purpose while others like the HY-MED (Neostar Medical Technologies Inc., Manchester GA) silicone catheters have no such holes and the suture should be tied around the Y-hub (Fig. 12.10).

By far, the most significant cause of catheter dysfunction is intraluminal or periluminal (fibrin sleeve) clot. Catheter dysfunction causes immense

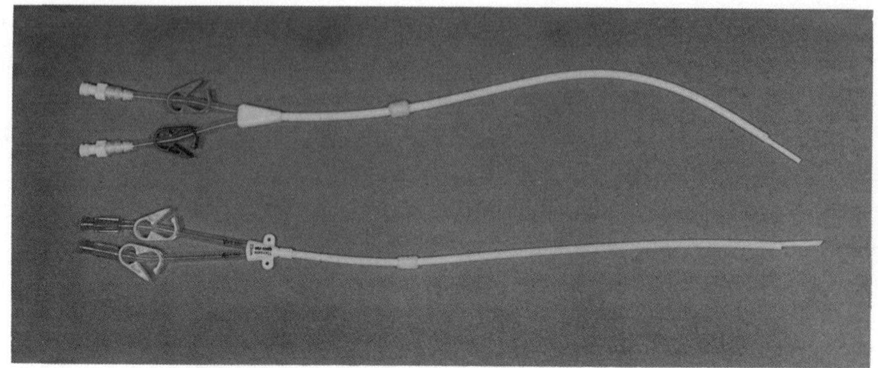

Fig. 12.10. Vas-Cath and HY-MED catheters.

problems for the dialysis patient ranging from delayed, missed or inadequate dialysis treatment to hospitalisation, emotional upheaval and medical risks. The need for a rapid, easily-performed and effective minimally-invasive procedure for restoring catheter function cannot be overstated. Any technique or treatment that can salvage these catheters will lead to considerable cost savings, preservation of central veins (by avoiding repeat cannulation or surgery) and avoidance of complex access surgery. A discussion of some of the commonly used treatment modalities follows.

12.7.2.1. *Thrombolysis*

Thrombolysis of dialysis access using fibrinolytic enzymes has been extensively studied during the last 25 years (68). Three fibrinolytic agents [streptokinase, tissue plasminogen activator (TPA) and urokinase] are available for this purpose. Streptokinase, due to its erratic response, allergic reactions and drug resistance problems, is unsatisfactory (65, 69, 70) as fibrinolysis may need to be repeated many times. Tissue plasminogen activator is expensive (66, 71) and does not achieve better results than the less expensive urokinase (68, 72). Urokinase has thus become the thrombolytic agent of choice. It offers the advantages over streptokinase of a direct mode of action, a shorter effective half-life and no antigenicity. Urokinase can therefore be used in cases of recurrent thrombosis (73). Thrombolytic therapy in one form or other is successful in most cases and should be the first-line therapy undertaken for non-positional catheter malfunction (65).

Urokinase is an enzyme that catalyses the conversion of plasminogen to plasmin, thereby lysing clot. Clot within a catheter lumen can usually be prevented by heparin but if it accumulates in the catheter, it can be dissolved by small amounts of urokinase (5000 IU/ml).

Catheter clearance with urokinase

One millilitre of urokinase (5000 IU/ml) is instilled into each port followed by an amount of saline to fill each lumen of the catheter. This is left in place for at least 20 minutes, after which haemodialysis is re-attempted. Instillation of urokinase in this fashion successfully restores function in 74–97% of cases (2, 65, 74).

Systemic urokinase

Experience with removed catheters show that intraluminal clots are easily sucked into a syringe. Such a clot should not interfere with blood flow because it can easily be aspirated or pushed into the venous system. For it to cause obstruction, the intraluminal clot has to be attached to a clot formed outside the catheter. Such clot is much bigger and cannot be aspirated (75). Urokinase locking of catheter lumen yields poor results in the long term (40–50% success rate) due to the fact that the external fibrin sleeve reaches the catheter tip in a short time (76). High-dose systemic urokinase works because it dissolves most of the external clot (76, 77).

If repeated urokinase lock fails to clear the catheter lumen, the patient is admitted for urokinase infusion. Using an infusion regimen of 20 000 units per lumen per hour for six hours, 55–79.5% of catheters are restored to successful function (36, 45). Results from centres where infusion is carried out over a longer duration with far less urokinase are being awaited.

Should this low to moderate dose of urokinase be unsuccessful in clearing catheter blockage, high-dose systemic urokinase should be considered. In the early nineties, Uldall (78) advocated the infusion of 125 000 units of urokinase through one lumen over two hours and then 125 000 units through the other lumen. Twardowski (76) describes another method of administering high-dose urokinase in cases where blood cannot be aspirated or flow is less than 100 ml/min. An amount of 250 000 IU urokinase is dissolved in 100 ml normal saline and infused over three hours through both lumen (125 000 IU over 1.5 hours consecutively through each lumen). Provided the patient is not receiving heparin, this treatment can be carried out safely on an outpatient basis. This treatment restores or improves flow in 89% of catheters (76) but dialysis shifts may have to be rearranged.

If dialysis can be started but with poor flows, dialysis with systemic heparinisation should be continued while 250 000 IU urokinase (dissolved in 100 ml normal saline) is infused through the venous port over three hours with a successful outcome in 94% of cases (76). Results of newer protocols that infuse a similar dose of urokinase over a much shorter period (e.g. 10 minutes) are being awaited. High-dose intradialytic urokinase is expensive but has the advantage of a high probability of success.

Contraindications to systemic urokinase

Contraindications to thrombolysis (76, 79) are shown in Table 12.1. While administration of urokinase is in progress, monitoring for adverse reactions should include pulse, respiration and blood pressure at 15- to 30-minute intervals. Patients are also observed for possible allergic reactions (skin rashes and bronchospasm) and bleeding.

12.7.2.2. *Low dose warfarin*

There is evidence to suggest that anticoagulation with warfarin may be helpful in patients with repeated catheter blockage. The idea to use warfarin as prophylaxis against catheter-related thrombosis came from experience with its use in intravenous hyperalimentation. The concept has now been tested in a prospective, open, randomised study where low-dose warfarin (1 mg/day) reduced the incidence of thrombosis in patients with indwelling

Table 12.1. Contraindications to systemic urokinase

Absolute contraindications
Active internal bleeding
Surgical procedure within 24 hours
Recent intracranial or intraspinal surgery, intracranial neoplasm, arteriovenous malformation, or aneurysm
Arterial puncture within 24 hours
Percutaneous biopsy within 48 hours
Recent severe trauma
Severe uncontrolled hypertension
Relative contraindications
Left heart thrombus
Subacute bacterial endocarditis
Catheter related sepsis (risk of bacterial shedding into the bloodstream)
Pregnancy
Cerebrovascular disease
Hemorrhagic retinopathy
In a patient in whom anticoagulation is contraindicated
Allergy
Recent surgery (other than intracranial or intraspinal)
Recent biopsy or puncture of non-compressible vessel

CVC (for infusion chemotherapy) without inducing a haemorrhagic state (80). In that study, 4 (9.5%) of 42 patients who received warfarin had venogram-documented thrombosis compared to 15 (37.5%) of 40 patients who received no warfarin ($P < 0.001$). This approach is likely to gain popularity in many dialysis units.

12.7.2.3. *Percutaneous fibrin sleeve stripping*

Blockage, the commonest reason for loss of catheter function, is thought to be due to encasement of catheter shaft by a fibrin sleeve (81, 82).

Autopsy dissections and fluoroscopic injections in patients dying with indwelling subclavian catheters have conclusively shown that totally circumferential fibrin sleeves are consistently and extensively formed on these catheters (81). Histologically, a fibrin sleeve is made up of fibrin and platelets with a few red blood cells and polymorphonuclear leucocytes (83). Catheter blockage can sometimes be managed by reversal of arterial and venous lines or thrombolytic therapy. If these measures fail, the catheter is usually removed. Recent reports of successful prolongation of catheter function by percutaneous fibrin sleeve stripping (PFSS) (84–86), have led to a change in approach to failing haemodialysis catheters.

Functional catheter failure is defined as inability to achieve blood flow rates of 200 ml/min or higher to allow for satisfactory completion of dialysis using a catheter that is confirmed to lie in the correct position with demonstrable one-way obstruction (36, 45, 84). The catheter is re-assessed fluoroscopically to confirm the tip position and exclude mechanical causes of malfunction such as kinking. Angiography is then performed through both lumens of the catheter using non-ionic contrast medium. Upon demonstration of a fibrin sleeve, a 6F vascular sheath is introduced into the femoral vein and a 6F goose-neck Amplatz snare catheter (Microvena Corporation, USA) advanced through it into the right atrium. A guidewire is then passed from above through the venous (distal) port of the catheter into the IVC where it is encircled by the snare. The snare is then advanced up over the catheter to the confluence of the brachiocephalic veins. Multiple-stripping passes are made by closing the snare around the catheter and gently pulling down over it. A final proximal port angiogram is obtained to assess contrast material jet.

There is a theoretical risk of embolisation of fibrin sleeve fragments to the lungs. Up to date, no symptomatic pulmonary embolism has been reported following PFSS (85). However, it is prudent to avoid this procedure in patients with a right-to-left cardiac shunt. PFSS is a simple procedure with a low complication rate that can be carried out as a day case. It provides a useful alternative to repeated urokinase infusions and extends the longevity of haemodialysis catheters. PFSS restored function in 79–98% of attempts in well-positioned catheters (84, 85). Overall catheter patency rate of 90% at six months and 81% at one year have been reported with the application of multiple PFSS procedures (84).

12.7.3. *Infection*

Catheter-related infection is responsible for great morbidity among haemodialysis patients and, in many centres, it is the second leading cause of catheter loss. Infection of CVC may present as exit-site sepsis, tunnel abscess, peripheral bacteraemia or metastatic sepsis. It is estimated that intravascular-device-related bacteraemia has a mortality rate of between 10% and 20% (87). Haemodialysis catheters are responsible for a higher systemic infection rate than other catheter types (88).

The sources of infection include: migration of microorganisms from the patient's skin or catheter hub; contamination of the catheter connectors or infused solutions; and secondary colonisation during bacteraemia from a remote source (7). Thrombogenic catheter material, breaks in sterile technique during catheter insertion, site of catheter insertion, heavy bacterial colonisation and prolonged duration of catheterisation are recognised risk factors for infection. Diabetes (2, 15), a history of bacteraemia and immunocompromised status (89) are also risk factors for catheter-related bacteraemia. The prevalence and consequence of intraluminal colonisation of central venous catheters is unknown. Both microbial slime and host fibrin and fibronectin contribute to the formation of the biofilm which is essential for adherence and maintenance of bacterial colonisation (90). A cross-sectional study carried out in the UK demonstrates that the vast majority of central venous haemodialysis catheters (95%) are colonised by bacteria which can develop into peripheral bacteraemia and severe systemic sepsis (91). In another report with a 55% colonisation rate, associated catheter-related septicaemia contributed to death in one patient (92).

Comparison of infection rates is difficult due to lack of a standardised definition of catheter-related infection. Infection rates have been calculated in many ways due to lack of a standardised index that incorporates all the significant aspects of catheter infection. Firstly, it has been expressed as a percentage of the total number of catheters (44) or patients (2, 30, 45). This fails to take into account either the number of patients with catheters or vice versa, and the duration and/or number of dialysis episodes. Inclusion of even one patient with repeated complications will skew the outcome for the group as a whole (50). Secondly, when expressed as a fraction of catheter removals due to infection (2), both the duration of cannulation and the number of successfully treated episodes without catheter removal are ignored. According to Lund *et al.* (45), the importance of the duration of catheterisation can be presented in three ways: expression of infection rate as the number of events that occur per 100 days at risk, life-table analysis, and time to first catheter infection. Life-table analysis is recommended for data from events that occur over time in populations that are at risk for complications in different time periods; it is also recommended whenever other unrelated events may remove individual patients from evaluation. A short time to first infection could indicate a breakdown in the sterile technique used during placement (44, 93). Infectious complications of CVC should therefore be reported in a standardised manner taking into consideration the number of patients, number of catheters and the duration of catheterisation, in other words, using the denominator of catheter days. Furthermore, results must indicate the type of catheters involved. Infection is less frequent in patients with cuffed, tunnelled catheters.

12.7.3.1. *Management of catheter-related infection*

Most guidelines recommend removal of the catheter in patients with clinical signs of systemic sepsis, persistent exit-site infection, tunnel infection or persistent fever (> 72 hours duration) or bacteraemia during antibiotic treatment (7, 89). Given the type of patients who use CVC for dialysis, the issues of catheter salvage and access site salvage have become increasingly important in recent times.

One approach combines intravenous antibiotic therapy and replacement of the catheter over a guidewire using the same venous access tract (94, 95). Using this approach, Carlisle *et al.* (94) salvaged 17 of 21 access sites in patients with cuffed, tunnelled jugular silastic catheters. Shaffer (95) successfully treated 10 patients with 13 bacteraemia episodes by means of catheter exchange, antibiotic therapy and preservation of the venous entry site.

Recent reports suggest that catheter-related infections can be successfully treated without the need for removing the catheter in 25 to 100% of cases

(44, 96, 97). The first study to look specifically at the issue of catheter salvage in dialysis patients with catheter-related bacteraemia reported a failure rate of 68% and advised that antibiotic therapy alone is insufficient (89). However, attempted salvage did not increase the risk of complications. Patients with polymicrobial infection, gram-negative or complicated staphylococcal bacteraemia require antibiotic therapy for more than two weeks (44, 89) and a lower threshold for catheter removal.

12.7.3.2. *Prevention*

Application of several simple measures may reduce the incidence of catheter-related infections (7). Such measures include: the use of chlorhexidene (instead of povidone–iodine) for skin disinfection prior to catheter insertion (98); application of silver-impregnated collagen cuff to the subcutaneous portion of the catheter at insertion (99); use of antiseptic-bonded catheters; application of antibiotic ointment to the exit site during dressing changes; and avoidance of occlusive dressings.

Although prophylactic antibiotic cover for catheter insertion is practised widely, its value in reducing infectious complications is debatable. There is a lack of randomised, controlled prospective study comparing antibiotics with placebo as prophylactic cover for CVC insertion for dialysis. However, such trials in patients undergoing chemotherapy for malignant disease (100) or on intravenous nutrition (101) failed to demonstrate any beneficial effect of vancomycin in reducing catheter-related sepsis.

12.7.3.3. *Antimicrobial coating of CVC*

It has been suggested that coating of intravascular catheters with antimicrobial agents can reduce the incidence of catheter-related infections. Agents used for bonding include sulfadiazine and chlorhexidene gluconate (87, 98), teicoplanin (102), tridodecylmethylammonium chloride either with vancomycin or minocycline, and rifampin (103, 104).

Previous animal experiments have shown that antibiotic or antiseptic coating reduces bacterial adherence and biofilm formation on intravascular catheters (105, 106). Human studies, while confirming that such coatings can reduce bacterial colonisation, have been rather inconclusive as to whether such an effect is translated into a reduction of infective complications in the clinical setting (87, 107–110). Two recent large randomised trials, using standardised definition of catheter-related infections and molecular typing techniques for confirming the source of infection, have assessed the impact of impregnated catheters in reducing catheter-related bloodstream infection (104, 111). Both studies confirm the beneficial effect of antimicrobial coating in reducing bacterial colonisation and incidence of catheter related bloodstream infection. Maki *et al.* (111) reported a 44% reduction in catheter colonisation [13.5 compared with 24.1 colonised catheters per 100 catheters; relative risk, 0.56 (95% CI, 0.36 to 0.89); $P = 0.005$] and a 79% reduction in catheter-related bloodstream infections [1.6 compared with 7.6 infections per 1000 catheter days; relative risk, 0.21 (CI, 0.03 to 0.95); $P = 0.03$] by using catheters coated with chlorhexidene and silver sulfadiazine. Similarly, Raad and co-workers (104), using catheters pre-treated with tridodecylmethylammonium chloride and coated with minocycline and rifampin, reported a reduction in colonisation [8 compared with 26 colonised catheters per 100 catheters; relative risk, 3.13 (95% CI, 1.66 to 5.88); $P = 0.001$] and catheter-related bloodstream infection (0 compared with 7.34 infections per 1000 catheter days; $P = 0.01$).

While none of these studies reported any adverse events or infection with resistant organisms associated with the use of coated catheters, the call for further evaluation (112) is particularly warranted in the haemodialysis setting. It is, for example, important to establish how long a coated catheter can resist microbial colonisation. If this technology proves applicable to haemodialysis catheters, it will lead to considerable savings in cost but its use must be as an adjunct to, rather than a substitute for, good aseptic practices (112).

12.7.4. *Central venous stenosis*

Although stenosis of central veins can occur following trauma or surgery, by far, the most important aetiological consideration is post-CVC insertion.

Catheterisation of the SCV is complicated by stenosis in 19–53% as compared with the IJV where the figure varies from 0–10% (27, 56, 113, 114). The prevalence of IJV stenosis is not known as venography is not usually performed to assess this site and fistula formation does not cause a subclinical stenosis to manifest. The aetiology of subclavian stenosis has not been fully worked out but as pointed out by Cimochowski and co-workers (27), it is likely to be multifactorial. Endothelial disruption at the puncture site causes flow disturbance, setting the stage for venous thrombosis. The SCV rises slightly to go over the first rib before dropping into the heart and so the catheter follows the same course with maximum torque at the rib. Another factor is the to and fro movement of the catheter tip in the atrium or superior vena cava with cardiac action (115, 116). These factors produce irritation of the vein wall with eventual ulcer or thrombus formation (56). Support for these theories is provided by the fact that most if not all the strictures in the SCV occur in the proximal third, the usual site of puncture and a segment in close relation to the first rib. In contrast, the IJV is usually punctured on the anterior surface in the middle or lower third where it is not torqued over any structure.

Other factors identified as playing important roles in the causation of vein stenosis include: the number of punctures per vein (27, 117); left-sided insertion due to a sharper curve taken by the catheter (113); catheter-related infection (118); the role of high venous pressure and turbulence induced by an arteriovenouis fistula (119); and the type of material the catheter is made of (120). Leakage of plasticisers from the catheter at the site of contact with the vessel wall is thought to induce fibrosis (121). Silicone catheters, thought to be more flexible than polyurethane and polyethylene catheters, cause less trauma to the vein.

Though the majority of SCV stenosis are severe (> 70%) (27), most patients are asymptomatic until an arteriovenous fistula is formed in the ipsilateral arm. The problem then presents with high urea recirculation, high venous dialysis pressure, arm oedema and, in advanced cases, distended subcutaneous venous collaterals of the upper half of the body (122). Before a fistula is created in the ipsilateral arm of a patient with a history of SCV catheterisation, patency of the SCV must be confirmed by imaging.

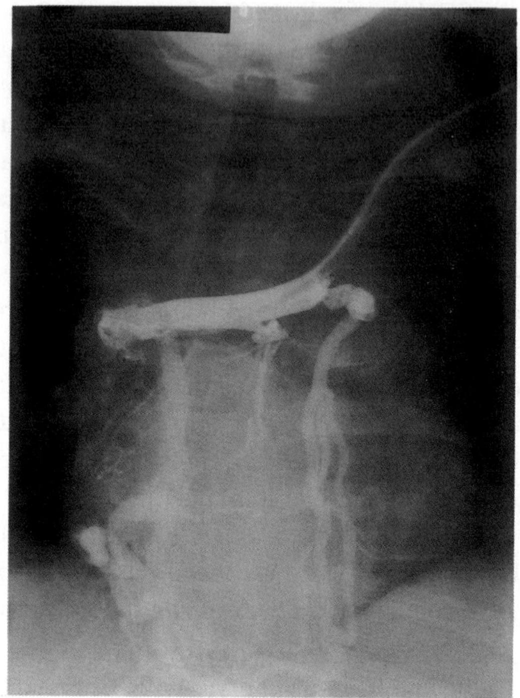

Fig 12.11. Occlusion of the superior vena cava after multiple CVC insertions.

Venography (Fig. 12.11) is the gold standard but Doppler ultrasound examination may also provide adequate evaluation of venous drainage from the limb (123).

Treatment is aimed at alleviating acute symptoms when present, preventing complications caused by thrombosis and minimising late sequelae and salvaging a fistula that is threatened. Arm elevation with or without anticoagulation and intravenous streptokinase have been shown to be beneficial in dealing with the acute symptoms (124). Percutaneous intervention with transluminal angioplasty is the preferred treatment for central venous stenosis. This has a reported initial failure rate of 30% (125) and a one year success rate of 35–36% (122, 126). Stenotic lesion can be classified by their behaviour following angioplasty. Elastic lesions (< 50% improvement) recur earlier than non-elastic (> 50% improvement) lesions (125). Stent placement combined with angioplasty is indicated in elastic central venous stenosis or if a stenosis recurs within three months after angioplasty. Repeat angioplasty for non-elastic stenosis offers long-term success rates close to those achieved by operative reconstruction (122, 125). Surgical treatment, indicated in patients with incapacitating venous hypertension of the upper extremities or head and neck, usually consists of bypass with externally reinforced PTFE grafts. Although surgery offers initial success and over 70% long-term patency rates (122), requirement for thoracotomy in patients with multiple co-morbidities makes it unattractive. Ligation of the arteriovenous fistula or graft is indicated in patients with incapacitating arm swelling or failed angioplasty due to impassable stenosis. Often that is all that is necessary.

12.8. Catheter Removal

The dacron or polyester cuff facilitates tissue in-growth. Once this has happened, the catheter must be removed surgically. Under local anaesthetic, the cuff is freed and the catheter pulled gently and smoothly. Sharp jerky motions could tear the catheter and should be avoided.

12.9. Long-Term Results

As shown in Table 12.2, the one-year cummulative catheter survival ranges from 47 to 93% (1, 2, 18, 36, 50) which compares favourably with the patency rates for secondary vascular access (50). Tesio twin catheters appear to achieve higher blood flow rates, lower complication rates and better survival. Because CVC are used for different purposes (temporary, prolonged temporary and permanent access), long-term results must be determined by patient subgroups. For example, 83 (99%) of 84 CVC inserted for prolonged temporary access functioned satisfactorily until removal, whereas 37 CVC intended for permanent access had a mean survival of 12.7 months (range, 3 weeks to 2.4 years) (65). It is also important to distinguish between catheter survival and technique survival. The former refers to the probable use-life of a single catheter while the latter refers to the probability of maintaining a

Table 12.2. Long-term results of CVC

Ref.	Catheters (Patients)	Mean Duration	Mean Flow (ml/min)	Clotting (%)	Infection (%)	Catheter Survival 1-yr (2-yr)
1	22 (18)	10 months	274	50	18.7	47 (41)
2	168 (131)	18.5 months	243	46	21	65
50	64 (51)	318 days				74 (43)
36	299 (194)	87 days		18.8	6.8	48
18	134* (108)	16 months	284	8.2	5.2	93 (91)
16	138* (129)	54 days	325	1.4	2.1	

* = Tesio catheter pairs

functioning vascular access by use of a permanent catheter. McLaughlin *et al.* (127) reported their results in this manner with one-year technique and catheter survivals of 81% and 48.7%, respectively.

From the foregoing, it is obvious why it is difficult to compare CVC with other secondary vascular access modalities. Tunnelled CVC provide a safe and effective long-term access for haemodialysis. Many renal units now rely increasingly on CVC for access.

12.10. Exciting Developments

The development of silicone elastomer (Solstice, Dow Corning Corp., Midland, Michigan) represented a significant advance in CVC technology. Biocompatible, softer, flexible and less thrombogenic than other polymers (128), it is thought to cause less vascular trauma. The search is still on for the ideal material for catheter formation.

Subcutaneous device

As percutaneous devices, CVC are constantly exposed to possible contamination leading to a high risk of infection. Generally, percutaneous devices tend to have poor patient acceptance due to restrictions they impose on patients' desire to shower, bath or swim. Subcutaneous vascular access ports have been successfully used in rhesus monkeys (129). Internal or external jugular veins were cannulated and the catheter attached to a subcutaneous port that was positioned on the back between the shoulder blades. The port/catheter system allowed easy serial blood sampling and intravenous drug administration on a daily basis with a reported mean functional life span of 243.61 days. Implanted ports (for example, BardPort™ Implanted Ports) have been developed to provide a reliable vascular access for patients who require long-term drug or fluid therapy (130). These ports which are made of biocompatible materials may consist of single- or dual-lumen, pre-connected or attachable catheter configurations. The application of this technology to human haemodialysis will reduce the risk of sepsis. It will also allow patients to enjoy improved personal hygiene and swim or participate in other water sports as desired, thereby improving the quality of life. A subcutaneous device, which provides vascular access through an implanted port, is presently under clinical trial (personal communication). A valved chamber allows percutaneous access through needle cannulas. The device can be rotated in the subcutaneous plane to vary the skin needling site. Initial experience indicates excellent patient and nursing acceptance. If long-term results are acceptable, this may offer a reasonable alternative to AV access.

Split-lumen catheter

KIMAL PLC has recently introduced the Ash Split Catheter which is said to combine the ease of insertion of a standard double-lumen catheter with the flow characteristics of a two-catheter system. The Ash Split Catheter, made of Bio-Flex™, features two discrete lumens joined at the Dacron cuff. This new catheter holds promise for the future.

Stabilising the catheter tip

To test out the hypothesis that central vein thrombosis is caused by catheter injury to the vein wall and that it can be reduced by stabilising the catheter tip, Kohler and Kirkman (115) studied CVC in a porcine model. Sealed silicon elastic catheters with or without stabilising loops were inserted via the EJV. They demonstrated that mural thrombosis at the tip of indwelling CVC is caused by chronic mechanical vessel wall injury and that it is prevented by stabilising the tip. The catheters have already been modified to use a retractable wire mechanism instead of a loop to ensure trouble free insertion and removal. This is a potentially significant development in the search for an ideal catheter.

An alternative to fluoroscopy

A new catheter lined with a wire allows positioning using a hand-held probe. The wire is connected to the probe which emits a magnetic field. When the probe is moved beyond the tip of the wire, a sudden electrical drop-off is sensed, allowing the operator to follow the tip's progression. Thus, the tip of the catheter can be placed in the right atrium, based on external landmarks. Adoption of this technology will allow more catheters to be inserted in the dialysis unit and therefore free up theatre and radiological space with significant cost savings.

12.11. The Ideal Haemodialysis Catheter

The ideal vascular access for haemodialysis should be longstanding, free of complications and easy to use. It must be designed to allow adequate blood flow (> 300 ml/min) with its tips staggered to minimise recirculation. The catheter material must be sturdy enough not to collapse when negative pressure is applied to suck in blood at high flows and should not mechanically induce haemolysis. It must also possess the following characteristics:

- No thrombogenicity
- No affinity for bacteria/possession of antibacterial cuff
- Easy care
- Kink free

- Suitable cross-sectional profile
- Radiopacity
- Ease of external fixation
- Availability of repair kits
- Availability of different sizes and lengths
- Minimal recirculation

Obviously, no such catheter exists and the ideal catheter remains firmly a dream. As highlighted above, industry working with clinicians and instrument engineers are making significant efforts towards the realisation of this dream.

12.12. Conclusion

The widespread use of central venous catheters has greatly increased the treatment options for patients with acute renal failure or ESRD. Compared to AV fistulae and AV grafts, haemodialysis catheters have significant drawbacks: the possibility of complications during insertion, unreliable function, the risk of infection or clotting with prolonged use, patient inconvenience and the likelihood of venous stenosis in some sites. As discussed above, the design and insertion techniques of these catheters have continued to evolve to overcome these problems. Tunnelled CVC now compare favourably with arteriovenous grafts and other secondary vascular accesses. The changing demographics of the dialysis population will ensure that CVC will have an expanding role. Consideration must be given to the overall clinical situation of the patient, experience and motivation of surgical, radiological or nephrological staff in deciding who inserts these catheters. Lessons from the past dictate that the right IJV is the preferred site for insertion of CVC.

The goal is to have a primary failure rate (proportion of catheters unable to deliver adequate blood flow during the first dialysis) of $< 5\%$ and a cumulative insertion complication rate of $\leq 2\%$ (53). The improving technology of CVC coupled with safer insertion techniques will ensure that these targets are met and, if possible, surpassed.

References

1. Shusterman, N.H., Kloss, K. and Mullen, J.L. (1988). Successful use of double-lumen, silicone rubber catheters for permanent haemodialysis access. *Kidney Int*, **35**, 887–890.
2. Moss, A.H. *et al.* (1990). Use of silicone dual lumen catheter with a Dacron cuff as a long term vascular access for haemodialysis patients. *Am J Kidney Dis*, **16**, 211–215.
3. Uldall, R. *et al.* (1993). A new vascular access catheter for haemodialysis. *Am J Kidney Dis*, **21**, 270–277.
4. Seghal, A.R. *et al.* (1998). Barriers to adequate delivery of haemodialysis. *Am J Kidney Dis*, **31**, 593–601.
5. Jean, G. *et al.* (1997). Central venous catheters for haemodialysis: Looking for optimal blood flow. *Nephrol Dial Transplant*, **12**, 1689–1691.
6. Twardowski, Z.J. (1998). High-dose intradialytic urokinase to restore the patency of permanent central vein haemodialysis catheters. *Am J Kidney Dis*, **31**, 841–847.
7. Fan, P.-Y. (1994). Acute vascular access: New advances. *Adv Ren Replace Ther*, **1**, 90–98.
8. Windus, D.W. (1993). Vascular access: A nephrologist's view. *Am J Kidney Dis*, **21**, 457–471.
9. Kumwenda, M.J., Wright, F.K. and Haybittle, D.J. (1996). Survey of permanent central venous catheters for haemodialysis in the UK. *Nephrol Dial Transplant*, **11**, 830–832.
10. Hoenich, N.A. and Donnelly, P.K. (1994). Technology and clinical application of large-bore and implantable catheters. *Artif Organs,* **18**, 276–282.
11. Walters, G.K. *et al.* (1997). A comparison of the dual lumen and coaxial catheters for temporary haemodialysis access. *Int J Artif Organs,* **20**, 208–212.
12. Atherikul, K., Schwab, S.J. and Conlon, P.J. (1998). Adequacy of haemodialysis with cuffed central-vein catheters. *Nephrol Dial Transplant*, **13**, 745–749.
13. Francis, D.M.A., Ward, M.K. and Taylor, R.M.R. (1982). Right-atrial catheters for long-term vascular access in haemodialysis patients. *Lancet,* **2**, 301–302.
14. Demers, H.G. *et al.* (1989). Soft right atrial catheter for temporary or permanent vascular access. *Dial Transplant*, **18**, 130–132+139.
15. Cappello, M. *et al.* (1989). Central venous access for haemodialysis using the Hickman catheter. *Nephrol Dial Transplant*, **4**, 988–992.

16. Canaud, B. *et al.* (1986). Internal jugular vein cannulation using 2 Silastic catheters: A new, simple and safe long-term vascular access for extracorporeal treatment. *Nephron*, **43**, 133–138.
17. Ringrose, T. (1998). What's new in vascular access? *Br J Renal Med*, **Spring**, 6–8.
18. Tesio, F. *et al.* (1994). Double catheterisation of the internal jugular vein for haemodialysis: Indications, techniques and clinical results. *Artif Organs*, **18**, 301–304.
19. Millner, M.R. *et al.* (1995). Tesio twin dialysis catheter system: A new catheter for haemodialysis. *AJR*, **164**, 1519–1520.
20. Prabhu, P.N. *et al.* (1997). Long-term performance and complications of the Tesio Twin Catheter system for haemodialysis access. *Am J Kidney Dis*, **30**, 213–218.
21. Shaldon, S., Chiandussi, L. and Higgs S.B. (1961). Haemodialysis by percutaneous catheterisation of the femoral artery and vein with regional heparinisation. *Lancet*, **2**, 857–859.
22. Copley, J.B. *et al.* (1984). Transabdominal angio-access catheter for long-term haemodialysis. *Ann Intern Med*, **100**, 236–237.
23. Erben, J. *et al.* (1969). Experience with routine use of subclavian vein cannulation in haemodialysis. *Proc Eur Dial Transplant Assoc*, **6**, 59–64.
24. Jones, C.E. and Walters, G.K. (1992). Efficacy of the supraclavicular route for temporary haemodialysis access. *South Med J*, **85**, 725–728.
25. Uldall, P.R. *et al.* (1979). A subclavian cannula for temporary vascular access for haemodialysis or plasmapheresis. *Dial Transplant*, **8**, 963–968.
26. Uldall, P.R. (1994). Subclavian cannulation is no longer necessary or justified in patients with end-stage renal failure. *Seminars Dial*, **7**, 161–164.
27. Cimochowski, G.E. *et al.* (1990). Superiority of the internal jugular over the subclavian access for temporary dialysis. *Nephron*, **54**, 154–161.
28. Bour, E.S. *et al.* (1990). Experience with the double lumen Silastic catheter for haemoaccess. *Surg Gyn Obstet*, **171**, 33–39.
29. Purchase, L. and Gault, M.H. (1991). Haemodialysis with a permcath kept open with streptokinase and later citrate in a heparin sensitive patient. *Nephron*, **58**, 119–120.
30. Dunn, J., Nylander, W. and Ritchie, R. (1987). Central venous dialysis access: Experience with a dual-lumen, silicone rubber catheter. *Surgery*, **102**, 784–789.
31. Dickson, C.S. *et al.* (1996). Placement of internal jugular vein central venous catheters: Anatomic ultrasound assessment and literature review. *Surgical Rounds*, **19**, 102–107.

32. Vanholder, R. and Ringoir, S. (1994). Vascular access for haemodialysis. *Artif Organs*, **18**, 263–265.
33. Defalque, R.J. (1974). Percutaneous catheterisation of the internal jugular vein. *Anesth Analg*, **53**, 116–121.
34. Lund, G.B., Trerotola, S.O. and Scheel, P.J. (1995). Percutaneous translumbar inferior vena cava cannulation for haemodialysis. *Am J Kidney Dis*, **25**, 732–737.
35. Lund, G.B. *et al.* (1990). Translumbar inferior vena cava catheters for long-term venous access. *Radiology*, **174**, 31–35.
36. Trerotola, S.O. *et al.* (1997). Outcome of tunnelled haemodialysis catheters placed via the right internal jugular vein by interventional radiologists. *Radiology*, **203**, 489–495.
37. McDowell, D.E. *et al.* (1993). Percutaneously placed dual-lumen silicone catheters for long-term haemodialysis. *Am Surg*, **59**, 569–573.
38. Silberzweig, J.E. and Mitty, H.A. (1998). Central venous access: Low internal jugular vein approach using imaging guidance. *AJR*, **170**, 1617–1620.
39. Rao, T.L.K., Wong, A.Y. and Salem, M.R. (1977). A new approach to percutaneous catheterisation of the internal jugular vein. *Anaesthesiology*, **46**, 362–364.
40. Khatri, V.P. and Espinosa, M.H. (1994). A safer technique of internal jugular venipuncture: Experience with 320 cases. *J Cardiothorac Vasc Anesth*, **8**, 663–667.
41. Apsner, R. *et al.* (1996). Alternative puncture site for implantable permanent haemodialysis catheters. *Nephrol Dial Transplant*, **11**, 2293–2295.
42. Teichgraber, U.K. *et al.* (1997). A sonographically guided technique for central venous access. *AJR*, **169**, 731–733.
43. Kwon, T.H., Kim, Y.L. and Cho, D.K. (1997). Ultrasound-guided cannulation of the femoral vein for acute haemodialysis access. *Nephrol Dial Transplant*, **12**, 1009–1012.
44. Swartz, R.D. *et al.* (1994). Successful use of cuffed central venous haemodialysis catheters inserted percutaneously. *J Am Soc Nephrol*, **4**, 1719–1725.
45. Lund, G.B. *et al.* (1996). Outcome of tunnelled haemodialysis catheters placed by radiologists. *Radiology*, **198**, 467–472.
46. Vanholder, R., Hoenich, N. and Ringoir, S. (1987). Morbidity and mortality of central venous catheter haemodialysis: A review of 10 years experience. *Nephron*, **47**, 274–279.

47. Vanherweghem, J. *et al.* (1986). Complications related to subclavian catheters for haemodialysis: Report and review. *Am J Nephrol*, **6**, 339–345.
48. Uldall, P.R. *et al.* (1993). A new vascular access catheter for haemodialysis. *Am J Kidney*, **21**, 270–277.
49. Sznajder, J.I. *et al.* (1986). Central vein catheterization: Failure and complication rates by three percutaneous approaches. *Arch Intern Med*, **146**, 259–261.
50. Mosquera, D.A., Gibson, S.P. and Goldman, M.D. (1992). Vascular access surgery: A 2-year study and comparison with the PermCath. *Nephrol Dial Transplant*, **7**, 1111–1115.
51. Held, P.J. *et al.* (1996). Executive summary In United States Renal Data System 1996 Annual Data Report. *Am J Kidney Dis*, **28** (3 Suppl 2), S12–S20.
52. Obrador, G.T. and Pereira, B.J.G. (1998). Early referral to the nephrologist and timely initiation of renal replacement therapy: A paradigm shift in the management of patients with chronic renal failure. *Am J Kidney Dis*, **31**, 398–417.
53. NKF-DOQI Clinical practice guidelines for vascular access. (1997). *Am J Kidney Dis*, **30**, S150–S191.
54. Bander, S.J. and Schwab, S.J. (1992). Central venous angioaccess for haemodialysis and its complications. *Semin Dial*, **5**, 121–128.
55. Collier, P.E. *et al.* (1998). Cardiac tamponad from central venous catheters. *Am J Surg*, **176**, 212–214.
56. De Moor, B., Vanholder, R. and Ringoir, S. (1994). Subclavian vein haemodialysis catheters: Advantages and disadvantages. *Artif Organs*, **18**, 293–297.
57. Flaccadori, E. *et al.* (1996). Cardiac arrhythmias during central venous catheter procedures in acute renal failure: A prospective study. *J Am Soc Nephrol*, **7**, 1079–1084.
58. DeMasi, R.J. *et al.* (1993). Vocal cord paralysis following placement of a double lumen haemodialysis catheter. *Dial Transplant*, **22**, 734–735.
59. Bambauer, R. *et al.* (1994). Complications and side effects associated with large-bore catheters in the subclavian and internal jugular veins. *Artificial Organs*, **18**, 318–321.
60. Barton, B.R., Hermann, G. and Weil, R. (1983). Cardiothoracic emergencies associated with subclavian haemodialysis catheters. *JAMA*, **250**, 2660–2662.
61. Fine, A., Churchill, D. and Gault, H. (1981). Fatality due to subclavian dialysis catheter. *Nephron*, **29**, 99–100.
62. Tapson, J.S. and Uldall, P.R. Fatal haemothorax caused by a haemodialysis catheter. *Arch Intern Med*, **144**, 1685–1687.

63. Wijeyesinghe, E.C.R. *et al.* (1983). Right atrial ball thrombus as a complication of subclavian catheter insertion for haemodialysis. *Int J Artif Organs*, **10**, 102–104.
64. Leblanc, M. *et al.* (1996). Central venous dialysis catheter dysfunction. *Adv Ren Replace Ther*, **4**, 377–389.
65. Suhocki, P.V. *et al.* (1996). Silastic cuffed catheters for haemodialysis vascular access: Thrombolytic and mechanical correction of malfunction. *Am J Kidney Dis*, **28**, 379–386.
66. Paulsen, D. *et al.* (1993). Use of tissue plasminogen activator for reopening of clotted dialysis catheters. *Nephron*, **64**, 468–470.
67. Ladocsi, L.T. and Cohn, J.D. (1995). An improved technique for the placement of haemodialysis catheters in the internal jugular vein. *J Am Coll Surg*, **180**, 343–345.
68. Beathard, G.A. (1994). The treatment of vascular access graft dysfunction: A nephrologist's view and experience. *Adv Ren Replace Ther*, **1**, 131–147.
69. Klimas, V.A. *et al.* (1984). Low dose streptokinase therapy for thrombosed arteriovenous fistulas. *Trans Am Soc Artif Organs*, **30**, 511–513.
70. McNamara, T.O. and Fischer, J.R. (1985). Thrombolysis of peripheral arterial and graft occlusions: Improved results using high-dose urokinase. *Am J Radiol*, **144**, 769–775.
71. Ahmed, A., Shapiro, W.B. and Porush, J.G. (1993). The use of tissue plasminogen acttivator to declot arteriovenous accesses in haemodialysis patients. *Am J Kidney Dis*, **21**, 38–43.
72. Roberts, A.C. *et al.* (1993). Pulse-spray pharmacomechanical thrombolysis for the treatment of thrombosed dialysis access grafts. *Am J Surg*, **166**, 221–226.
73. Kumpe, D.A., Cohen, M.A.H. and Durham J.D. (1992). Treatment of failed haemodialysis access sites: Comparison of surgical treatment with thrombolysis/angiopasty. *Semin Vasc Surg*, **5**, 118–127.
74. Schwab, S.J. *et al.* (1988). Prospective evaluation of a Dacron cuffed haemodialysis catheterfor prolonged use. *Am J Kidney Dis*, **11**, 166–169.
75. Twardowski, Z.J. (1995). Percutaneous blood access for haemodialysis. *Semin Dial*, **8**, 175–186.
76. Twardowski, Z.J. (1998). High-dose intradialytic urokinase to restore the patency of permanent central vein haemodialysis catheters. *Am J Kidney Dis*, **31**, 841–847.
77. Twardowski, Z.J. (1996). What is the role of permanent central vein access in haemodialysis patients? *Semin Dial*, **9**, 394–395.

78. Uldall, R. *et al.* (1993). Maintaining the patency of double-lumen silastic jugular catheters for haemodialysis. *Int J Artif Organs*, **16**, 37–40.
79 Valji, K. *et al.* (1991). Pharmaco-mechanical thrombolysis and angioplasty in the management of clotted haemodialysis grafts: Early and late clinical results. *Radiology*, **178**, 243–247.
80. Bern, M.M. *et al.* (1990). Very low doses of warfarin can prevent thrombosis in central venous catheters: a randomised prospective trial. *Ann Intern Med*, **112**, 423–428.
81. Hoshal, V.L., Ause, R.G. and Hoskins, P.A. (1971). Fibrin sleeve formation on indwelling subclavian central venous catheters. *Arch Surg*, **102**, 353–358.
82. Cardella, J.F., Lukens, M.L. and Fox, P.S. (1994). Fibrin sheath entrapment of peripherally inserted central catheters. *JVIR*, **5**, 439–442.
83. Ahmed, N. and Payne, R.F. (1976). Thrombosis after central venous catheterisation. *Med J Aust*, **1**, 217–220.
84. Crain, M.R. *et al.* (1996). Fibrin sleeve stripping for salvage of failing haemodialysis catheters: Technique and initial results. *Radiology*, **198**, 41–44.
85. Rockall, A.G. *et al.* (1997). Stripping of failing haemodialysis catheters using the Amplatz gooseneck snare. Clinical Radiology, **52**, 616–620.
86. Johnstone, R.D. *et al.* (1999). Percutaneous fibrin sleeve stripping of failing haemodialysis catheters. *Nephrol Dial Transplant*, **14**, 688–691.
87. Pemberton, L.B. *et al.* (1996). No difference in catheter sepsis between standard and antiseptic central venous catheters. *Arch Surg*, **131**, 986–989.
88. Gosbell, I.B. *et al.* (1995). Infection associated with central venous catheters: A prospective survey. *Med J Aust*, **162**, 210–213.
89. Marr, K.A. *et al.* (1997). Catheter-related bacteraemia and outcome of attempted catheter salvage in patients undergoing haemodialysis. *Ann Intern Med*, **127**, 275–280.
90. Raad, I.I. (1994). The pathogenesis and prevention of central venous catheter-related infections. *Middle East J Anesthesiol*, **12**, 381–403.
91. Dittmer, I.D. *et al.* (1989). Bacterial colonisation and peripheral bacteraemia associated with central venous haemodialysis catheters: A cross-sectional study. *Nephrology*, **3**, 557–561.
92. Almirall, J. *et al.* (1989). Infection of haemodialysis catheters: Incidence and mechanisms. *Am J Nephrol*, **9**, 454–459.
93. Shaffer, D. *et al.* (1992). The use of Dacron cuffed silicone catheters as long-term haemodialysis aaccess. *ASAIO J*, **38**, 55–58.

94. Carlisle, E.J. *et al.* (1991). Septicaemia in long-term jugular haemodialysis catheters: Eradicating the infection by changing the catheter over a guidewire. *Int J Artif Organs*, **14**, 150–153.
95. Shaffer, D. (1995). Catheter-related sepsis complicating long-term, tunnelled central venous dialysis catheters: Management by guidewire exchange. *Am J Kidney Dis*, **25**, 593–596.
96. Capdevila, J.A. *et al.* (1993). Successful treatment of haemodialysis catheter related sepsis without catheter removal. *Nephrol Dial Transplant*, **8**, 231–234.
97. Wheat, L.J. *et al.* (1979). Serologic diagnosis of access device-related staphylococcal bacteraemia. *Am J Med*, **67**, 603–607.
98. Maki, D.G., Ringer, M. and Alvarado, C.J. (1991). Prospective randomized trial of povidone–iodine, alcohol, and chlorhexidine for prevention of infection associated with central venous and arterial catheters. *Lancet*, **338**, 339–343.
99. Darouiche, R.O. and Raad, I.I. (1997). Prevention of catheter-related infections: The skin. *Nutrition*, **13**, 26S–29S.
100. Ranson, M.R. *et al.* (1990). Double-blind placebo controlled study of vancomycin prophylaxis for central venous catheter insertion in cancer patients. *J Hosp Infect*, **15**, 95–102.
101. McKee, R. *et al.* (1985). Does antibiotic prophylaxis at the time of catheter insertion reduce the incidence of catheter related sepsis in intravenous nutrition? *J Hosp Infect*, **6**, 419–425.
102. Bach, A. *et al.* (1996). Retention of the antibiotic teicoplanin on a hydromer-coated central venous catheter to prevent bacterial colonisation in postoperative surgical patients. *Intensive Care Medicine*, **22**, 1066–1069.
103. Thornton, J., Todd, N.J. and Webster, N.R. (1996). Central venous line sepsis in the intensive care unit. A study comparing antibiotic coated catheters with plain catheters. *Anaesthesia*, **51**, 1018–1820.
104. Raad, I. *et al.* (1997). Central venous catheters coated with minocycline and rifampin for the prevention of catheter-related colonisation and bloodstream infections: A randomised, double-blind trial. *Ann Intern Med*, **127**, 267–274.
105. Bach, A. *et al* (1994). Prevention of bacterial colonisation of intravenous catheters by antiseptic impregnation of polyurethane polymers. *J Antimicrob Chemother*, **33**, 969–978.
106. Greenfield, J.I. *et al.* (1995). Decreased bacterial adherence and biofilm formation on chlorhexidine and silver sulfudiazine-impregnated central venous catheters implanted in swine. *Crit Care Med*, **23**, 894–900.

107. Ciresi, D.L. *et al.* (1996). Failure of antiseptic bonding to prevent central venous catheter-related infection and sepsis. *Am Surg*, **62**, 641–665.
108. Kamal, G.D. *et al.* (1991). Reduced intravascular catheter infection by antibiotic bonding: A prospective, randomised, controlled trial. *JAMA*, **265**, 2363–2368.
109. Tennenber, S. *et al.* (1997). A prospective randomised trial of an antibiotic- and antiseptic-coated central venous catheter in the prevention of catheter-related infections. *Arch Surg*, **132**, 1348–1351.
110. George, S.J., Vuddamalay, P. and Boscoe, M.J. (1997). Antiseptic-impregnated central venous catheters reduce the incidence of bacterial colonisation and associated infection in immunocompromised transplant patients. *Eur J Anesthesiol*, **14**, 428–431.
111. Maki, D.G. *et al.* (1997). Prevention of central venous catheter-related bloodstream infection by use of an antiseptic-impregnated catheter: A randomised, controlled trial. *Ann Intern Med*, **127**, 257–266.
112. Pearson, M.L. and Abrutyn, E. (1997). Reducing the risk for catheter-related infections: A new strategy. *Ann Intern Med*, **127**, 304–306.
113. Schillinger, F. *et al.* (1991). Post catheterisation vein stenosis in haemodialysis: Comparative angiographic study of 50 subclavian and 50 internal jugular accesses. *Nephrol Dial Transplant*, **6**, 722–724.
114. Stalter, K.A., Stevens, G.F. and Sterling, W.A. (1986). Late stenosis of the subclavian vein after haemodialysis catheter injury. Surgery, **100**, 924–927.
115. Kohler, T.R. and Kirkman, T.R. (1998). Central venous catheter failure is induced by injury and can be prevented by stabilising the catheter tip. *J Vasc Surg*, **28**, 59–66.
116. Williams, E.C. (1990). Catheter-related thrombosis. *Clin Cardiol,* **13**, V134–136.
117. Vanherweghem, J.L. *et al.* (1986). Subclavian vein thrombosis: A frequent complication of subclavian cannulation for haemodialysis. *Clin Nephrol*, **26**, 235–238.
118. Hernandez, D. *et al.* (1993). Subclavian catheter related infection is a major risk factor for the late development of subclavian vein stenosis. *Nephrol Dial Transplant*, **8**, 227–230.
119. Barret, N. *et al.* (1988). Subclavian stenosis: A major complication of subclavian dialysis catheters. *Nephrol Dial Transplant,* **3**, 423–425.
120. Beenen, L. *et al.* (1994). The incidence of subclavian vein stenosis using silicone catheters for haemodialysis. *Artif Organs,* **18**, 289–292.

121. Ponz, E. *et al.* (1990). Mechanism of haemodialysis-associated subclavian vein stenosis. *Nephron*, **56**, 227–228.
122. Wisselink, W. *et al.* (1993). Comparison of operative reconstruction and percutaneous balloon dilatation for central venous obstruction. *Am J Surg,* **166**, 200–205.
123. Surratt, R.S. *et al.* (1991). The importance of preoperative evaluation of the subclavian vein in dialysis access planning. *Am J Roentgenol*, **156**, 623–625.
124. Korzets, A. *et al.* (1991). Subclavian vein stenosis, permanent cardiac pacemakers and the haemodialysed patient. *Nephron*, **58**, 103–105.
125. Kovalik, E.C. *et al.* (1994). Correction of central venous stenoses: Use of angioplasty and vascular Wallstents. *Kidney Int*, **45**, 1177–1181.
126. Glanz, S. *et al.* (1988). Axillary and subclavian vein stenosis: Percutaneous angioplasty. *Radiology*, **168**, 371–373.
127. McLaughlin, K. *et al.* (1997). Long-term vascular access for haemodialysis using silicon dual-lumen catheters with guidewire replacement of catheters for technique salvage. *Am J Kidney Dis*, **29**, 553–559.
128. Welch, G.W. *et al.* (1974). The role of catheter composition in the development of thrombophlebitis. *Surg Gynecol Obstet*, **138**, 421–424.
129. Wojnicki, F.H., Bacher, J.D. and Glowa, J.R. (1994). Use of subcutaneous vascular access ports in rhesus monkeys. *Lab Anim Sci*, **44,** 491–494.
130. Bard Access Systems. (1997). BardPort™ implanted ports. Website: http://www.bardacess.com/implanted.htm.

CHAPTER 13

ACCESS FOR PAEDIATRIC PATIENTS

Oswald N Fernando FRCS, FRCSEd
Transplantation Unit
Royal Free Hospital
Pond Street
London NW3 2QG, UK

Vascular access plays a significant role in the management of infants and children who not only suffer from end-stage renal failure but require continued access for purposes of drug therapy and repeated blood tests (1). For many years, the combination of small vessels and large-diameter catheters limited the efforts of clinicians. In recent years, a variety of catheters of smaller diameter have become available. In particular, the introduction of siliconised catheters has allowed long-term placement of catheters. This chapter will highlight the indications for the placement of such catheters and other methods of blood access in paediatric patients. It will also consider other special indications for access in children (2).

13.1. Investigations and Preparation

A clear understanding of the anatomy of the blood vessels in children is an absolute prerequisite for any surgeon who undertakes vascular access procedures in paediatric patients. While experience in adult surgery is useful, an appreciation of the special needs of children is vital to obtaining success. Not only is it necessary to understand the anatomy, it is also necessary to understand that anomalies can occur and also to appreciate that blood vessels in children are finer and much smaller than in adults.

It is often necessary to carry out investigative procedures in these patients prior to surgical exploration. These include Doppler ultrasound, and angiography of the arteries and veins to assess patency. The availability of MRI scanning to assess the major blood vessels such as aorta and cava and also the neck veins has been studied in several centres. It is clearly important that these studies are matched by careful clinical examination and mapping of the blood vessels in a particular region. Of special importance is a consideration of whether previous procedures have been carried out.

Recognition of allergy to certain materials that are used in prostheses has become important in the last few years. Latex allergy is especially important and the addition of latex to any prosthesis has to be noted prior to implantation in a particular recipient.

13.2. Indications

At its simplest level, it may be necessary for the purpose of haemodynamic monitoring and this is useful in the Intensive Care Unit. Central venous access provides a channel to measure central venous pressure and it also provides access for the infusion of different forms of medications. In the majority of cases, these devices are placed percutaneously.

Along with the development of sophisticated vascular access techniques, the indications for their use have increased. While repeated phlebotomy can be carried out, it may be necessary to infuse drugs centrally. Chemotherapeutic agents for patients with cancer range from therapy for short to repeated periods of treatment. During periods of remission, vascular access lines will be useful for blood sampling and supplemental infusions of blood or blood products. Parenteral nutrition is often necessary in children who have low birthweight and those patients who have malignancies and are receiving chemotherapy. Conditions associated with alimentary type of anomalies or hypermetabolic states with limited oral intake may provide a further indication for the use of vascular access for these patients.

By far, the most important requirement for vascular access is haemodialysis. Acute renal failure may be managed with temporary vascular lines but any form of prolonged dialysis, such as patients awaiting a cadaveric renal transplant, will require permanent vascular access. The different options available will be discussed in the next section but clearly the need for a reliable vascular access in these children cannot be overemphasised.

13.3. Vascular Access Devices

A whole variety of vascular access devices and methods are currently available (Table 13.1). Each manufacturer claims advantage with respect to a particular requirement and in fact for the longevity and durability of each device. The clinician presented with these devices is therefore faced with the difficult decision of selection (3).

Table 13.1. Methods and devices available.

1. Arteriovenous shunts
2. Internal arterial venous shunts (grafts)
3. Central access lines
4. Arteriovenous fistulae
5. Peritoneal dialysis

The choice of vascular access is dependent primarily on the age and size of the patient, the duration for which the therapy is required and the flow rates that need to be achieved. The initial assessment, therefore, is made by determining the length of time the device is anticipated to be required. The methods used for short-term requirements differ considerably from those requirements for long-term periods (more than a few weeks). The devices used in the paediatric patients need to be of appropriate size and length to match each individual patient (4, 5).

13.4. Short-Term Vascular Access

Short-term access is used when the duration of treatment is from several days up to two weeks. Indications in these situations include dialysis for acute renal failure, plasmapheresis for immune complex diseases, TPN, antibiotic and pharmacologic therapy, and intensive haemodynamic monitoring. In very small children, umbilical artery and vein catheters are commonly used. However, these are inadequate for procedures such as dialysis or plasmapheresis. Insertion of these catheters is performed at the bedside and the catheter tip is positioned at the level of the thoracic spine above the origin of the major blood vessels or below the level of the renal vessels. Complications have been reported with both high- and low-level placements. These include thrombosis of aorta and visceral vessel embolisation with extremity or visceral ischaemia, bowel perforation and vascular perforation with intraperitoneal haemorrhage. Other complications of these manoeuvres have included arterial aneurysm and aortic coarctations. Thrombosis remains the most significant risk with these catheters and it should be clearly understood that their placement is only up to a maximum of seven to ten days.

Peripheral intravenous catheters can be used but are unsatisfactory for any use beyond three to five days. Infection and phlebitis are commonly seen following these.

13.4.1. *Peripherally-inserted central catheters*

These are now commonly used and as the skills of radiologists have improved, the technique has gained popularity. They have decreased the necessity for cutdowns and intra-operative catheter placement. The peripherally-inserted central catheter is a radio-opaque, flexible silastic tube that is inserted percutaneously through a peripheral vein and threaded centrally. These catheters come in a variety of sizes and the smallest size is 23-gauge and can withstand flow rates of up 125 ml per min. Slightly larger catheters can tolerate flow rates of 250 ml per min and the catheters can remain in place for several weeks.

The insertion of peripheral central catheters is performed in the X-ray department under radiological control. The cephalic or basilic veins and veins of the antecubital fossa are generally used but the external jugular or subclavian veins can also be utilised. Strict asepsis needs to be observed and very careful placement of the tip of the catheter is achieved by radiography. Flushing of the catheter with heparinised saline is important. These catheters need to be secured, and sterility and patency maintained. These are safe in neonates as well as in older children. The rate of phlebitis is significantly lower than in peripheral catheters and they can be used for periods of up to a month.

13.4.2. *Percutaneous central venous lines*

The technology relating to these catheters has improved in the last five years. Uncuffed catheters are made of plastic or silastic. Their use is intended for one to four weeks, development of smaller and more flexible catheters has allowed safe use in neonates and small children (6, 7). Double-lumen catheters are available and can be used for infusions, dialysis, temporary dialysis and blood sampling. The softer catheters are clearly advantageous in that they can be used for longer periods. The sites of insertion for these catheters are the subclavian, internal jugular and femoral veins. Heavy sedation or general anaesthesia is required and the patient will require aseptic conditions of an operating theatre or aseptic radiology room. All technical details regarding insertion of central catheters will be discussed in a later section.

13.4.3. *External arterial venous shunts*

The original technique was introduced by Scribner and Quinton (8) in 1960. While these were used for many years, their popularity has since been reduced with the advent of newer methods of blood access. Although they remained the mainstay of dialysis access in children for many years, there are very few commercial devices available for external arterial venous shunts. Arteriovenous shunts require careful planning and even more careful management. Anticoagulation is required to ensure the continued patency of the blood vessel but it became clear over the years that many children would require revision and a prograde movement of the vessel tip higher up in the arm or leg to continue with haemodialysis. This had limitations as it was not possible to progress beyond the lower margin of the popliteal triangle in the leg and above the elbow in the arm. They left unsightly scars in many children and often led to considerable developmental difficulties because of the requirement to use an end artery and vein. In fact, the mere presence of a external device in a limb incapacitated children and restricted their activities as far as sports and swimming were concerned. Other complications that occurred with these devices included bleeding, skin necrosis and infection.

13.5. Long-Term Vascular Access Devices

Long-term vascular access devices are necessary when access is required for longer than four or five weeks. The available devices include cuffed catheters, made of silastic. They may be either single- (Hickman catheter) or a double-lumen [Perm Cath (Quinton Instruments Comp., Seatle WA)]. For purposes of efficient haemodialysis, it is essential to have two ports. Therefore, if single lumen catheters are used, it will be necessary to have two lines at different end steps. The majority of catheters used for children are the double-lumen catheters and these can either be of oval or circular configuration. Catheters with circular side-by-side cross-sectional configuration achieve higher and more reliable blood flows. All long-term catheters need to have a cuff to prevent accidental dislodgement of the catheter and also to prevent infection from progressing along the line on the catheter. Double-lumen catheters come in a variety of sizes from 8 to 11.5F and with varying lengths (12, 18 and 24 cm) from cuff to tip. In addition to this, catheters may be pre-curved or straight. Cuffed catheters are essential for long-term use and, if looked after carefully, can remain and function for prolonged periods.

13.5.1. *Central access devices*

These form the major proportion of the devices used in children. It is an absolute pre-requisite that central blood vessels are investigated for patency especially if previous procedures have been carried out.

13.5.1.1. *Technique of percutaneous insertion*

Skilled radiologists have now become extremely proficient at the insertion of percutaneous central catheters (9). Older children and adolescents require generous sedation but younger children will require general anaesthesia. The patient is placed supine with a small shoulder pad to help extend the neck. The head is rotated slightly to the opposite side and a Trendelenburg position will help to distend the vein.

The operator palpates for the artery and vein in the neck and sometimes introduces a small needle to locate the exact position of the vein. A larger needle attached to a 10-ml syringe is then inserted into the jugular vein. Following aspiration of blood, a guidewire is passed centrally. Often a minor cardiac irregularity indicates that the guidewire is in the correct position. The position of the wire is further confirmed by radiological screening.

An incision is made by the wire to permit the cuffed catheter to be placed in a subcutaneous tunnel and drawn in for insertion into the vein. The cuff of the catheter must be positioned at least 2 cm away from the exit point. Once the line is primed with saline, a dilator with a sheath is passed over the guidewire. The wire and dilator are removed and the catheter is passed centrally via the peel-away sheath. It is important to place the catheter accurately to allow maximum flows. Similar procedures may be used for the subclavian vein and the femoral vein.

13.5.1.2. *Technique of surgical placement*

Catheters are placed via a cutdown onto the internal or external jugular vein. The femoral vein may also be used although this is clearly associated with more complications in smaller children. The cuff should be at least 2 cm from the skin entry site as this distance will allow tissue in-growth and prevent accidental dislodgement. If catheters are placed in the groin, the exit should be on the abdominal wall above the diaper line.

13.5.1.3. *Tesio twin single-lumen catheters*

Over the last few years, two single-lumen catheters have been used in the major blood vessels. These catheters (Tesio) are placed in a step-like fashion and provide adequate dialysis. It is reported that the complication rates are reduced and the catheters may be introduced percutaneously through sheaths. While two catheters can be introduced into a large jugular vein in an adult, similar placement in a small child may be difficult. These catheters have a one-year survival of 30%. A close collaboration between the surgeon and radiologist is crucial for successful placement of these lines.

13.5.2. *Arteriovenous fistulae*

The Cimino–Brescia subcutaneous arteriovenous fistulae are constructed when prolonged dialysis or cadaveric transplantation is anticipated (10, 11). They are suitable for children weighing over 25 kg. The rule for fistula formation should be distal before proximal. Modern microsurgical techniques will help the creation of these fistulae as the blood vessels used are of small calibre and are more prone to vascular spasm. Patients who have had previous steroid therapy have tissues that are more fragile.

The creation of a fistula in a small child has been reported but the continuing use of these fistulae requires education and training of nursing personnel. One must consider that even if a fistula can be created, it is sometimes almost impossible to educate a small child to remain still with two needles stuck in his arm. Methods of creating an arteriovenous fistula have been described elsewhere in this book. It must be remembered that the non-dominant arm should always be used where possible. Gentle handling of tissues is crucial to achieving success.

13.5.3. *Internal AV shunts (grafts)*

While the use of vascular prosthetic grafts is popular in the United States for adults, their use in small children has been questioned. Wide variations occur in the patency rates. The majority are placed in the upper arm but there are reports of unusual sites such as axillo-femoral, brachio-jugular and aorto-caval bridge grafts. Part of the problem lies in the necessity to needle the site on a regular basis and, in addition, the known complications of thrombosis and stenosis make their use more difficult. Adolescents will however be able to make use of this device.

13.6. Complications

Infection, thrombosis, malfunction, migration and vessel perforation have all been reported (Table 13.2) (12, 13). The most common complications of blood access are malfunction and thrombosis. While it is crucial to assess the procoagulant status of all children undergoing vascular access, it is equally important to assess whether any factors such as deficiencies in coagulation are present. Malfunction may often be related to thrombosis but the unrecognised angulational kinking of a catheter can sometimes account for poor function. Circular catheters are more prone to angulation as opposed to the "double-D" configuration or oval-shaped catheters. Malfunction is also related to the unwitting migration of a catheter caused by an infant who

Table 13.2. Complications of central venous catheters

1. Mechanical: Pneumothorax/haemothorax/haemorrhage
2. Arterial injury and false aneurysm
3. Stenosis
4. Air embolism
5. Catheter damage/angulation
6. Superior vena cava thrombosis
7. Infection
8. Lymphatic fistula
9. Cardiac perforation

may pull or dislodge a catheter. Critical placement and secure attachment are imperative in preventing this hazard. Confirmation of the position of the tip of the catheter by radiology is mandatory. Vessel perforation and cardiac penetration have been described especially with the use of stiff or hard catheters and even the use of a Seldinger technique has not prevented this complication (13).

13.6.1. *Complications of AV fistulae*

Thrombus formation remains a major problem in arterial venous fistulae. Fistulography and urgent thrombectomy may be required for treatment of this complication. Lymphatic oedema and monomegaly have been reported and occasionally, children will develop a swollen painful hand if a side-to-side fistula is created and there is venous congestion in the distal run-off. Congestive cardiac failure has also been reported with a large fistula or two simultaneously functioning fistulae. False aneurysms may develop in children who have repeated venepuncture in the same position. Occasionally, severe bleeding has been reported following dialysis.

13.6.2. *Treatment of complications*

The use of interventional radiology has facilitated the management of complications of vascular access. Obstruction of fistulae and internal bridge grafts can be defined by careful arteriography or venography and any stenosis may be dilated with balloon angioplasty. Placement of stents in a stenotic segment can also be used to prolong the life of a fistula. Adequacy of dialysis may thus be ensured and any further obstructive venous complications or thrombosis may be prevented. Addition of aspirin and dypiramadole to the anticoagulant regime will often help in the survival of these grafts.

13.7. Peritoneal Dialysis

Peritoneal dialysis via a Tenchkoff catheter is currently the primary method of dialysis in very small children (14). It is especially useful in children who are small and those who live far away from dialysis centres. Adolescent children also adapt well to this method of dialysis but do require constant supervision and careful management. Success depends on a collaborative home environment and the ability to be trained in sterile techniques. In addition to this, carers must be able to rapidly identify the complications. The use of this method of dialysis has greatly reduced the need of vascular access for small children.

Two configurations of peritoneal dialysis catheters are available; one relies on the use of a curved catheter while the other relies on a straight catheter. Both require cuffs and a long-embedded tunnel to prevent infection from the skin into the peritoneal cavity.

Peritoneal dialysis catheters can be placed percutaneously or with the help of a laparoscope (15). We have found that carrying out an omentectomy helps in maintaining these catheters for long periods. Should facilities for laparoscopic insertion or percutaneous insertion not be available, it is possible to do a small laparotomy, carry out the omentectomy and place the peritoneal dialysis catheter in the appropriate position. In doing an omentectomy, it is crucial that all the omental blood vessels are carefully ligated as patients may have profuse haemorrhage.

The two cuffed catheters are placed in position with one cuff residing within the rectus sheath and the other in the subcutaneous tissue at least 4 to 5 cm proximal to the exit site. Some catheters are manufactured with a pre-curved angulation (see Fig. 15.2) and these may facilitate the placement of a catheter within a long tunnel. Where catheters are placed in very small children, it is important that the exit site is placed well above the nappy line.

13.8. The Role of Nursing Staff in the Management of Vascular Access

The role of nursing staff in the management of vascular access and peritoneal dialysis in children is extremely important. The most common reason for hospitalisation is early thrombosis and infection. Training in strict hygiene and aseptic techniques by the nursing staff will help in obtaining longevity in these devices. It is therefore mandatory that all haemodialysis nurses have a full training in the management of these catheters.

13.9. The Role of the Radiologist

Radiology has played an increasing role in the management of vascular access in children. The use of sophisticated equipment and the excellent skills of the interventional radiologist have changed the emphasis from traditional surgical methods to percutaneous placement. A close collaboration between physicians, surgeons and radiologists has resulted in improved results and management of complications. Investigation and assessment of blood vessels either by ultrasonography, MRI scanning and direct vision with angiography have clearly helped in the decision-making process and planning of vascular access in small children. In addition to this, radiologists have become expert at placement of catheters. These may be done in the angiography suite or with guidance from ultrasonography. Percutaneous Seldinger-technique-guided placement of catheters have helped in producing a non-invasive technique of blood access.

In addition to help with placement of catheters, the initial management of complications has fallen into the hands of the radiologists. Stenosis can be treated with balloon angioplasty and thrombectomy can be used to remove early thrombi. Occasionally, a catheter may break and be dislodged in the vascular system. Radiologists have now become skilled in retrieving these pieces with loops and snares.

References

1. Kapoian, T. and Sherman, R.A. (1997) A brief history of vascular access for hemodialysis: An unfinished story. *Seminars in Nephrology*, **17**, 239–245.
2. Alexander, S.R. *et al.* (1993) Maintenance dialysis in North American children and adolescents: A preliminary report. North American Pediatric Renal Transplant Cooperative Study (NAPRTCS). *Kidney Int*, **43**, S104.
3. Bell, P.R.F. (1984) Haemodialysis in infants. *Br J Hosp Med*, **32**, 168.
4. Newman, B.M. *et al.* (1986) Percutaneous central venous catheterization in children: First-line choice for venous access. *J Pediatr Surg*, **21**, 685.
5. Schustermann, N.H. *et al.* (1989) Successful use of double lumen, silicone rubber catheter for permanent haemodialysis. *Kidney Int*, **35**, 887–890.

6. Kon, V. *et al.* (1981) Short- and long-term dialysis in ambulatory pediatric patients by subclavian catheter. *Kidney Int*, **21**, 171.
7. Lally, K.P. *et al.* (1992) Use of a subclavian venous catheter for short- and long-term hemodialysis in children. *J Pediatr Surg*, **22**, 840.
8. Buselmeier, T.J. (1971) Arteriovenous shunts for pediatric haemodialysis. *Surgery*, **70**, 638.
9. Dolcourt, J.L. and Bose, C.L. (1982) Percutaneous insertion of Silastic central venous catheters in newborn infants. *Pediatrics*, **70**, 484.
10. Brescia, M.J. and Cimino, J.E. (1966) Chronic Hemodialysis using venepuncture and a surgically created arteriovenous fistula. *N Engl J Med*, **275**, 1089–1092.
11. Burkhart, H.M. and Cikrit, D.F. (1997) Arteriovenous fistulae for hemodialysis [Review; 35 refs]. *Seminars in Vascular Surgery*, **10**, 162–165.
12. Feldman, H.I., Kobrin, S. and Wasserstein, A. (1996) Hemodialysis vascular access morbidity [editorial]. [Review; 88 refs]. *J Am Soc Nephrol*, **7**, 523–535.
13. Barton, B.R., Hermanm G. and Weil, R. (1983) Cardiothoracic emergencies associated with subclavian haemodialysis catheters. *JAMA*, **250**, 2260.
14. Orkin, B.A. *et al.* (1983) Continuous ambulatory peritoneal dialysis in children. *Arch Surg*, **118**, 1398–1402.
15. Brunk, E. (1985) Peritoneoscopic placement of a Tenchkoff catheter for chronic peritoneal dialysis. *Endoscopy*, **17**, 186–188.

CHAPTER 14

RADIOLOGY OF ACCESS

Aghiad Al-Kutoubi MD, FRCR, DMRD
Department of Diagnostic Radiology
American University of Beirut Medical Centre
Beirut, Lebanon

14.1. Pre-Operative Venography

With the increasing use of central venous catheters, the incidence of venous stenosis, particularly in the subclavian and innominate veins, is becoming more of a problem (1). Pre-operative venography of the upper limbs is therefore recommended especially when there is clear history of previous line insertion and previous failed fistula surgery. Detailed views of the peripheral veins must always be accompanied by views of the central veins. A small needle (typically 21G or 19G) is placed in a vein in the dorsum of the hand and contrast-injected with multiple views to identify all the veins, particularly the superficial ones such as the cephalic. A tourniquet can be used to improve the visualisation of the central veins. Digital subtraction angiography (DSA) is a useful method for the assessment of the subclavian and innominate veins and should be routinely used.

14.2. Fistulography

Imaging of dialysis fistulas requires a meticulous approach and attention to detail as well as knowledge of the type of surgery used to provide access because there are numerous alternatives and permutations (2). Whether it is a venous fistula or a synthetic one, careful demonstration of the full "loop" to include the inflow, outflow as well as the conduit is an essential requirement, otherwise, problems may be overlooked or missed. The technique varies depending on the type of fistula.

14.2.1. *Arteriovenous fistulas*

The referring physician should provide details of the type and site of the fistula, particularly if several procedures had been undertaken. The patient is positioned on a fluoroscopy or angiography table, and careful palpation and examination of the fistula are performed to establish the anatomy and presence or otherwise of a thrill or buzz at the anastomosis. Inspection and palpation of the draining vein will help decide on the puncture point. The venous side of the fistula is cannulated as close to the anastomosis as possible. A butterfly needle or a cannula is used, depending on the site of the fistula and the size of the vein. We prefer to use 18G needles/cannulas to ensure sufficient flow of contrast within the fistula as well as the possibility of threading a guidewire for catheterisation should this prove necessary. If DSA is performed, then dilute contrast of a density of 140–200 is used in aliquots of 20 ml injected firmly to achieve a constant column of contrast. Views are obtained to outline the venous outflow from the anastomosis site to the superior vena cava (SVC). Any areas of narrowing or dilatation are assessed with extra views in different projections. Once the venous anatomy is established, attention is directed at the arteriovenous anastomosis itself. A narrow pneumatic pressure cuff linked to a calibrated pressure gauge is then applied to the limb central to the anastomosis and cannula, and inflated until the thrill is no longer palpable. Contrast is then injected and multiple views taken until the anastomosis is adequately profiled (Fig. 14.1).

Studies of the arterial side may not be necessary if the venous side is occluded. The images are then reviewed and printed with and without subtraction to identify the bone anatomy and establish landmarks. A decision on suitable intervention is made before the cannula is removed to avoid repeated punctures as much as possible.

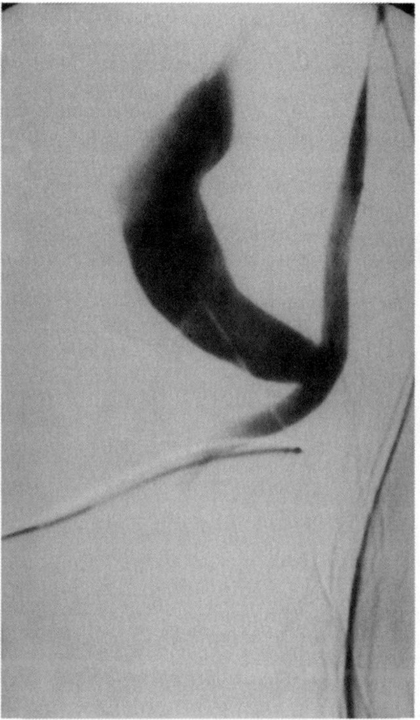

Fig. 14.1. The arteriovenous anastomosis of a brachial fistula is profiled after a tourniquet is applied over the lower arm.

14.2.2. *Synthetic fistulas*

Clinical examination of the fistula is the first step to define the anatomy although some of the loops, particularly in the lower limbs, may prove difficult to feel, in which case, fluoroscopy to identify the support rings or ultrasound guidance may be necessary to perform the puncture. The loop which is generally made from expanded polytetrafluoroethylene (PTFE) is punctured in the direction of the arterial anastomosis and as close to it as possible using an 18G needle. Contrast is used to image the arterial and venous anastomoses in different projections, if necessary, followed by imaging of the proximal veins to the SVC or IVC. Pneumatic cuffs or tourniquets are

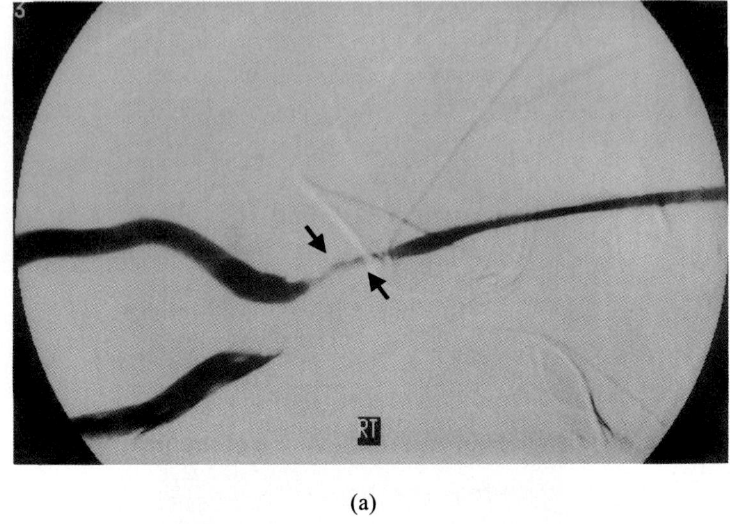

(a)

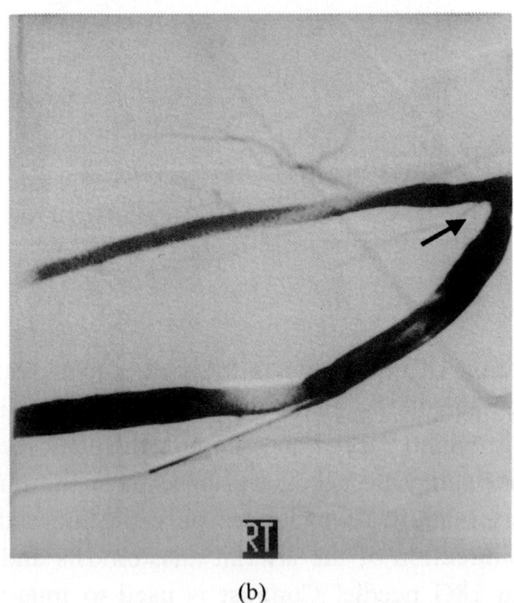

(b)

Fig. 14.2. PTFE loopogram. (a) A stenosis is seen at the venous anastomosis (arrows). (b) The arterial anastomosis is widely open (arrow).

seldom of any benefit in improving imaging in this situation and introduction of a catheter towards or through the arterial anastomosis may prove necessary in a small percentage of cases. As mentioned above, the use of DSA makes it easier to achieve good images of the central veins in particular (Fig. 14.2).

14.2.3. *Central venous catheters*

Malfunctioning catheters may require imaging assessment. Plain films such as chest X-rays are sufficient to evaluate the position of the catheter. Contrast injections in all lumens of the catheter are required for full assessment of the flow pattern and to establish whether the malfunction is indeed due to thrombus, fibrin sheath or some other cause. Rapid sequence conventional fluoroscopy films are probably better than DSA in this situation as the thin fibrin sheath may be overlooked on the subtracted images.

14.2.4. *Peritoneal dialysis catheters*

Plain films will establish the position of the catheter and show any kinks or migration. Contrast injections may be used to confirm loculation of the peritoneal fluid, but in cases of persistent leaks, a CT scan study of the abdomen after installation of 400–500 ml of dilute contrast (typically 5–10%) into the peritoneal cavity is the method of choice for demonstrating the site and extent of the leak.

14.3. Magnetic Resonance Angiography (MRA)

Although MRA has been successfully used in the assessment of vascular problems, the role of this technique in the imaging of fistulas remains unexplored due mainly to the complexity and variability of the venous anatomy as well as the susceptibility of MRA to artefacts from surgical clips at the fistula or turbulence from the irregular flow producing false positive or negative results. However, it may be useful for the odd patient who is allergic to iodine-based contrast media and in whom the clinical presentation is more suggestive of a central venous problem (3).

14.4. Ultrasound

Doppler examination of the fistula is an excellent non-invasive method for the assessment of flow and stenoses, particularly in synthetic grafts where the problem is usually in the immediate vicinity of the loop and is generally confined to a single vessel (4). It may also be used in the documentation of the result of intervention on focal lesions.

The complexity of venous drainage in arteriovenous fistulas, however, may result in confusion and false results. It is also difficult to assess the central veins with this method.

14.5. Management of Complications

It is probably best to discuss complications from the point of view of the technique used for dialysis.

14.5.1. *Fistula-based access*

The complications can be divided into three sections.

14.5.1.1. *Venous thrombosis*

This is a common presentation occurring in 44–70% of patients (5) and is often preceded by signs of slowing flow and elevated pressures during dialysis. The thrombosis may occur within the draining vein itself in cases of direct arteriovenous fistulas or in the PTFE loop when this is present. Either way, it is important that the thrombosis is recognised and treated as quickly as possible. Treatment methods vary and several new techniques have emerged in the last few years in an effort to improve the results and reduce the complications. There is probably no role for simple aspiration of the thrombus as it seldom yields a good clinical result. The two main radiological options are thrombolysis and mechanical thrombectomy.

Thrombolysis

For thrombolysis to be effective, a cannula or a small catheter is inserted into the thrombosed fistula and directed towards the venous end. A slow-dose regimen is preferable in a manner similar to that used in peripheral arterial thrombolysis (6). Alternatively, mechanical thrombolysis using pulse spray techniques may be used (7, 8, 9). Urokinase or recombinant tissue

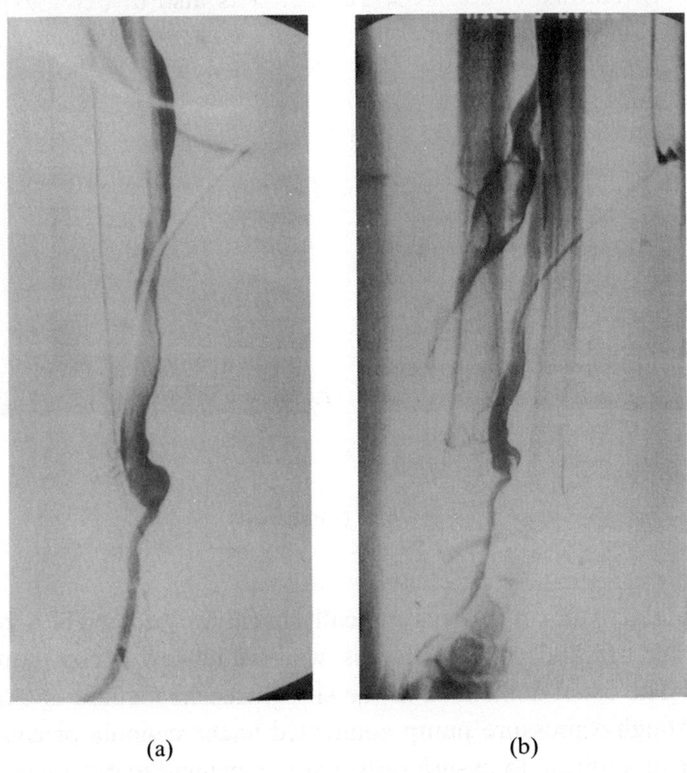

(a) (b)

Fig. 14.3. (a) Fistulogram demonstrates thrombosis of the venous limb of a radial arteriovenous fistula. The arterial limb has been punctured. (b) Following slow infusion of rTPA, some opacification of the thrombus-filled vein is seen. (c) At the end of the treatment, there is good filling of the draining vein. (d) Image at elbow demonstrates severe stenosis of the draining vein which has caused the thrombosis.

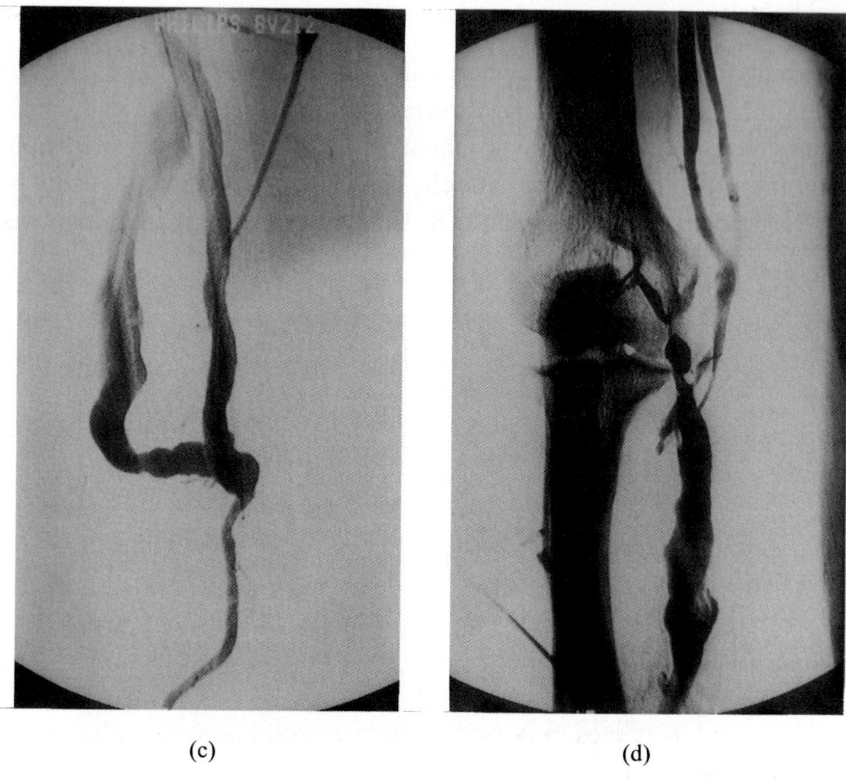

(c) (d)

Fig. 14.3. (*Continued*)

plasminogen activator (rTPA) is typically used. We prefer rTPA because of its short half-life and ease of use as well as its lower cost compared to urokinase. The usual dose used in the slow infusion method is 0.5–1 mg/hr infused through a pressure pump connected to the cannula or catheter with continuous monitoring to ensure satisfactory flow and to detect any bleeding which may occur (Fig. 14.3).

A common drawback with thrombolysis relates to the fact that most patients will have undergone "needling" of the fistula a few hours before they are referred to the radiology department. This results in bleeding from the site of the needle as lysis of the clot occurs. This can occur in the arteriovenous and synthetic fistulas and will often necessitate the termination of thrombolysis. It is also worth remembering that thrombolysis, particularly in the elderly, is associated with a higher risk of bleeding intracranially and at sites of a recent surgery. So, even with the low peripheral dose, thrombolysis should not be used if there is a history of peptic ulceration, recent stroke or recent major surgery.

Mechanical thrombectomy

These devices were developed to overcome the difficulties which were encountered with thrombolysis. The rationale of the treatment is that some form of mechanical "maceration" of the thrombus takes place resulting in free flow through the fistula. The maceration effect is achieved through rotational movement of a moving part at the tip of the catheter which is connected to a pressure source (either gas or saline injected under pressure). The mechanical movement as well as the vortex effect cause the clot to fragment and change into microparticles which can be aspirated through an "exhaust" lumen within the catheter in some designs, or diluted as the particles mix with the flowing blood in other designs (Figs. 14.4 and 14.5).

Many of the patients probably suffer from small pulmonary emboli but the majority of these are subclinical and are of no significance (10). Successful declotting in 89% of the cases has been reported in PTFE grafts as well as arteriovenous fistulas, particularly with the Hydrolyser catheter which requires saline to create the vortex and includes an exhaust lumen to get rid of the particles (11). The catheter should be introduced through a vascular sheath in the thrombosed graft and then advanced as the saline is injected to achieve clearance of the thrombus. This is documented by taking views of the loop and the arterial and venous ends. If a stenosis is demonstrated, then this could be treated before the sheath is removed. Many patients will complain of some discomfort when these devices enter the vein, which probably relates to the pressure of the vortex on the wall of the vein.

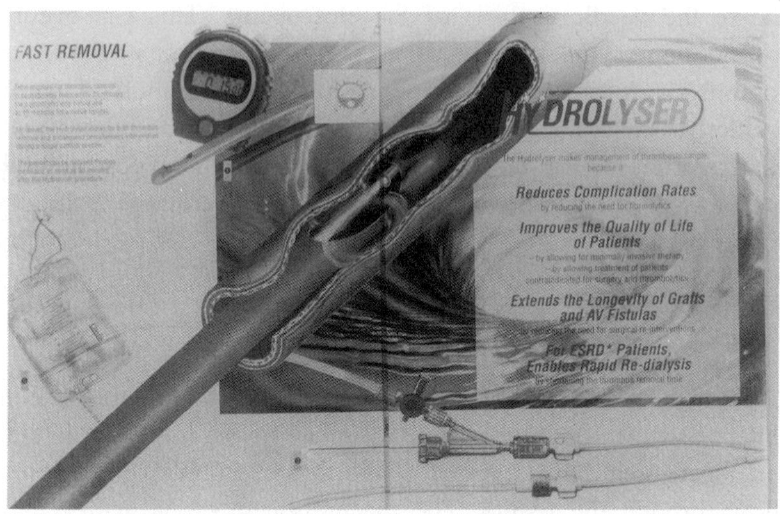

(a)

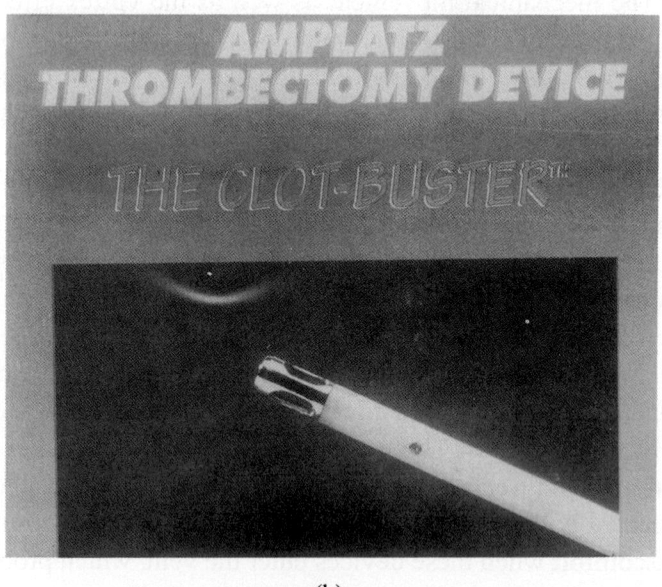

(b)

Fig. 14.4. (a) The Hydrolyser catheter and (b) the Amplatz clot-buster.

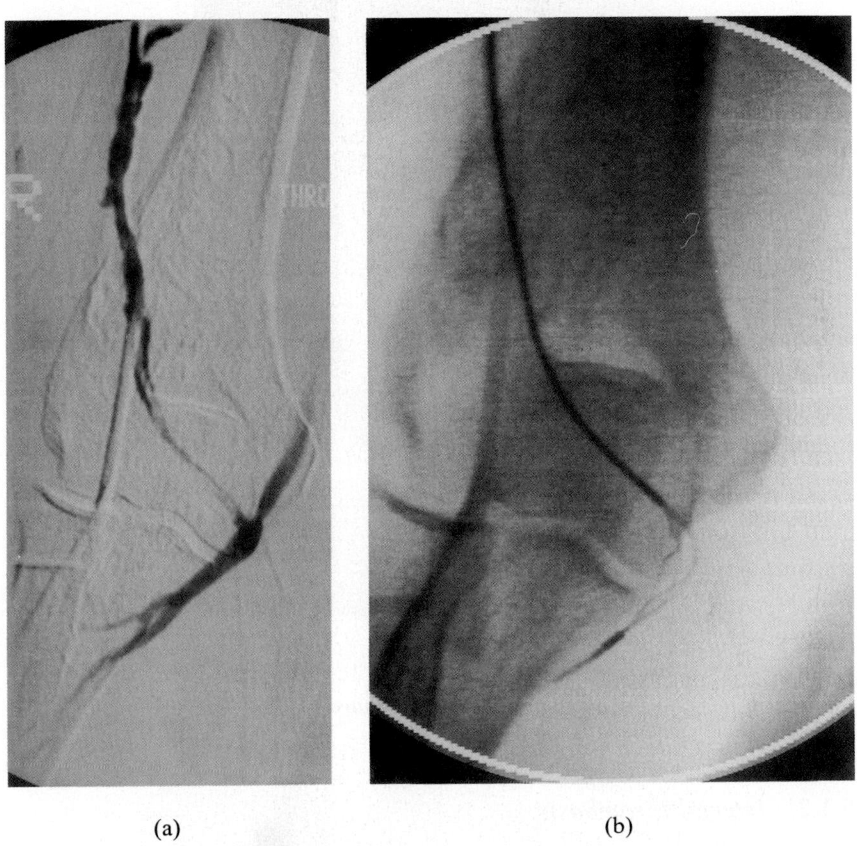

(a) (b)

Fig. 14.5. (a) Fistulogram showing thrombus within the venous side of a brachial fistula. (b) The Arrow–Trerotola PTD basket thrombectomy device through the fistula. (c) Fistulogram after thrombectomy showing a patent vein.

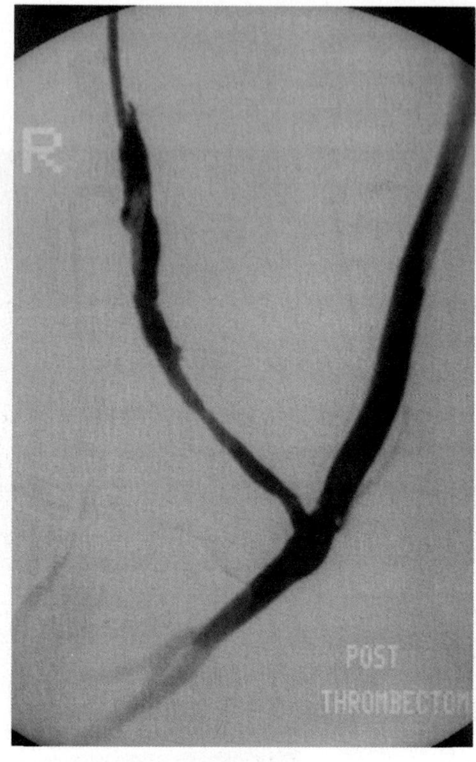

(c)

Fig. 14.5. (*Continued*)

14.5.1.2. *Arterial thrombosis*

This is much less common than venous thrombosis and is commonly associated with occlusion of the venous limb. In many instances, arterial surgery may be the only option but thrombolysis can be used effectively if the amount of thrombus is small. The artery may be approached through a femoral puncture or, anatomy permitting, through a direct puncture of the brachial or even the radial artery. The regimen used is similar to that used in the venous system with the same precautions and contraindications (Fig. 14.6).

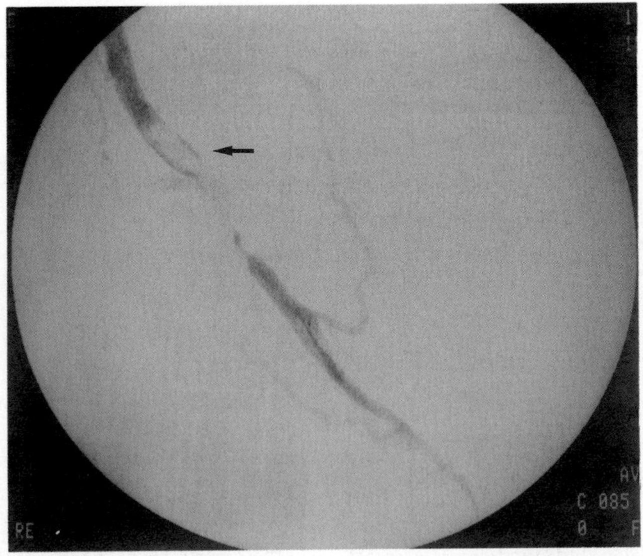

Fig. 14.6. Fistulogram at the elbow level. The vein is completely occluded. There is a significant stenosis at the site of the anastomosis with thrombus within the distal brachial artery appearing as a filling defect (arrow).

14.5.1.3. *Venous stenosis*

This is the commonest cause of late fistula failures and is often preceded by elevation in the venous dialysis pressure (VDP) (12). The stenosis may be at the anastomosis, at the site of needling or in the central veins. The treatment of venous stenoses is often disappointing as the restenosis rate is quite high irrespective of the treatment method used. The rate of patency following angioplasty is at best about 40% at one year (13–15). Metal stents have been used to improve the long-term results both in the peripheral veins and in the central stenoses (16–19). The initial results were promising but the long-term results indicated a high incidence of restenosis due to intimal hyperplasia, either secondary to the needle trauma in the peripheral veins or previous central line insertions in the central veins (Figs. 14.7–14.10).

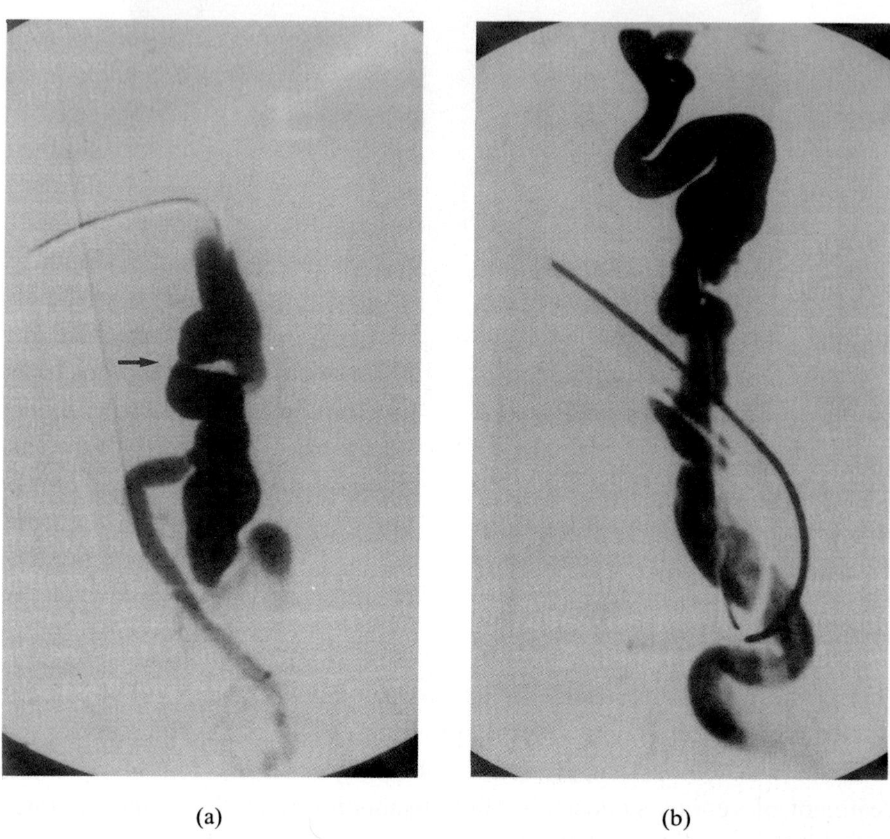

(a) (b)

Fig. 14.7. (a) Fistulogram demonstrates a fairly tight stenosis (arrow) in the draining vein of a radial fistula approximately 2 cm from the anastomosis. (b) Following angioplasty, the vein is widely patent.

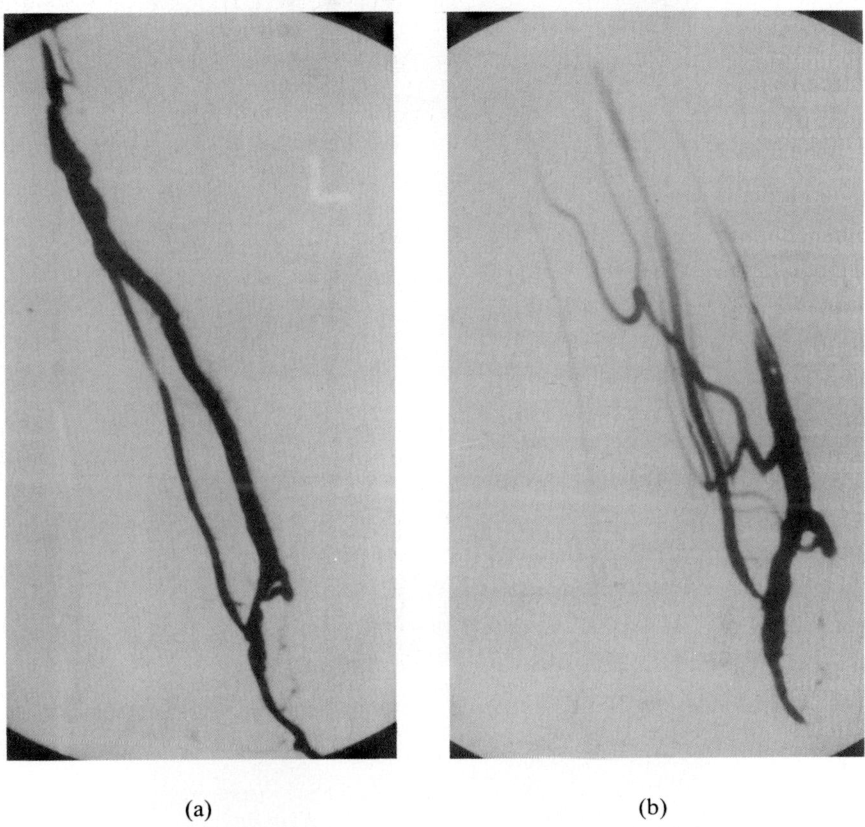

(a) (b)

Fig. 14.8. (a) Venous stenosis just beyond the anastomosis. (b) After angioplasty, the vein is widely patent.

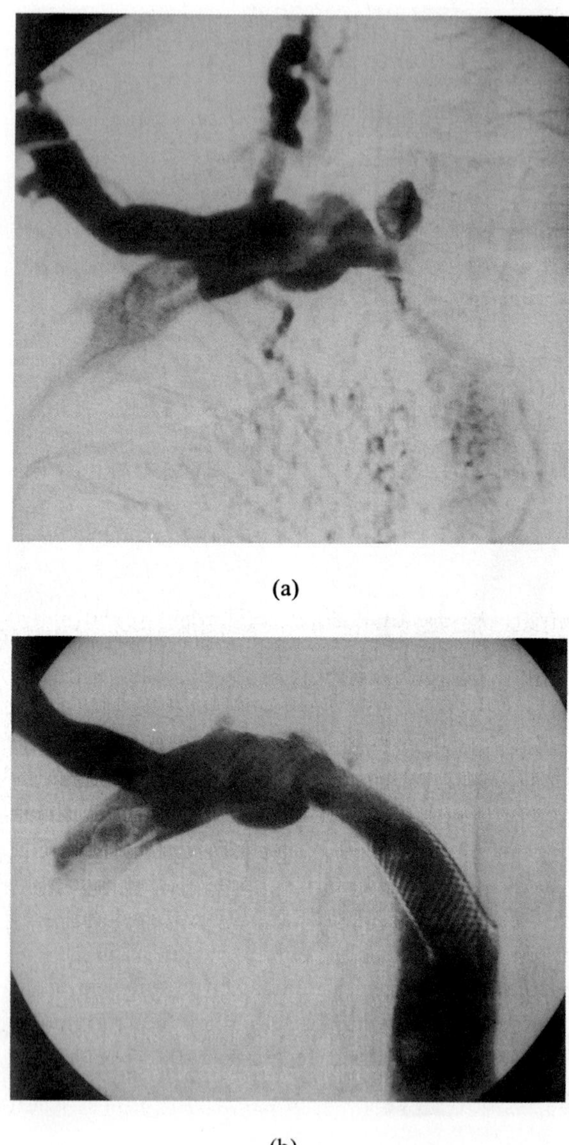

Fig. 14.9. (a) Near total occlusion of the right innominate vein with some filling of collateral veins. (b) After recanalisation and stenting with a Wallstent, the lumen is nearly normal.

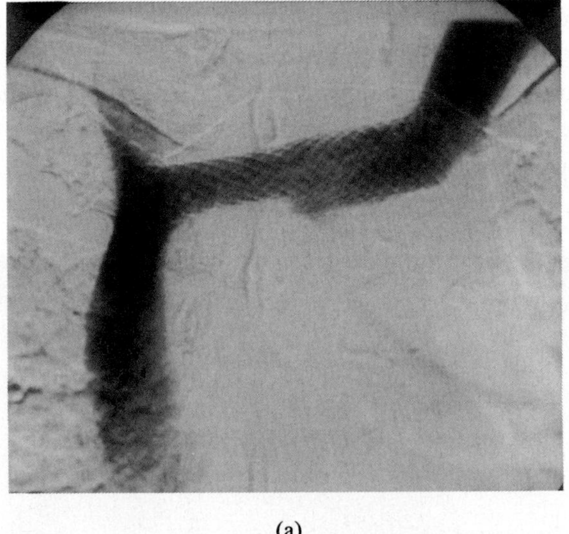

(a)

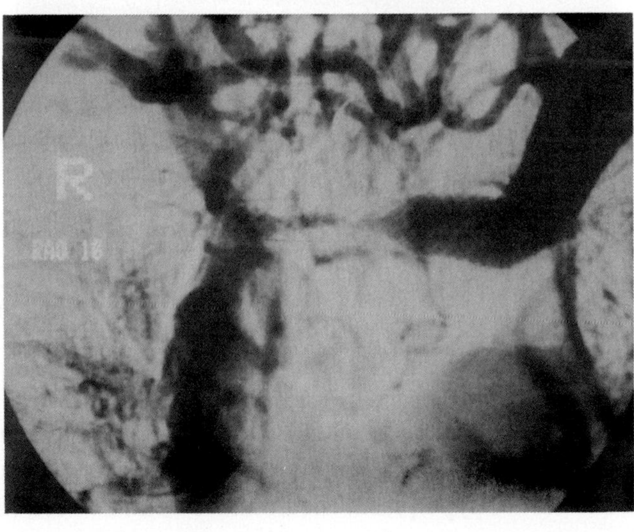

(b)

Fig. 14.10. (a) Stent has been placed in the left innominate vein at the site of a short-segment occlusion and the results are excellent. (b) A few months later, a severe stenosis has developed inside the stent at the site of the initial lesion, probably due to intimal hyperplasia.

There is also the fact that stents placed in peripheral veins restrict the use of that vein for needling which may be a major disadvantage if this was the only available or developed vein. Complications of stents include thrombosis, migration, infection and disintegration of the struts due to continuous movement or compression (20) (Fig. 14.11).

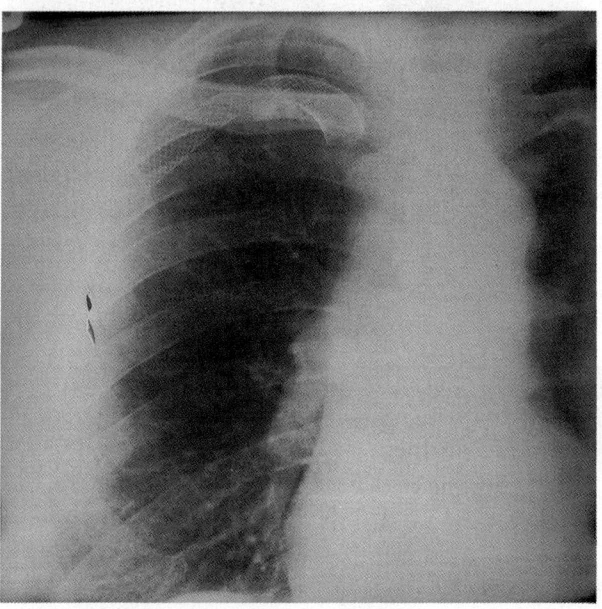

Fig. 14.11. Disruption of the struts of a Wallstent due to compression between the clavicle and first rib.

Atherectomy devices like the Simpson Atherocath have been used to shell out intimal hyperplasia at the site of restenosis but without lasting effect as further intimal hyperplasia accumulates within weeks (21). A recent balloon catheter with cutting blades (Fig. 14.12) has been developed to achieve a deep cut into the intimal hyperplasia, therefore resulting in a wider lumen (22).

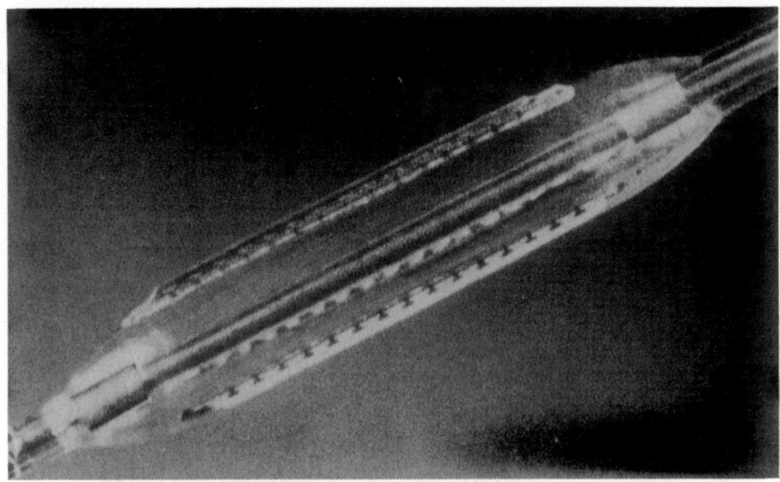

Fig. 14.12. Cutting blades are placed on the surface of the balloon catheter to achieve deep cuts into the intimal hyperplasia.

14.5.2. *Catheter-based access*

The complications are further divided into those that are related to the catheter insertion technique and those that are related to the catheter itself.

14.5.2.1. *Complications of catheter insertion*

A large number of complications have been reported in the literature. Many of these are related to the approach used for placement of the initial catheter. In experienced hands, the complications are rare but there is little doubt that care and meticulousness, in addition to the use of modern imaging and particularly ultrasound guidance (Fig. 14.13), will avoid many of the complications that are listed below.

Arterial injury

This occurs either to the subclavian, the carotid or one of their smaller branches. Injury to the femoral arteries may also occur in lower limb access.

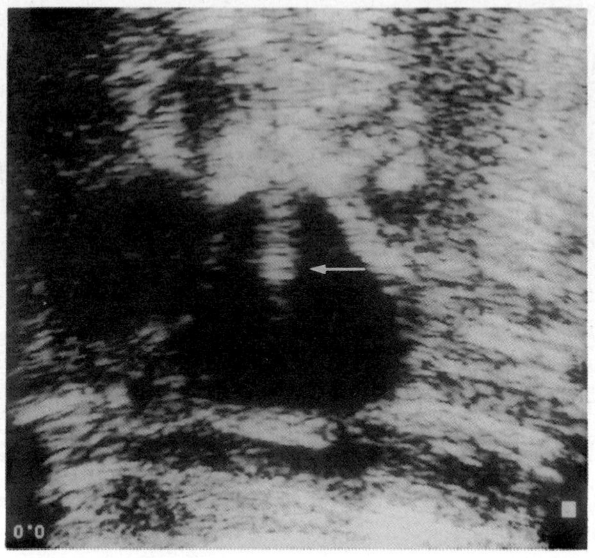

Fig. 14.13. Puncture of the jugular vein under ultrasound guidance. The linear reflection of the needle tip is clearly seen within the jugular vein (arrow).

The injury may take the form of dissection of the artery, formation of a false aneurysm or an arteriovenous fistula.

Arterial dissection does not usually require intervention and is best treated conservatively unless it results in acute ischaemia in which case it can be treated either surgically or by means of placement of an arterial stent at the site of the dissection if it is accessible

False aneurysms and AV fistulas may require surgery but it is generally possible to treat the injury percutaneously and through the placement of embolisation material, such as metal coils in the false aneurysm or at the fistula (Fig. 14.14).

Pneumothorax

The treatment is by either aspiration if the pneumothorax is small or insertion of a tube if it is causing embarrassment to the respiratory system.

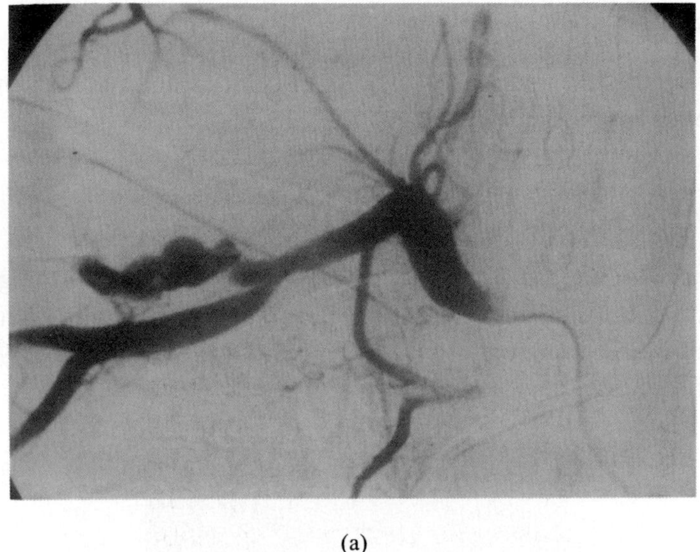

(a)

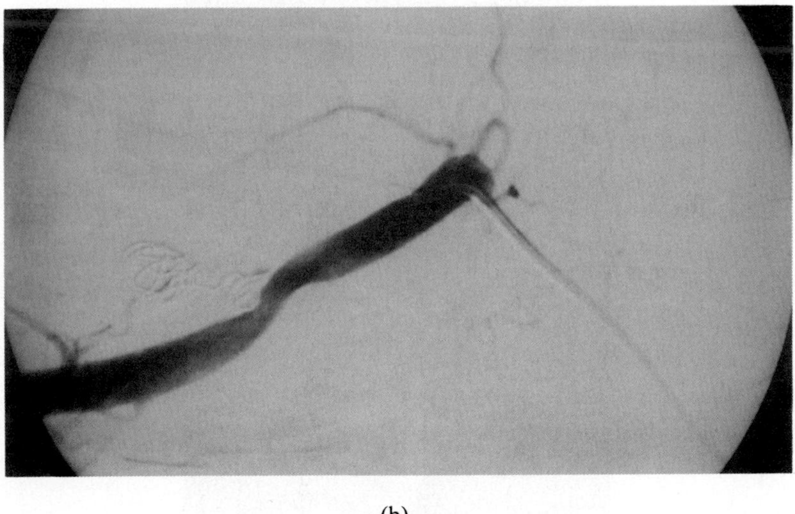

(b)

Fig. 14.14. (a) Arteriogram demonstrating a false aneurysm of the right subclavian artery from the insertion of a subclavian venous line. (b) After the placement of coils the aneurysm no longer fills.

Nerve injury

This may be the most serious of the complications and requires immediate referral to an appropriate specialist.

Lost wire/sheath

Loss of the wire into the vein through carelessness can lead to migration of the wire into the heart or lungs, or perforation of a major vein. In the majority of cases, it is possible to retrieve the wire percutaneously through the use of a snaring device (Figs. 14.15 and 14.16).

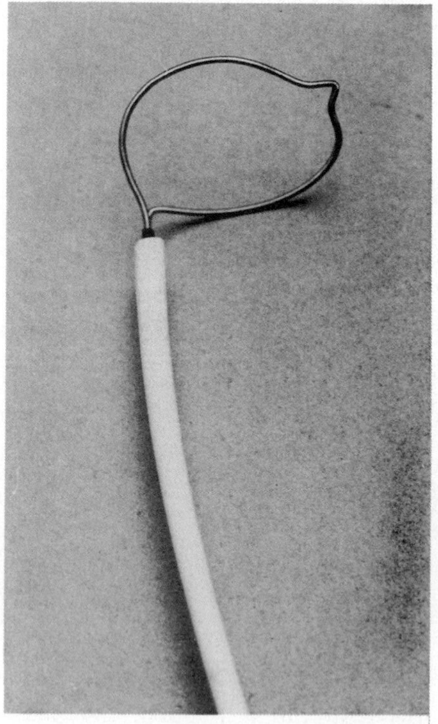

Fig. 14.15. The Amplatz goose-neck snare.

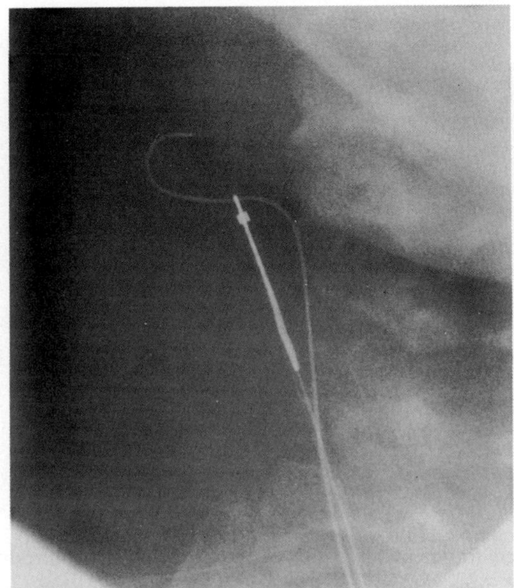

Fig. 14.16. A retained wire in the jugular vein is snared by the Amplatz snare prior to removal via the femoral vein.

A rarer but potentially more difficult problem is the separation of part of the peel-away sheath that is used to insert the catheter inside the patient. The sheath should always be inspected after removal to ascertain that it has come out whole, and if not, an attempt is made to remove it percutaneously (Fig. 14.17).

Air embolism

This should be a rare complication if care is taken during the insertion of the catheter through the sheath to prevent air entry by using a clamp on the sheath and tilting the table head down. Air tends to accumulate in the pulmonary outflow tract. Should it happen, the head is tilted down and enough time is allowed for the air to disperse, with monitoring of the amount of air on screening (Fig. 14.18).

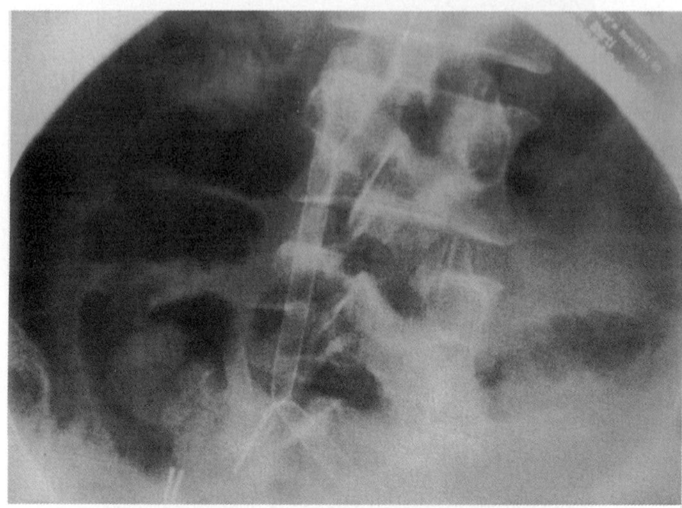

Fig. 14.17. A retained fragment of a peel-away sheath used to introduce the hemodialysis catheter is snared and pulled down to the iliac vein. The fragment is removed through a 12F sheath inserted in the femoral vein.

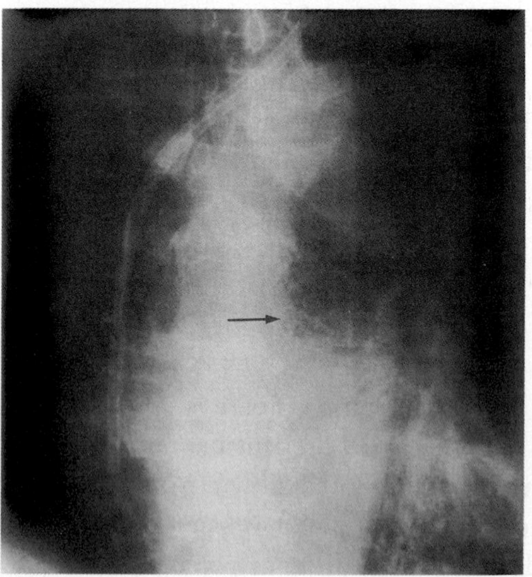

Fig. 14.18. Air within the pulmonary outflow tract (arrow).

The patient may complain of pleural pain and suffer a drop in blood pressure but the amount of air aspirated is seldom sufficient to cause tamponade of the pulmonary artery, in which case, more drastic resuscitation measures may be needed.

Haematoma formation (mediastinum/retroperitoneum)

Many of the renal patients have abnormal clotting. Therefore, they are more liable to excessive bleeding from puncture sites, particularly if arterial injury were to occur. It is important that these complications are suspected and documented early so that supportive measures can be undertaken.

14.5.2.2. *Catheter-related complications*

Poor function

This is commonly due to thrombus at the tip of the catheter, probably related to fibrin sheath formation around it, and is encountered in about 50% of patients (5, 23, 24). Contrast studies will show a filling defect at the tip of one or both the arterial and venous lumens of the catheter, and on occasions, a lucent sleeve may be seen surrounding the catheter (Fig. 14.19).

The classical treatment of this complication has been with urokinase injected into the affected lumen. The success rate is variable and frequently short-lived (25). More recently, stripping of the fibrin sheath percutaneously has proved to be an effective and fairly safe procedure which helps to maintain the function of the catheter for a good length of time provided that the catheter is not kinked or malpositioned. Success rates of around 70% have been reported (24, 26) (Fig. 20). Replacement of the catheter may be the best and most cost-effective option however (27).

Infection

The rate of infection of venous catheters is similar, whether the catheter is inserted in the operating theatre or in the radiology department, and is

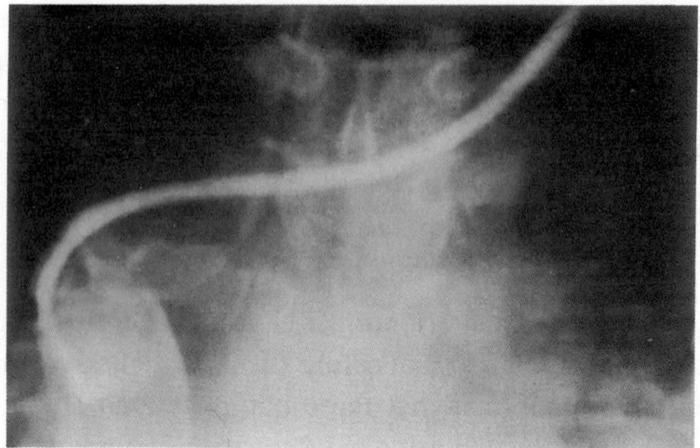

Fig. 14.19. Lineogram demonstrating irregular flow from the tip of the catheter in keeping with thrombus formation. A lucent sleeve surrounding the distal part of the catheter probably represents a fibrin sheath.

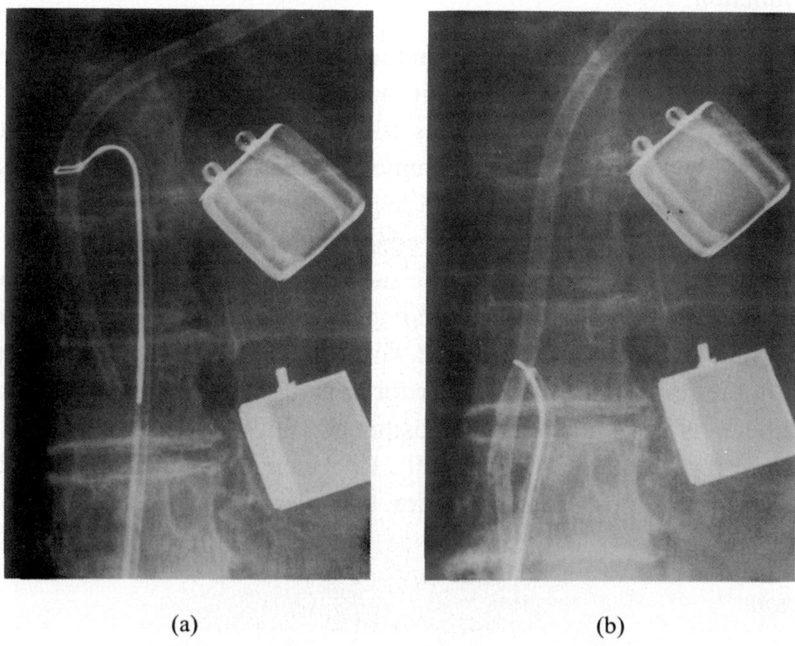

(a) (b)

Fig. 14.20. (a) and (b) A snare is advanced over the dialysis catheter, tightened and pulled down to "strip" the fibrin sheath off.

probably more related to the daily handling of the catheters rather than the insertion method (28, 29). The only treatment is the removal or replacement of the catheter after an appropriate period of antibiotic treatment. To avoid contamination, it is preferable not to use the same entry point for the new catheter.

Kinks and compression

Compression of the catheter is sometimes observed from the muscles and other soft tissues at the site of the subcutaneous tunnel (Fig. 14.21) and is frequently observed on the post-procedure erect chest X-ray even though the position might appear satisfactory when the patient is in the supine position during the insertion procedure. This may result in malfunction of the catheter and should be corrected before the patient leaves the department in which the insertion took place. Gentle blunt dissection is usually sufficient to remove the compression, but in some instances, the catheter may be damaged from the compression and may need to be changed.

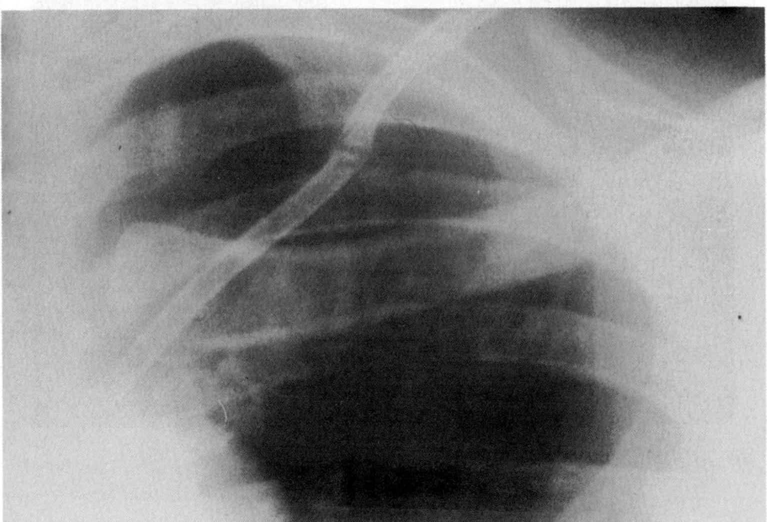

Fig. 14.21. A severe kink is seen in the tunnelled catheter in the subcutaneous tunnel. This was only apparent on the erect film taken after the insertion of the catheter.

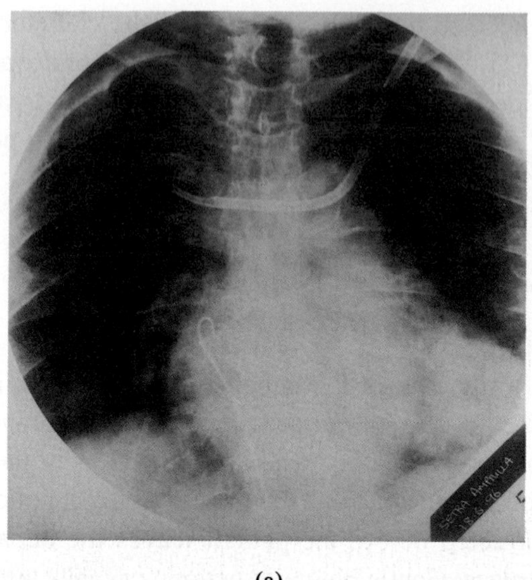

(a)

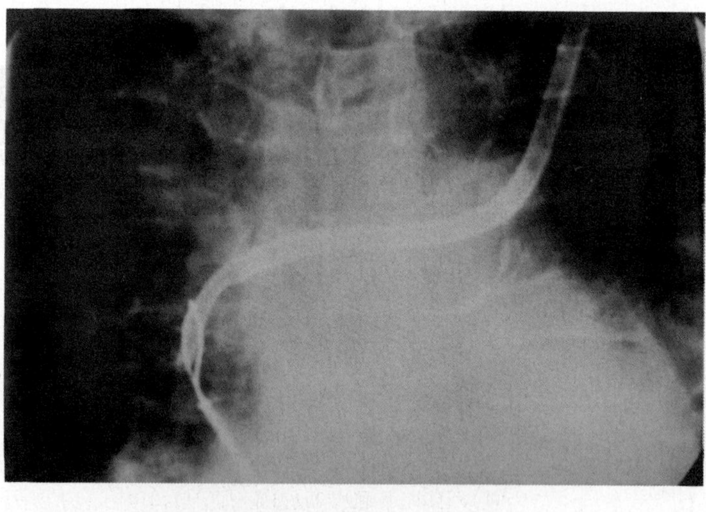

(b)

Fig. 14.22. (a) The catheter has been introduced through the left jugular vein and its tip has entered the origin of the azygos vein. (b) The tip of the catheter is pulled down to the lower SVC.

Malposition and migration

Insertion of the catheter from the left jugular or subclavian may result in its tip lying at the junction of the innominate veins or even protruding into the origin of the azygos vein, leading to malfunction. It should be possible to reposition the catheter tip through the use of a snare from the femoral route combined with loosening of the subcutaneous tunnel to allow for advancement of the catheter into the SVC (Fig. 14.22).

Longer catheters should be used from the left side to avoid this complication. Migration into a different vein, such as the jugular, when a catheter is inserted from the subclavian is sometimes seen and is correctable by the same snaring technique (Fig. 14.23).

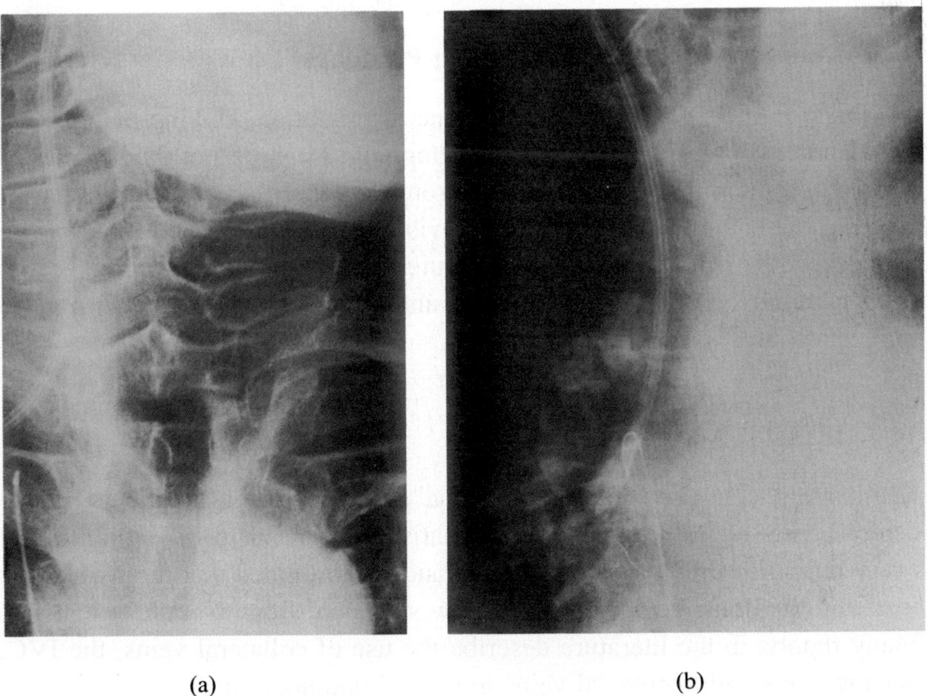

(a) (b)

Fig. 14.23. (a) The tip of the subclavian venous catheter has migrated into the jugular vein. (b) The catheter was repositioned.

Erosion of vascular structures

Long-term catheters may erode through the wall of the vein or SVC into the perivascular space or surrounding structures. This is a very rare complication which necessitates removal of the catheter, and sometimes, surgical treatment of the site of perforation.

Venous thrombosis/stenosis

This was discussed above. The only important consideration here is that intervention on the vein may necessitate the removal of the existing catheter, and therefore, loss of access. Careful planning of the procedure will avoid this problem.

14.6. Management of Malfunctioning Peritoneal Dialysis Catheters

Malfunction of PD catheters may be due to migration, kinking or blocking of sideholes by omental in-growth or wrapping. Migration of the catheter or occlusion can sometimes be corrected through the insertion of a stiff guidewire to bring the catheter down into the pelvis or open the blocked lumen (30). More recently, percutaneous use of "cauterisation" through the catheter was used to remove in-growing omentum and restore the function with good immediate success (31).

14.7. Difficult Access

Maintaining central venous access is a problem in some patients whose central veins have thrombosed. Alternative sites need to be explored and every possible large vein should be considered provided that the size of the vein and the drainage pattern will allow sufficient flow to achieve dialysis. Many reports in the literature describe the use of collateral veins, the IVC, hepatic veins and intercostal veins in such situations (32).

14.7.1. *Inferior vena cava (IVC) access*

It is possible to puncture the IVC using fluoroscopy guidance but we prefer to image the IVC with CT scanning first to establish the anatomy. CT guidance is then used to plan the route to avoid puncturing other organs such as the colon or kidneys A thin needle is used to access the IVC below the renal veins at about the level of L3 and then a small stiff wire is introduced. If a mobile fluoroscopy system is available in the CT room, then the procedure can be completed there. Alternatively, the patient can be transferred in the prone position to the appropriate fluoroscopy room with the wire in position and fluoroscopy is used to exchange the thin wire for a stiff wire that would support the insertion of a peel-away sheath through which a single- or a dual-lumen catheter can be inserted. The amount of movement in the lumbar region is considerable which makes fixation of the line difficult, particularly if the muscle mass is thin (Fig. 14.24). Therefore,

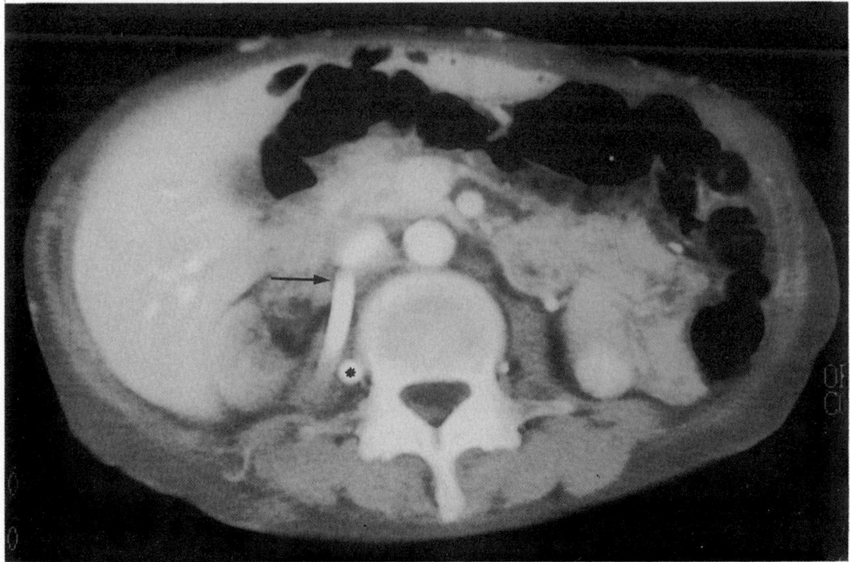

Fig. 14.24. CT scan of the abdomen showing the route of catheter entry into the IVC (arrow). Note the thin psoas muscles, and the proximity of the kidneys and lumbar veins (asterisk) to the puncture route.

it is advisable to advance the line as high in the venous system as possible, preferably in the SVC if it is patent (Fig. 14.25) as migration of the line is very common (Fig. 14.26). Good results and low complications have been reported with this approach (33).

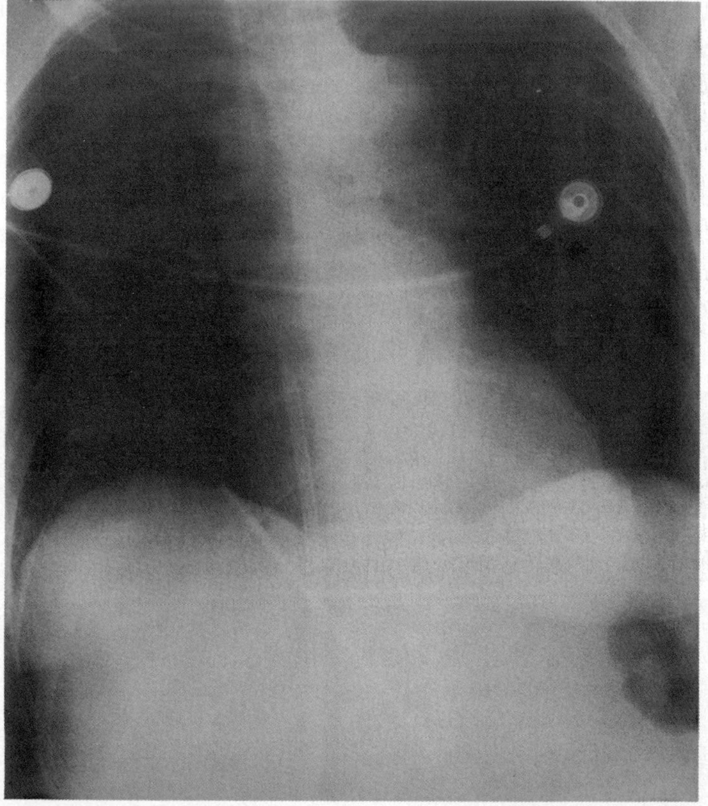

Fig. 14.25. The tip of the catheter is placed in the region of the lower SVC.

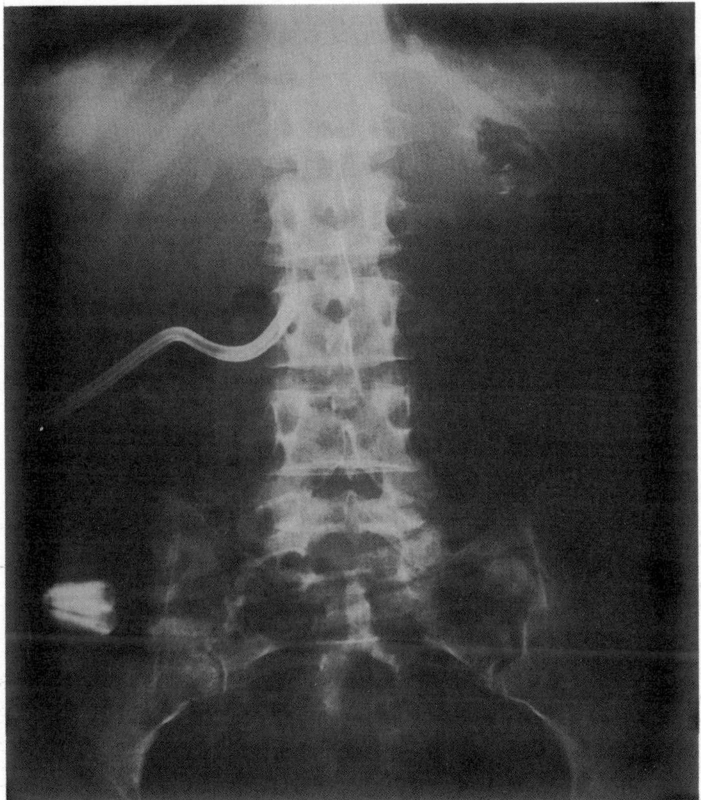

Fig. 14.26. The catheter has retracted almost completely out of the IVC due to lumbar movement.

14.7.2. *The azygos system*

This is commonly enlarged in patients with chronic renal disease because of the fluid overload. It is also an important collateral pathway for the drainage of the upper and lower parts of the body in cases of IVC or SVC obstruction. There are commonly communications with the external jugular which can be used to place a wire into the azygos over which the catheter can be advanced. Access is easier in cases of patency of the innominate veins as the origin of the azygos is usually patent and can be easily entered (Figs. 14.27

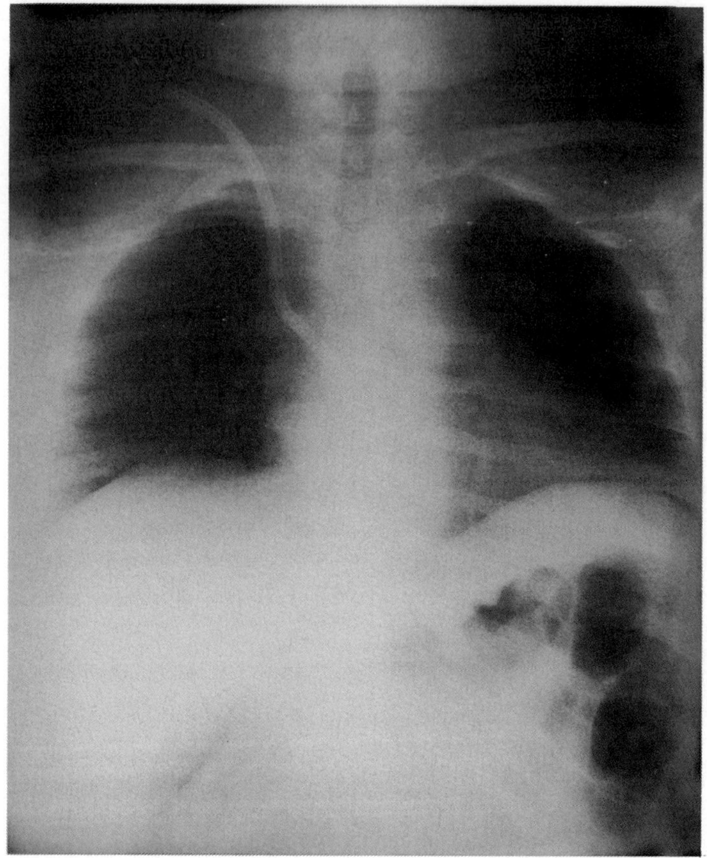

Fig. 14.27. The tip of the catheter is in the azygos vein. Note the medial course of the lower end of the catheter.

and 14.28). A large azygos vein is capable of tolerating high flow with drainage occurring through the paravertebral plexus and through the renal veins as well as the intercostal veins, but the normal blood flow through it is sluggish resulting in thrombus formation within weeks and sometimes days. Nevertheless, this route is worth considering if there is no other access.

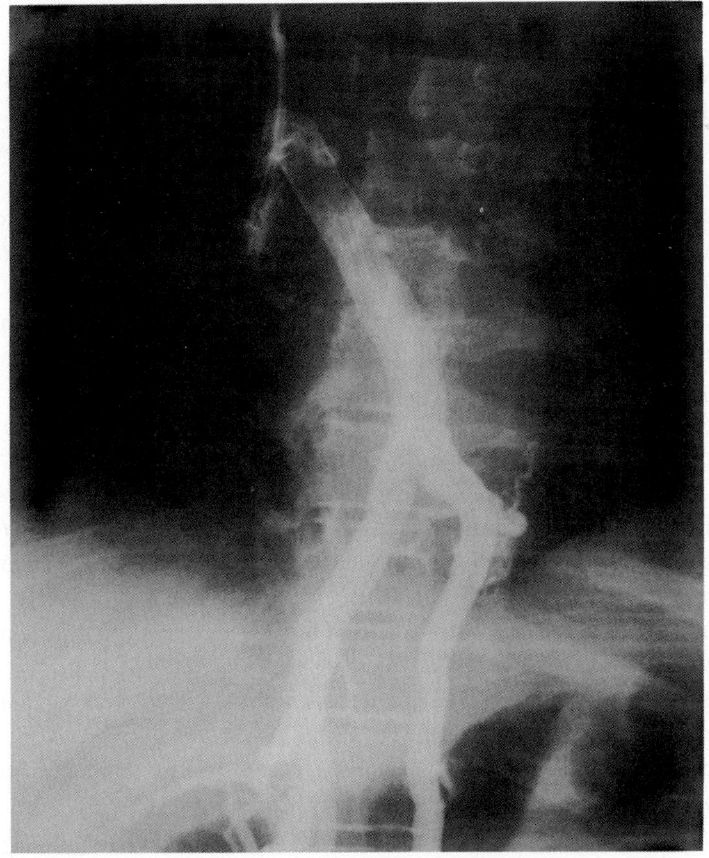

Fig. 14.28. Contrast injection into the azygos vein. The drainage eventually occurs through the lumbar plexus which communicates with the renal vein.

References

1. Surratt, R.S. *et al.* (1991). The importance of preoperative evaluation of the subclavian vein in dialysis access planning. *AJR*, **156**, 623–625.
2. Koo Seen Lin, L.C. and Burnapp, L. (1996). Contemporary vascular access surgery for chronic haemodialysis. *J Roy Coll Surg Edinburgh*, **41**, 164–169.
3. Prince, M.R., Grist, T.M. and Debatin, J.F. (1997). *3D Contrast MR Angiography*, pp. 135–150. Springer, Germany.

4. Strauch, B. *et al.* (1992). Forecasting thrombosis of vascular access with doppler color flow imaging. *Am J Kidney Dis*, **19**, 554–557.
5. Hodges, T.C. *et al.* (1997). Longitudinal comparison of dialysis access methods: Risk factors for failure. *J Vasc Surg*, **26**, 1009–1019.
6. Berridge, D.C. *et al.* (1995). Thrombolysis in arterial graft thrombosis. *Eur J Vasc Endovasc Surg*, **9**, 129–132.
7. Beathard, G.A., Welch, B.R. and Maidment, S.J. (1996). Mechanical thrombolysis for the treatment of thrombosed hemodialysis access grafts. *Radiology*, **200**, 711–716.
8. Verta, M.J. Jr. (1993). Endovascular salvage of failing PTFE grafts. In *Vascular Access for Hemodialysis* — III, (eds. M.L. Henry and R.M. Feguson), p. 201, Precept Press Inc: USA.
9. Koo Seen Lin, L.C., Taube, D.E. and Al-Kutoubi, A. (1995). Acute management of subclavian vein thrombosis. *Br J Surg*, **82**, 1140.
10. Swan, T.L. *et al.* (1995). Pulmonary embolism following hemodialysis access thrombolysis/thrombectomy. *J Vasc Interventional Radiol*, **6**, 683–686.
11. Overbosch, E.H. *et al.* (1996). Occluded hemodialysis shunts: Dutch multicenter experience with the Hydrolyser catheter. *Radiology*, **201**, 485–488.
12. Schwab, S.J. *et al.* (1989). Prevention of hemodialysis fistula thrombosis. Early detection of venous stenoses. *Kidney Int*, **36**, 707–711.
13. Glanz. S. *et al.* (1984). Dialysis access fistulas: Treatment of stenoses by transluminal angioplasty. *Radiology*, **152**, 631–642.
14. Kanterman, R.Y. *et al.* (1995). Dialysis access grafts: Anatomic location of venous stenosis and results of angioplasty. *Radiology*, **195**, 135–139.
15. Schwab, S.J. *et al.* (1987). Transluminal angioplasty of venous stenoses in polytetrafluoroethylene vascular access grafts. *Kidney Int*, **32**, 395–398.
16. Quinn, S.F. *et al.* (1995). Percutaneous transluminal angioplasty versus endovascular stent placement in the treatment of venous stenoses in patients undergoing hemodialysis: Intermediate results. *J Vasc Interventional Radiol*, **6**, 851–855.
17. Vorwek, D. *et al.* (1995). Venous stenosis and occlusion in hemodialysis shunts: Follow-up results of stent placement in 65 patients. *Radiology*, **195**, 140–146.
18. Gibson, M. and Al-Kutoubi, A. (1994). Endovascular stents in subclavian and innominate vein stenosis. *J Interventional Radiol*, **9**, 113–120.
19. Turmel-Rodrigues, L.A. *et al.* (1997). Wallstents and Craggstents in hemodialysis grafts and fistulas: Results for selective indications. *J Vasc Interventional Radiol*, **8**, 975–982.

20. Rhodes, A.I., Gibson, M. and Al-Kutoubi, A. (1997). Mechanical disruption of a Wallstent in the subclavian vein between the clavicle and first rib. *J Interventional Radiol*, **12**, 151–153.
21. Vorwerk, D. and Guenther, R.W. (1990). Removal of intimal hyperplasia in vascular endoprostheses by atherectomy and balloon dilatation. *AJR*, **154**, 617–619.
22. Vorwerk, D. *et al.* (1996). Hemodialysis fistulas and grafts: Use of cutting balloons to dilate venous stenoses. *Radiology*, **201**, 846.
23. Bolz, K.D. *et al.* (1995). Catheter malfunction and thrombus formation on double-lumen hemodialysis catheters: An intravascular ultrasonographic study. *Am J Kidney Dis*, **25**, 597–602.
24. Rockall, A.G. *et al.* (1997). Stripping of failing haemodialysis catheters using the Amplatz gooseneck snare. *Clin Radiol*, **52**, 616–620.
25. Shrivastava, D. *et al.* (1994). Salvage of clotted jugular vein hemodialysis catheters. *Nephron*, **68**, 77–79.
26. Crain, M.R. *et al.* (1996). Fibrin sleeve stripping for salvage of failing hemodialysis catheters: Technique and initial results. *Radiology*, **198**, 41–44.
27. Merport, M. *et al.* (2000). Fibrin sheath stripping versus cathether exchange for the treatment of failed tunnelled haemodialysis catheters: Randomised clinical trials. *JVIR*, **11**, 1115–1120.
28. Trerotola, S.O. *et al.* (1997). Outcome of tunnelled hemodialysis catheters placed via the right internal jugular vein by interventional radiologists. *Radiology*, **203**, 489–495.
29. Duszak, R. Jr. *et al.* (1998). Replacement of failing tunnelled hemodialysis catheters through pre-existing subcutaneous tunnels: A comparison of catheter function and infection rates for *de novo* placements and over-the-wire exchanges. *J Vasc Interventional Radiol*, **9**, 321–327.
30. Degesys, G.E. *et al.* (1985). Tenchkoff peritoneal dialysis catheters: The use of fluoroscopy in management. *Radiology*, **154**, 819–820.
31. Lim, S.J. *et al.* (1998). Recanalization of obstructed Tenchkoff peritoneal dialysis catheter: Wire/stylet manipulation combined with endoluminal electro-cauterization. *Cardiovasc Interventional Radiol*, **21**, 435–438.
32. Andrews, J.C. (1994). Percutaneous placement of a Hickman catheter with the use of an intercostal vein for access. *J Vasc Interventional Radiol*, **5**, 859–861.
33. Lund, G.B., Terotola, S.O. and Scheel, P.J. Jr. (1995). Percutaneous translumbar inferior vena cava cannulation for hemodialysis. *Am J Kidney Dis*, **25**, 732–737.

CHAPTER 15

PERITONEAL DIALYSIS ACCESS

Murat A Akyol MD, FRCS
Scottish Liver Transplant Unit
Royal Infirmary of Edinburgh
1 Lauriston Place
Edinburgh EH3 9YW, UK

15.1. Introduction

Peritoneal dialysis is a well-established form of renal replacement therapy used by almost half of the patients on dialysis in the United Kingdom and an estimated 120 000 patients (around 15% of the dialysis population) worldwide (1). Improvement in our understanding of fluid and solute transport across the peritoneal membrane, refinement in dialysis techniques and advances in surgery for peritoneal access have all contributed to the current success of peritoneal dialysis.

15.2. Peritoneal Dialysis Catheters

15.2.1. *Catheter design*

The prototype of current peritoneal dialysis catheters was developed in the early 1970s by Tenckhoff (2). There have been many variations in catheter design since then but the standard Tenckhoff catheter remains the most commonly used catheter throughout the world.

Peritoneal dialysis catheters consist of an intraperitoneal, a subcutaneous and an external portion.

The intraperitoneal segment has multiple perforations to allow inflow and outflow of fluid. Most of the variations in catheter design are due to the differences in this intraperitoneal segment. The original Tenckhoff catheter consists of a straight intraperitoneal segment with multiple perforations (Fig. 15.1). A common variation is a coiled intra-abdominal portion with a longer perforated segment potentially improving the flow rate and reducing the tendency for catheter migration and omental wrapping. Other variations to the intra-abdominal segment of catheters include silicon discs used in the Toronto Western Hospital catheter (Fig. 15.1) and a T-fluted design of the Ash catheter. All these design features have been developed to improve

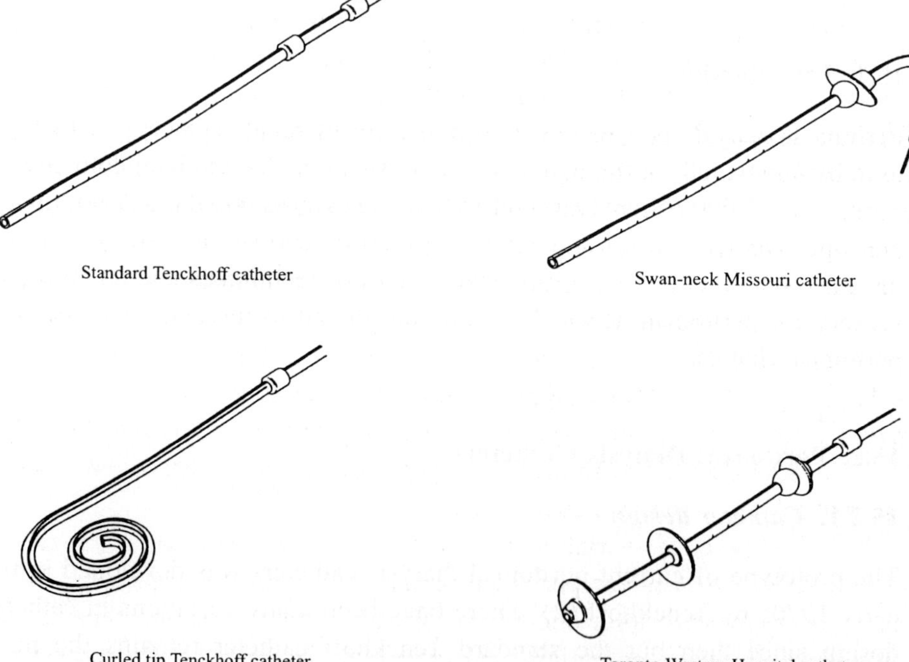

Fig. 15.1. Illustration of some of the commonly used peritoneal dialysis catheters. [Reprinted with permission from Gokal *et al.* (5); copyright©1998 International Society for Peritoneal Dialysis]

dialysate flow and reduce propensity of the catheter to migrate or to be wrapped by omentum. There is no evidence demonstrating clear superiority of any particular catheter design over the others.

15.2.2. *Catheter materials*

Most peritoneal dialysis catheters are manufactured from silicon rubber (silastic), which is soft and flexible through a wide range of temperatures and is biologically inert. An exception is the Cruz catheter (3) which is made of polyurethane. Polyurethane possesses a number of theoretical advantages. Its ability to bond with other materials allows such catheters to have built-in adapters. This eliminates the problem of reduced lumen size at the level of the adapter and the wear and tear of the catheter segment mounted over the adapter. In addition, the greater strength of polyurethane compared with silicon (approximately ten-fold difference in tensile strength) allows polyurethane catheters to have thinner walls and larger lumens which increases the flow rate. Polyurethane is also thermoplastic which allows it to conform to the shape and configuration of surrounding tissues at body temperature, thereby reducing tissue stress. The increased flow rate made possible by the use of polyurethane reduces the time required to do fluid exchanges and may increase patient compliance. Furthermore, in automated peritoneal dialysis, the time spent during inflow and outflow relative to total cycle time is also reduced and the dialysis efficiency is enhanced because of the longer effective dwell times (4). Concern has been expressed about potential hydrolysis of the polyurethane surface with cracking of the catheter material especially when exposed to polyethylene glycol or alcohol (5). Large-scale experience with the use of polyurethane catheters has only been reported from one dialysis centre (3, 4).

A further potential variation in catheter material is the development of silver-coated catheters for peritoneal dialysis (6). Recent technology with pulsed laser ion beam application has allowed surface coating of soft catheters with silver, thereby potentially imparting to the catheter the powerful antibacterial properties of silver. Initial experience in animal models suggest that silver-coated catheters do have a higher infection resistance and similar biocompatibility compared with uncoated silastic catheters. Theoretical concerns about long-term side effects and silver toxicity remain, and there is very little clinical experience with the use of these new silver-coated catheters.

15.2.3. *Subcutaneous section*

The subcutaneous segment of peritoneal dialysis catheters usually include one or two Dacron cuffs which allow tissue in-growth, reduce the propensity for fluid leak and act as barriers against infection. In addition to the number of cuffs (one versus two), a further variation of the subcutaneous section of the catheter is in its shape. Catheters with straight subcutaneous segments are implanted in an arcuate subcutaneous tunnel directing the catheter exit site laterally or caudally. A permanent bend in the subcutaneous section of the catheter (such as the swan-neck Tenckhoff catheter, the Moncrief–Popovich catheter and the Cruz catheter — Fig. 15.2) eliminates the resilience force or "shape memory" of straight catheters, thereby reducing the tendency for the migration of the intra-abdominal section out of the pelvis. The anchorage of catheters is provided by incorporation of subcutaneous tissues into the cuffs. Some catheters such as the Toronto Western Hospital or the Missouri catheters have a flange and bead in place of a simple Dacron deeper cuff, which strengthens the anchorage of the catheter to the abdominal wall and reduces the risk of fluid leakage. Because of these design features (for instance, the slanted flange of the Missouri catheter), some catheters need to be mirror images of each other designed for right- and left-sided subcutaneous tunnels and are not interchangeable. There is little evidence from controlled trials that these design features of the subcutaneous portion of various catheters impart a significant advantage, except the difference between one cuff and two cuffs. In a number of studies (7–9), the single-cuffed catheters were shown to be associated with a shorter time to the first peritonitis episode, more exit site complications and shorter survival times compared with double-cuffed catheters.

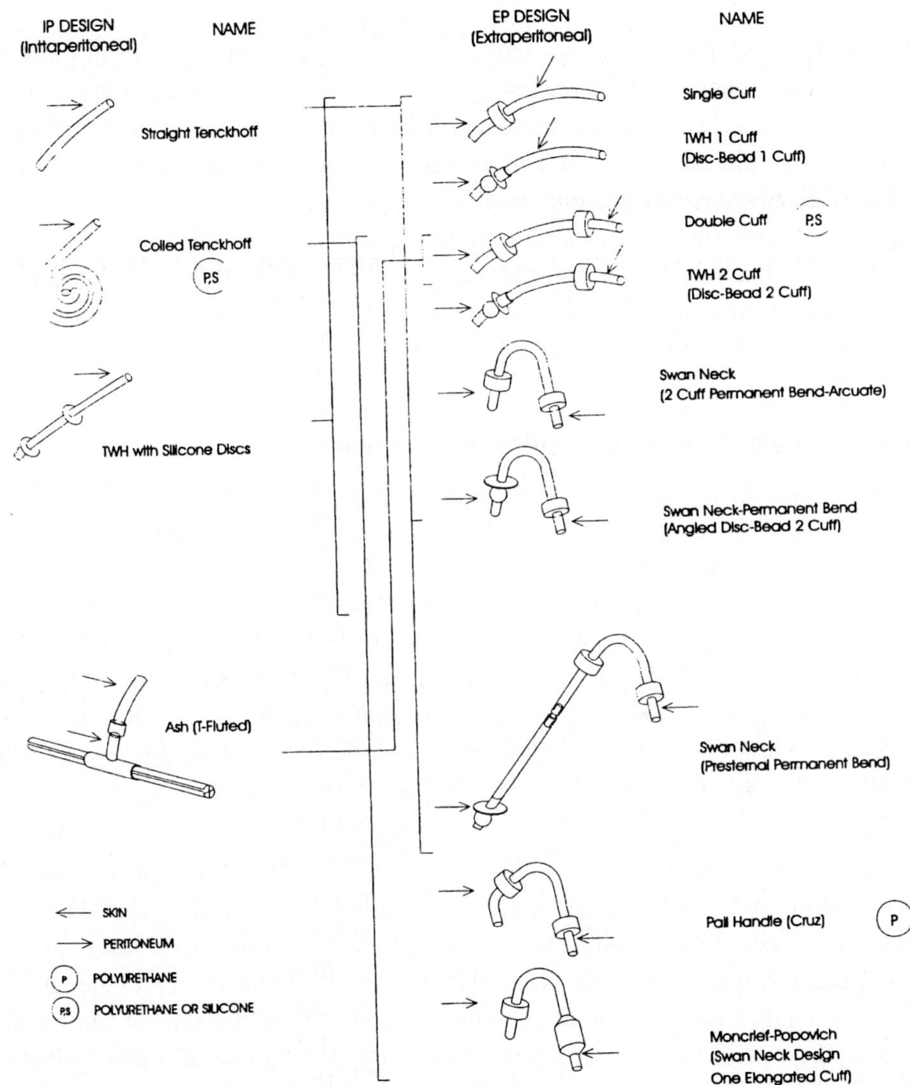

Fig. 15.2. Currently available chronic peritoneal catheters showing combinations of intraperitoneal and extraperitoneal designs. [Reprinted with permission from Gokal *et al.* (5); copyright© 1998 International Society for Peritoneal Dialysis]

15.3. Insertion of Peritoneal Dialysis Catheters

Insertion of peritoneal dialysis catheters can be performed on an in-patient or out-patient basis, with or without general anaesthesia using a percutaneous, open surgical or laparoscopic approach. Whichever technique is used, catheter insertion should be undertaken in an operating room under standard sterile surgical conditions. Technically, none of the procedures involved are complex but all require meticulous attention to detail. The establishment of safe, permanent and reliable peritoneal access on a consistent basis relies upon experienced surgeons (or physicians) who have a regular commitment to the care of patients with chronic renal failure.

15.3.1. *Preparation prior to catheter insertion*

Patients should be fully counselled about the options for dialysis and the procedures involved in dialysis access surgery including potential complications. Once the choice for chronic peritoneal dialysis is made, the patient should be assessed for evidence of abdominal wall herniae or any other source of potential infection in the abdominal wall. If present, herniae should be repaired at the time of catheter insertion. The location of the exit site should also be carefully determined and marked on the skin prior to insertion. The exit site should be marked with the patient seated upright and should avoid the belt line as well as abdominal folds or previous scars. For nasal carriers of *Staphylococcus aureus*, eradication therapy has been shown to reduce the incidence of exit site infections (9). Appropriate bowel preparation and emptying the bladder immediately before the procedure are also required. Trials about use of prophylactic antibiotics have given conflicting results (5,9–11). Nevertheless, most of the direct evidence as well as a large body of indirect evidence concerning the use of antibiotic prophylaxis prior to surgical interventions suggest the use of a prophylactic antibiotic. The antibiotic chosen should have anti-staphylococcal activity and should be given prior to the procedure and perhaps repeated once or twice after catheter insertion. In practice, a cephalosporin antibiotic or flucloxacillin is most commonly used. Vancomycin is effective but should not be used routinely to avoid development of resistant microorganisms.

15.3.2. *Catheter insertion techniques*

15.3.2.1. *Open surgical insertion*

This is the most common method of insertion for chronic peritoneal dialysis. It is possible to perform the procedure under local infiltration anaesthesia in thin patients but general anaesthesia is preferred. Of the many different skin incisions that can be used, a small vertical paramedian incision is frequently used (10), although a small transverse incision is equally good. This allows the deeper cuff of the catheter to lie within the well-padded and vascular rectus muscle. The posterior rectus sheath fused with the peritoneum provides an easily identifiable and strong layer securing water tightness. The small opening in the peritoneum through which the intra-abdominal portion of the catheter enters the peritoneal cavity is secured with a purse-string suture. The deeper cuff of the catheter is always left in the pre-peritoneal space and not within the peritoneal cavity. The subcutaneous portion of the catheter is then tunnelled towards the predetermined exit site. A distance of at least 3 cm should be left between the subcutaneous cuff and the exit site, and the catheter should be directed laterally or preferably caudally at the exit site. It has been shown that cranially-directed exit sites are associated with a higher incidence of infection (7, 11). Another important consideration is to make a small exit site that snugly fits around the catheter. It is best to avoid sharp tunnelling devices to minimise the risk of bleeding in the subcutaneous tunnel. Catheters must not be tunnelled by introducing a surgical instrument from the skin surface into the subcutaneous tunnel. This practice results in an inappropriately large exit site allowing free movement of the catheter and increases the risk of colonisation and subsequent exit site infection. The catheter should be checked at the time of insertion for patency, ensuring free flow in both directions.

Some authors have recommended partial omentectomy at the time of catheter insertion in an attempt to reduce the incidence of omental wrapping of the intra-abdominal portion of the catheter (12). Very few units have adopted this relatively invasive approach. A further modification reported is the initial subcutaneous embedding of the external portion of the catheter (13). The rationale for this is to enable uncontaminated wound healing and better tissue in-growth into the cuffs, resulting in reduced incidence of tunnel infections or fluid leaks. In practice, the technique is unduly complex, resulting in a high incidence of abdominal wall complications such as haematoma, seroma, infection and catheter occlusion.

15.3.2.2. *Percutaneous insertion*

Percutaneous insertion of peritoneal dialysis catheters can be performed either using a Trocar or a modification of the Seldinger technique using a "peel apart" sheath. Percutaneous insertion is commonly performed under local anaesthesia and the introduction of the Trocar or guidewire into the peritoneal cavity is often at the midline. Percutaneous insertion techniques are not suitable for obese patients or for those who may have intra-abdominal adhesions. Percutaneous insertion has not been compared with other catheter insertion techniques in a controlled trial. Some authors have reported satisfactory results with percutaneous insertion (14). There is concern about the potential for early fluid leak with the percutaneous technique. Other mechanical complications such as outflow failure and potentially serious complications such as intra-abdominal viscus perforation or haemorrhage have also been reported. Percutaneous catheter insertion should not be used for patients who are required to commence dialysis soon after catheter insertion because of the greater probability of leak.

15.3.2.3. *Laparoscopic insertion*

This is a more recent development using minimally invasive surgical techniques. It usually requires general anaesthesia. The traditional advantages of minimally invasive surgery, such as reduced post-operative pain and early recovery, are less important since the alternative open surgical approach is also "minimally invasive". In experienced hands, it provides results comparable to the open surgical approach (15, 16). It may have the additional advantage of being able to visualise the peritoneal cavity in patients who may have intra-abdominal adhesions. It is possible to perform this procedure using a mini-camera of 3 mm diameter. Technical considerations dictate that the entry site of the catheter into the peritoneal cavity is usually in the lower abdomen or the suprapubic region. This results in the exit site being higher in the abdominal wall and makes it difficult to fashion a caudally-directed exit site. It is also more difficult to use double-cuffed catheters. The peritoneum deep to the deeper cuff is not secured with a purse-string suture during laparoscopic insertion, which may potentially result in a higher incidence of fluid leaks, especially if the catheter is used early.

15.4. Complications

These can be classified into infective and non-infective complications.

15.4.1. *Infective complications*

- Exit site infections
- Tunnel/wound infections
- Peritonitis

15.4.1.1. *Exit site infection*

Acute exit site infection is defined as a purulent and/or blood-stained discharge from the exit site which may or may not be associated with erythema, oedema, tenderness and overgrowth of granulation tissue. The presence of pain or crusting of the exit site alone is not indicative of infection. Almost all exit sites are colonised by bacteria within a few weeks of catheter insertion. Positive cultures from normal-looking exit sites indicate colonisation, not infection. Cultures should only be taken for abnormal-looking exit sites.

Staphylococcus aureus is responsible for the large majority of exit site infections. *Staphylococcus epidermidis* is an infrequent cause. *Pseudomonas aeruginosa*, other gram-negative bacilli and fungi are rare causes of exit site infection (5).

Diagnosis of an exit site infection is an indication for antibiotic therapy. The choice of antibiotic should be determined by culture and sensitivity results. The commencement of antibiotic should not be delayed and an initial antibiotic active against gram-positive organisms should be commenced while bacteriology results are being awaited. Topical treatment such as hypertonic saline dressings or other antiseptic solutions, cauterisation of granulation tissue and exteriorisation, and shaving of the infected superficial cuffs may also be useful on selected cases. There are no conclusive data on the choice and duration of antibiotic therapy or the route of administration. In general, topical antibiotics are not recommended and prolonged systemic antibiotics may be necessary until the exit site appears normal. Persistent or recurrent exit site infections, especially when there is associated peritonitis caused by the same organism, require catheter removal.

15.4.1.2. *Tunnel infection*

Acute tunnel infections (sometimes associated with wound infection) are soft tissue infections of the subcutaneous tunnel presenting as a tender, swollen, erythematous tunnel. Chronic tunnel infections may be clinically less obvious, presenting as intermittent or chronic purulent blood-stained discharge expressed by pressure over the subcutaneous cuff or tunnel. Bacteriology is similar to that of exit site infections but the risk for peritonitis is greater. Initial management of tunnel infections also requires antibiotic therapy. Their principles of therapy are similar to those of exit site infections. However, the risks of peritonitis are higher, requiring a lower threshold for catheter removal. In cases of low-grade chronic tunnel infections (or exit site infections) without peritonitis, it may be possible to replace the catheter rather than removal and delayed re-insertion.

15.4.1.3. *Peritonitis*

This is the most common complication of peritoneal dialysis leading to discontinuation of dialysis, at least temporarily. The mechanisms of infection, clinical presentation, bacteriology and management principles of peritoneal-dialysis-associated peritonitis are different from peritonitis (encountered in general surgical practice) associated with viscus perforation.

Increasing knowledge of CAPD peritonitis and advances in the fluid delivery systems and connectors have led to a gradual decrease in the incidence of peritonitis throughout the 1980s. The disconnect system (also known as the Y system) allows the peritoneal effluent to be drained after the connection is made with a new bag and any touch contamination can be flushed out before new fluid enters the peritoneal cavity. This allows the patient not to carry the empty bag and transfer set between exchanges and is associated with a lower incidence of infection (17). With the increasing use of disconnect systems, many centres now achieve peritonitis rates of one episode every 2–3 patient years or better (18).

Peritonitis associated with peritoneal dialysis manifests itself as a cloudy effluent which contains a high number of white blood cells. Abdominal pain and clinical signs of peritoneal irritation are almost always present but do not usually precede the detection of a cloudy effluent. Seventy five percent of peritonitis episodes are caused by gram-positive organisms. *Staphylococcus epidermidis* is more common than *Staphylococcus aureus* by a ratio of 2:1 (1). The presence of other bacterial growth in cultures should alert the clinicians of the probability of the gastro-intestinal tract being the primary source of infection. The diagnosis of peritonitis associated with gram-positive organisms is not an immediate indication for laparotomy. Intravenous or intraperitoneal antibiotic therapy is commenced without delay and is guided by bacterial sensitivity results. Quantitative assessment of white blood cell content of the effluent is a useful guide to the progression of disease or response to therapy. Prolonged episodes of peritonitis with no response to antibiotic therapy, clinical signs of deteriorating sepsis or mixed bacterial growth, gram-negative organisms or fungi in the dialysis effluent, are indications for early surgical intervention. Persistent staphylococcal peritonitis is often caused by colonisation of the catheter or persistent infection in the exit site/tunnel. In such cases, removal of the catheter followed by appropriate antibiotic therapy and a temporary period on haemodialysis is usually required.

If the episode of peritonitis has necessitated catheter removal, a further catheter may be re-inserted once antibiotic therapy has been discontinued and the patient has remained free of any clinical sign of infection for at least 48 hours.

15.4.2. *Non-infective complications*

- Complications of the insertion procedure
  — Inadvertent injury to intra-abdominal viscera
  — Haemorrhage (intra-abdominal or abdominal wall)
  — Incisional hernia
- Specific "dialysis-related" complications
  — Obstruction to flow through catheter
  — Fluid leaks
  — Erosion of catheter into viscera
  — Sclerosing peritonitis

15.4.2.1. *Complications of the insertion procedure*

Appropriate patient selection and preparation including bladder emptying and bowel preparation combined with competent catheter insertion techniques should ensure that these complications occur very rarely. Inadvertent injury to intra-abdominal viscera or haemorrhage may be more likely to occur with percutaneous or laparoscopic insertion methods but there is no reported data to substantiate these theoretical concerns.

15.4.2.1. *Specific "dialysis-related" complications*

(i) *Obstruction*

Lack of flow through the catheter suggests catheter malposition or migration, catheter blockage or omental wrapping around the intra-abdominal portion of the catheter. Complete obstruction preventing inflow is rare and suggests kinking of the subcutaneous segment of the catheter or blockage of the catheter with intraluminal debris, clot or fibrin. Outflow obstruction is the most common mechanical problem associated with peritoneal dialysis catheters. It is often due to enwrapment of the intra-abdominal segment of the catheter by omentum. Incorrectly positioned catheters, displacement of the catheter tip out of the pelvis or entrapment in peritoneal pockets as a result of adhesions can also present as one-way (outflow) obstruction. Abdominal radiography to assess the position of the catheter can be misleading, especially if a single film is taken in one plane. If catheter malposition is shown on radiography it may be possible to "float" such catheters back into the pelvis. The abdomen is filled with 2–3 litres of dialysate and the patient is positioned such that the displaced tip of the catheter is initially submerged in the fluid. Changing the patient's position may then free the catheter and allow it to return to the pelvis. Successful repositioning of displaced catheters by manipulation with a stiff wire or by laparoscopic means has also been reported (19).

(ii) *Fluid leaks*

External fluid leaks, presenting as dialysis fluid appearing through the exit site or wound, occur early and indicate inadequate sealing of the site of entry of the catheter into the peritoneal cavity. Interruption of peritoneal dialysis for a few weeks may allow sealing of the leak. Persistent or major leaks require surgical correction.

Fluid leaks can also be internal, presenting as hydrothorax or swelling of the external genitalia. Such leaks usually present later. Swelling of the genitalia very often indicates the presence of a patent processus vaginalis. Occasionally, minor leaks into the abdominal wall can also track down into the region of the groin. If there is doubt about the diagnosis, "herniography" by means of intra-abdominal radiological contrast medium or radioisotope establishes the diagnosis. A patent processus vaginalis requires surgical correction. Fluid leaks into the thoracic cavity almost always occur as a result of multiple small communications through the diaphragm. Laparotomy or thoracotomy in an attempt to repair the diaphragm in such cases is a major surgical procedure, is rarely successful and should not be attempted. Thoracoscopic talcum pleurodesis is a simple and less invasive procedure which may be adequate in sealing the diaphragmatic leak. Persistent hydrothorax may require discontinuation of peritoneal dialysis.

(iii) *Erosion of catheter into viscera*

This is a rare complication of peritoneal dialysis catheters left dormant in the abdominal cavity. Gradual erosion of the catheter, usually into the intestine, may result in peritonitis or a more insidious development of an inflammatory mass.

(iv) *Sclerosing peritonitis*

This is a very rare and ill-understood complication associated with peritoneal dialysis. Repeated exposure of the peritoneal membrane to chlorhexidene or repeated/persistent low-grade peritoneal infection has been implicated in its pathophysiology. The intestine is seen to be completely wrapped with a firmly adherent fibrous membrane at the time of laparotomy in advanced cases. The condition is usually progressive and fatal with bowel obstruction. There have been reports of stabilisation of the process and long-term survival with immunosuppressive therapy and occasional successful surgical removal of the membrane (20).

15.5. Removal of Peritoneal Dialysis Catheters

Successful transplantation, infective complications and obstruction are the most common indications for dialysis catheter removal.

The practice of some units is to leave the peritoneal dialysis catheters *in situ* at the time of renal transplantation until stable allograft function has been obtained. This approach may allow continuation of peritoneal dialysis in the post-transplant period should there be a prolonged delay in the onset of allograft function. Alternatively, peritoneal dialysis catheters may be removed at the time of renal transplantation. In the uncommon eventuality of early transplant failure, transplant nephrectomy is indicated and a new peritoneal dialysis catheter can be inserted at the time of transplant nephrectomy.

Removal of peritoneal dialysis catheters for non-infective complications or following successful transplantation involves a minor surgical procedure which requires dissection of the cuff(s) subcutaneously. Removal of catheters for infective complications requires greater consideration and expertise in terms of timing and surgical approach. Catheter removal for exit site or tunnel infections, or repeated episodes of *Staphylococcus epidermidis* infections, involves a similar minor procedure requiring mobilisation of the subcutaneous cuff(s) without any exploration of the peritoneal cavity (21). Clinically severe peritonitis, especially when associated with gram-negative bacteria, mixed organisms or fungi requires a different approach. In such cases, a laparotomy which allows adequate exploration of all four quadrants of the abdominal cavity and a thorough abdominal lavage is indicated.

References

1. Gokal, R. and Mallick, N. P. (1999). Peritoneal dialysis. *Lancet*, **353**, 823–828.
2. Tenckhoff, H. *et al.* (1973). Chronic peritoneal dialysis. *Proc Eur Dial Transpl Assoc*, **10**, 363–370.
3. Cruz, C. (1992). Are the new peritoneal dialysis catheters better than the Tenckhoff catheter? *Semin Dial*, **5**(3), 202–204.
4. Cruz, C. (1997). The Cruz catheter and its functional characteristics. *Peritoneal Dial Int*, **17** (suppl 2), 5146–5148.
5. Gokal, R. *et al.* (1998). Peritoneal catheters and exit-site practices: Towards optimum peritoneal access: 1998 update. *Peritoneal Dial Int*, **18**, 11–33.
6. Dasgupta, M.K. (1997). Silver-coated catheters in peritoneal dialysis. *Peritoneal Dial Int*, **17** (suppl 2), 5142–5145.
7. Warady, B.A., Sullivan, E.K. and Alexander, S.R. (1996). Lessons from the peritoneal dialysis patient database: A report of the North American paediatric renal transplant co-operative study. *Kidney Int*, **49** (suppl 53), 568–571.
8. Honda, M. *et al.* (1996). The Japanese National Registry Data in paediatric CAPD patients: A report of the study group of paediatric conference. *Peritoneal Dialy Int*, **16**, 269–275.
9. Cruz, C. (1996). Implantation techniques for peritoneal dialysis catheters. *Peritoneal Dialy Int*, **16** (suppl 1), 5319–5321.
10. Eklund, B.H. (1995). Surgical implantation of CAPD catheters: Presentation of midline incision — Lateral placement method and a review of 110 procedures. *Nephrol Dial Transpl*, **10**, 386–390.

11. Golper, T.A., Brier, M.E. and Bunke, M. (1996). Risk factors for peritonitis in long-term peritoneal dialysis: The Network 9 peritonitis and catheter survival studies. *Am J Kidney Dis*, **38**, 428–436.
12. Nicholson, M.L. *et al.* (1990). Factors influencing peritoneal catheter survival in continuous ambulatory peritoneal dialysis. *Ann Roy Coll Surg England*, **72**, 368–372.
13. Prischl, F.C. *et al.* (1997). Initial subcutaneous embedding of the peritoneal dialysis catheter — A critical appraisal of this new implantation technique. *Nephrol Dial Transpl*, **12**, 1661–67
14. Ates, K. *et al.* (1997). A comparison between percutaneous and surgical placement techniques of permanent peritoneal dialysis catheters. *Nephron*, **75**, 98–99.
15. Ash, S.R., Handt, A.E. and Bloch, R. (1983). Peritoneoscopic placement of the Tenckhoff catheter: Further clinical experience. *Peritoneal Dial Bull*, **3**, 8–12.
16. Copley, J.B. *et al.* (1996). Peritoneoscopic placement of Swan neck peritoneal dialysis catheters. *Peritoneal Dial Int*, **16** (suppl 1), 5330–5332.
17. Oreopoulos, D.G. (1998). A backward look at the first 20 years of CAPD. *Peritoneal Dial Int*, **18**, 360–362.
18. Gokal, R. (1996). CAPD overview. *Peritoneal Dial Int*, **16** (suppl 7), 513–518.
19. Jones, B. *et al.* (1998). Tenckhoff catheter salvage by closed stiff-wire manipulation without fluoroscopic control. *Peritoneal Dial Int*, **18**, 415–418.
20. Smith, L. *et al.* (1997). Sclerosing encapsulating peritonitis associated with continuous ambulatory peritoneal dialysis: Surgical management. *Am J Kidney Dis*, **29**, 456–460.
21. Hakim, N.S. and Matas, A.J. (1995). Technique of removal of the PD catheter. *J Am College Surg*, **180**(3), 350–352.

CHAPTER 16

NURSING CARE OF PATIENTS WITH DIALYSIS ACCESS

Joanne Emery RGN
Haemodialysis Unit
Lister Hospital
Stevenage SG1 4AB, UK

It has been widely documented that problems with renal patients' dialysis access account for the majority of their hospital admission episodes. Nursing staff are in a prime position to prevent, detect, identify problems that may occur with their patients' access and to initiate appropriate interventions. Ideally, this will result in the early recognition and treatment of problems in an attempt to avoid or minimise further surgical intervention (1).

The aims of this chapter are to:

- identify specific nursing care required by those individuals with dialysis access
- discuss cannulation methods
- explore access assessments.

16.1. Arteriovenous Fistulae

Twardowski (2) noted that the preservation of arteriovenous fistulae and grafts were dependent upon three factors: the quality of the patient's blood vessels used; the surgical technique; and the way in which the fistula or graft is cannulated.

However, as high-flux dialysis is now more widespread, there is a subsequent demand for blood flows in excess of 400 ml/min. Therefore, the fistula should not only be patent but have the capability of delivering the required blood flow to enable an adequate haemodialysis treatment (3) on an ongoing basis without the need for regular, expensive intervention (4).

16.1.1. *Nursing care for pre-arteriovenous fistula creation*

Prior to the surgical procedure for arteriovenous access creation, the patient will undergo assessment by the surgeon. The usual sites for arteriovenous fistula (AVF) or synthetic graft insertion are:

- radial artery to cephalic vein in the forearm
- brachial artery to cephalic vein in the upper arm
- straight graft between the radial artery and the cephalic or basilic vein
- loop graft between the brachial artery and the basilic vein
- loop graft between the femoral artery and the saphenous vein in the upper thigh.

Surgical assessment and type of access are considered in more detail in the respective chapters of this book. For the purpose of this chapter, it will be assumed that the AVF is in the patient's arm. However, the same principles could be applied to those patients with upper thigh grafts.

Once the site of the proposed access is determined, it is imperative that the vessels are protected from trauma to maximise the potential for success.

Education is an important aspect of the patient's pre-operative preparation. The following advice is given:

- removal of constrictive items from their identified limb, for example, a wrist watch prevents potential damage to the veins caused by their compression and the patient becomes accustomed to not wearing such items on their proposed fistula arm.
- avoidance of venepuncture on their identified limb prevents scarring, stenosis, thrombosis, haematoma formation and/or compression of the intended fistula veins.

As the patient already has renal impairment, careful monitoring of biochemical and haematological results are necessary. Hyperkalaemia, severe uraemia and symptomatic anaemia will require correction prior to anaesthesia and the nursing staff may need to plan for the patient's dialysis prior to surgery. On the morning of their surgery, diabetic patients will require regular blood sugar determinations, a reduction in their insulin dosage if insulin dependent and commencement of a 5% dextrose intravenous infusion (5).

16.1.2. *Nursing care for post-AVF creation*

The operative procedure can be performed under local anaesthetic, regional block or general anaesthetic. The anaesthetic chosen is dependent upon the type of AVF to be created and on the patient's general condition. The patient may attend as a day case or require overnight hospitalisation.

As with any AVF or synthetic graft insertion, the specific post-operative nursing care involves monitoring for the patency of the fistula, observing for signs of infection or other complications, and optimising the potential development of the fistula vessel.

Post-operative monitoring of the patient should be conducted at frequent intervals and the time duration between assessments extended as the patient's clinical condition dictates.

- The fistula arm must be kept warm; the patient's forearm can be loosely wrapped with cotton wool wadding to promote vasodilation and prevent occlusion of the vessel due to vasoconstriction and thrombus formation.
- Ensure that no constrictive clothing or dressings are applied to the fistula arm to prevent a compromised blood flow through the fistula and therefore reduce the risk of thrombus formation.
- Observe surgical site for significant blood loss and refer to the surgeons for prompt intervention.
- Monitor the strength of the palpable thrill and listen with a stethoscope for the arteriovenous bruit to detect reduction in blood flow-through or thrombosis within the fistula vessel.
- Check the colour, degree of swelling and temperature of the fingers to detect steal syndrome.

- Monitor the degree of sensation of the fingers to detect nerve damage and/ or steal syndrome,
- Monitor for hypotension as reduction in blood flow or fall in blood volume can lead to thrombus formation.
- Elevate the affected limb to promote venous return, reduce oedema and prevent compression of the fistula veins whilst facilitating wound healing.
- Observe for signs of infection both locally at the wound site and systemically.

Educating the patient regarding the ongoing care of his/her fistula is very important. This promotes the continued development of the fistula vessels, monitoring of the fistula development itself, the prevention of potential causes of fistula occlusion and knowledge of when to seek prompt nephrological opinion should any problems occur. Patients should be instructed as follows:

- monitor the bruit strength at least four times per day, and if this deteriorates, promptly contact the nephrology department
- keep the fistula arm warm
- observe the fistula for signs of infection and if present promptly contact the nephrology department
- avoid sleeping on the fistula arm
- not to wear constrictive clothing, wrist watches or bracelets on the fistula arm
- not to permit blood pressure readings on the fistula arm
- avoid carrying heavy items with the fistula arm
- from ten days post-operatively, exercise the fingers and fistula arm to promote arterial blood flow and venous return.

The shorter the patient's admission, the less time nursing staff have to teach the patient about caring for his/her fistula. In view of this, many renal units have developed written instructions for their patients and families to assist in this education process.

Uraemia causes platelet dysfunction and this may contribute to delayed wound healing and so should be considered when planning the removal of sutures or staples. Sutures are usually removed between two and three weeks following surgery. Ideally, absorbable stitches should be used.

An AVF will take several weeks to develop but should be ready for cannulation between six to eight weeks (4), an artificial graft should be ready from four weeks. However, the wishes of the access surgeon may influence the time frame to first cannulation.

16.1.3. *Cannulation techniques*

Prior to AVF needle insertion, the patient's fistula vessel should be assessed (see Sec. 16.1.5), skin prepared and sites anaesthetised as desired by the patient.

Skin preparation involves educating the patient to wash their AVF with soap, rinse and dry the area thoroughly. Further preparation is as advised by the hospital infection control team. Most units swab the injection sites with 70% alcohol and allow them to dry immediately before anaesthetic injection or needle insertion.

Anaesthesia of the sites can be achieved in two ways:

- local anaesthetic injection (the patient should be warned that this injection does sting)
- topical anaesthetic (either a cream or spray)

Patients are encouraged to relax their arms and avoid unnecessary arm movement. The needles are sited so that patient movement does not compromise their position. Therefore, the antecubital fossa should be avoided and needles must be securely taped to avoid displacement.

In patients with synthetic jump grafts, their native vessels should be cannulated if possible as this will preserve the life of their grafts.

Krönung (6) identified that plastic deformation of native AVF occurred following repeated punctures in limited areas. These puncture areas become dilated, resulting in turbulent blood flow and differing pressures within the vessel. This will subsequently lead to stenosis development at the dilated vessel margins [Figs. 16.1(a) and 16.1(b)].

Krönung (6) described three approaches to fistula cannulation:

- area puncture
- rope-ladder puncture
- buttonhole puncture or constant-site puncture

Figure 16.1 diagrammatically represents these approaches.

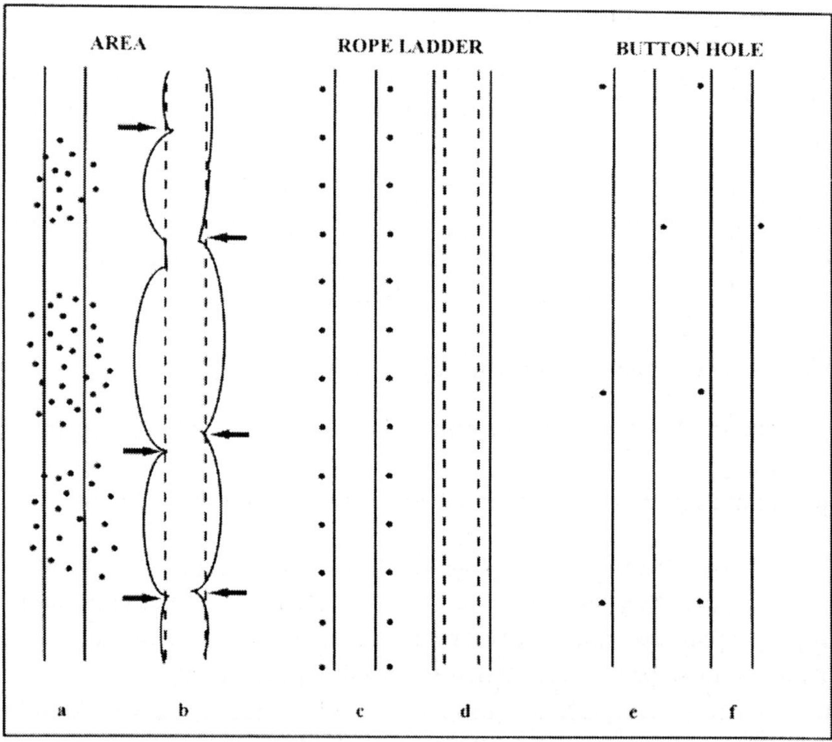

Fig. 16.1. Needling puncture as identified by Krönung (6) (→ denotes stenotic development).

Area puncture, as already identified, contributes to progressive dilation and stenotic development which can jeopardise the life of the fistula and therefore should be avoided for long-term practice. However, generalised dilation may be required initially for new, narrow fistula vessels to facilitate subsequent cannulation.

Rope-ladder puncture involves systematic cannulation at evenly spaced distances along the length of the fistula vessel. This produces slight dilation along the length of the fistula [Figs. 16.1(c) and 16.1(d)] whilst the density of punctures in any one area avoids excessive dilation and aneurysm formation. This approach is favoured for synthetic graft cannulation (7) to prevent graft degeneration and in moderately developed fistula vessels. Care needs to be taken with this approach as it can degenerate into area puncture if not monitored closely.

Buttonhole puncture is the cannulation of the fistula vessel directly through the previous needle site. When the site is exactly punctured, there is no dilatory effect [Figs. 16.1(e) and 16.1(f)]. Twardowski (2) advocated a single experienced needler to initially establish these sites as each needler will use the same direction, angle and depth during needle insertion. Once established, the site can be needled by less experienced staff. Obviously, this single-needler approach may have limitations in some renal units, for example, depending upon dialysis days and staff rosters.

The buttonhole puncture method enables easier and quicker cannulation; desensitisation of the site, hence less pain; and more reliable needle insertion, thus reducing extravasation and haematoma formation (2). There is conflicting advice regarding an increased incidence of infection with this technique, and so this warrants further investigation.

Some staff have reservations about cannulating through the still-healing site and will puncture an area as close to the previous needle site as possible, without exactly puncturing the initial site. Over time, this will result in dilation of the fistula vessel, a tendency to aneurysm formation and bleeding from around the fistula needle. If two to three puncture sites are identified for each of the arterial and venous needles, it would enable alternation of these sites at each dialysis and facilitate the staff's puncture technique.

Buttonhole puncture is advocated in large fistula vessels where no dilation is necessary and also in vessels with short available areas for cannulation.

No two AVF are alike, and depending upon the size and availability of a vessel for cannulation, a different puncture approach may need to be adopted for both the arterial and venous needle sites. For example, from the anastomosis, a large vein that deepens beneath the skin, with branches therefore becoming narrower and further along the forearm resurfaces, may benefit from buttonhole puncture of the arterial site as the vessel is large but less available due to its depth, whilst rope-ladder puncture is more appropriate for the venous sites due to the shallower but narrow vein.

Needle direction needs consideration. Retrograde insertion (the needle faces into the blood flow) optimises available blood flow particularly in small fistula vessels, is commonly used at the arterial site (Fig. 16.2) and lengthens the distance between blood outflow and inflow.

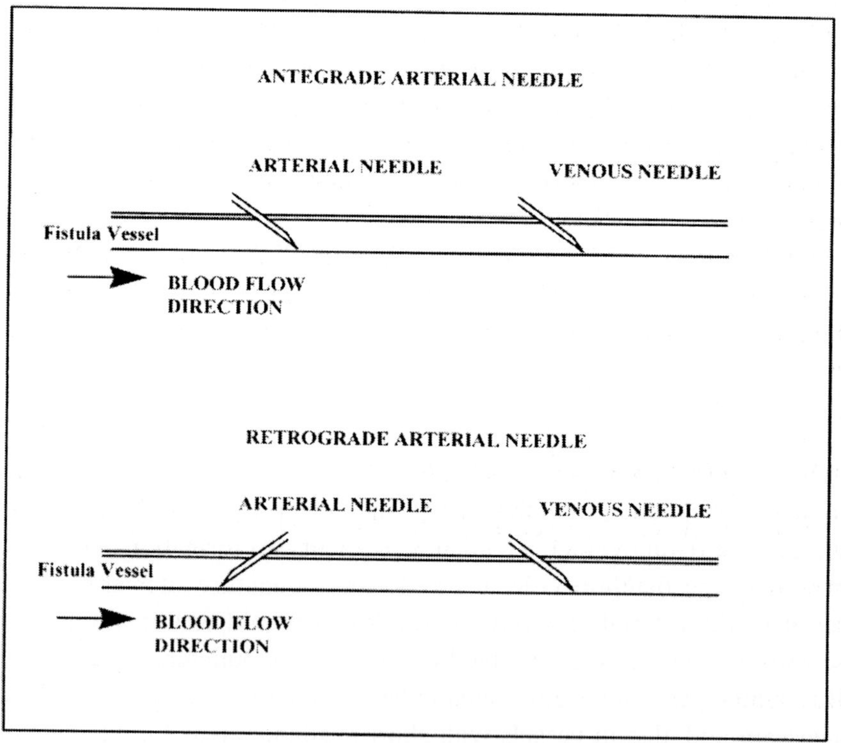

Fig. 16.2. Direction of fistula needles.

However, retrograde positioning can contribute to blood flow turbulence, haematoma formation following needle extraction and may require a longer period of pressure sustained at the needle site to arrest bleeding. The drawbacks of retrograde insertion need to be weighed against the benefit of optimising blood flow.

Antegrade insertion (Fig. 16.2) is utilised for venous needles and in those patients having a single-needle dialysis. The needle is positioned with the blood flow, thus minimising turbulence and reducing recirculation (see Sec. 16.1.4). A high degree of recirculation will result in deterioration of dialysis efficiency and the patient will eventually become under-dialysed.

An invaluable tool to assist staff with needle placement is a diagram or map of the patient's AVF. Needle position, direction and cannulation approach can all be identified, along with areas of the AVF that should be avoided, for example, where the vessel is too deep for reliable cannulation.

A patient with a new AVF and a vascular access catheter could initially have one needle only inserted to reduce the likelihood of cannulation problems. Once this site is established, the second needle is introduced.

The education of staff and patients who cannulate is extremely important to facilitate effective and reliable cannulation, to prevent the development of cannulation induced problems and to preserve the life of the AVF or synthetic graft.

16.1.4. *Complications associated with AVF*

The main complications associated with AVF and synthetic grafts are:

- thrombosis
- stenosis
- infection
- steal syndrome
- aneurysm

Each of these complications will now be explored in more depth.

16.1.4.1. *Thrombosis*

Thrombosis of the fistula vessel results in severe deterioration or loss of blood flow through the fistula. Nursing staff are unable to palpate the bruit and adequate blood flow for dialysis is not achievable.

Early thrombosis may be attributed to failure of the operative technique; hypotension; and compression of the fistula, for example, constrictive clothing or haematoma formation.

Later thrombosis is usually as a result of the development of a stenosis or narrowing of the fistula vessel (8). However, trauma to the vessel due to poor needling technique, the use of large bore fistula needles, and excessive compression of the fistula vein following needle removal may contribute to the development of a thrombus.

It is worth noting that a high ultrafiltration rate coupled with a low blood flow during haemodialysis will greatly increase the blood viscosity within the extracorporeal circuit from the dialyser to the patient, and so should be avoided.

Treatment for a fistula thrombosis consists of surgical or radiological intervention. In view of this, the patient may require a vascular access catheter for haemodialysis until permanent functioning access is established.

A patient that exhibits increased coagulability or repetitive thrombosis may benefit from oral anticoagulant therapy in an attempt to preserve the life of the AVF.

16.1.4.2. *Stenosis*

Stenosis of an AVF is not clinically obvious until it is well progressed. Narrowing of the vessel occurs due to intimal proliferation or thickening of the vein lining. This is attributed to high blood flow turbulence and vessel cannulation. An advanced stenosis may present as reduced dialysis adequacy. However, other causes for under-dialysis warrant investigation.

Subtle changes that nursing staff can detect early and act upon may prevent the patient becoming under-dialysed. A simple technique to assess for stenosis involves the compression of the AVF or synthetic graft between the two dialysis needles during dialysis at the prescribed blood flow rate whilst simultaneously observing the arterial and venous pressures (9):

- both pressures remain unchanged; this indicates a fully functioning fistula, blood flow is not compromised by a stenosis
- arterial pressure falls; this indicates an inflow stenosis, distal to the "arterial" needle position
- venous pressure rises; this indicates an outflow stenosis, proximal to the "venous" needle position

Problems achieving prescribed blood flow rates and progressive cannulation difficulties encountered by experienced nursing staff justify further investigation, as an inflow stenosis may be present.

Performing determinations of the degree of recirculation can help to identify access flow problems and therefore stenosis development. The recirculation of blood directly from the venous needle into the arterial needle reduces effective urea clearance and dialysis adequacy. It occurs when the two dialysis needles are sited too close together or when the extracorporeal blood flow rate is greater than the inflow into the fistula, causing a retrograde blood flow from the venous needle into the arterial needle in order to sustain the extracorporeal flow (4,8,10).

Percentage recirculation can be calculated using the following formula:

$$\frac{\text{Peripheral urea} - \text{Arterial (inflow) urea}}{\text{Peripheral urea} - \text{Venous (outflow) urea}} \times 100$$

For reliable results, the arterial and venous samples are taken simultaneously, the extracorporeal blood flow is immediately reduced to 50 ml/min for 30 seconds and then the peripheral sample is taken from the arterial inflow blood port. The three blood samples are taken 15 to 30 minutes into the patient's dialysis session.

Other methods of measuring recirculation involve thermodilution and optical density techniques. Although these methods require dialysis machine modifications, they do allow a series of tests during one treatment session with immediately available results.

A series of recirculation estimations in one treatment allows for the tests to be performed at different extracorporeal blood flow rates. It should be noted that percentage recirculation will increase with higher blood flow rates and as blood volume reduces. Windus and Delmez (3) advocated the regular use of recirculation studies to identify the development of AVF flow problems. They identified greater than 15% recirculation at 400 ml/min extracorporeal blood flow as an indication for further investigation.

Other indicators of outflow stenosis are prolonged bleeding from needle puncture sites, the development of a shrill bruit or change in bruit on auscultation with a stethoscope along the full length of the fistula vessel and gradual elevation in venous pressure.

Nursing staff need to be aware that there are many variables that can have an effect on venous pressure readings: extracorporeal blood flow; hypotension; displacement of the needle; needle lumen size; obstruction of the arterial or venous lines; coagulation within the extracorporeal circuit; and AVF flow problems.

Schwab (4) advocated the use of protocols for measuring venous pressures to enable a controlling factor for the aforementioned variables and facilitate the comparison of results. For conventional haemodialysis, he advocated:

- use of 16-gauge needle
- extracorporeal blood flow 200 ml/min
- measure venous dialysis pressure during the first 30 minutes of treatment
- if the venous pressure is > 150 mmHg for three consecutive dialysis sessions, a stenosis is indicated and further investigations are necessary.

And for high efficiency haemodialysis, he recommended:

- use of 14- or 15-gauge needle
- extracorporeal blood flow 200 ml/min
- measure venous dialysis pressure during the first ten minutes of treatment
- if the venous pressure is > 100 mmHg for three consecutive dialysis dialysis sessions, a stenosis is indicated and further investigations are necessary.

Obviously, for high efficiency or high flux dialysis, higher blood flows are required and the patient's prescription dialysis should be established once the ten minutes has elapsed and the pressure has been recorded.

Once the nursing staff suspect stenosis development, referral for other investigations can be sought. These investigations may include fistulagrams, or if available, colour flow Doppler ultrasounds, followed by an access clinic assessment with the access surgeon. If a stenosis is affecting the function of an AVF or synthetic graft, then further surgical intervention is necessary.

16.1.4.3. *Infection*

Nurses must observe the patient's AVF or synthetic graft for any signs of infection. The presence of infection requires prompt treatment as it can seriously compromise the patient's general health. Indeed, infection of a synthetic graft usually necessitates its removal.

Patients need to be educated and encouraged to report any of the following:

- erythema
- pain
- heat radiation
- swelling
- exudate from puncture sites
- systemic pyrexia

Careful skin preparation prior to cannulation is important and the patient should be encouraged to leave puncture site dressings intact for 24 hours.

If infection is evident, a medical assessment is necessary as the patient will require broad spectrum antibiotic cover until sensitivities are known. An infected area should never be cannulated as this could spread a local infection systematically and cause a septicaemia.

16.1.4.4. *Steal syndrome*

Steal syndrome, as mentioned earlier, is a reduction in blood flow to the hand caused by the diversion of arterial blood through the fistula or graft anastomosis. The patient's hand or fingers will feel cold, clammy, look pale and may be painful.

These symptoms may worsen as the fistula matures and when the patient is dialysing. Nurses can help to alleviate mild symptoms by lowering of the patient's arm during haemodialysis to facilitate gravitational blood flow and keeping their hand warm during dialysis to promote vasodilation, for example, by wearing a glove. However, as the patient is at risk of ischaemic skin ulceration, it is appropriate to refer them to the surgeon for further assessment. In cases of severe steal syndrome, the AVF may need to be ligated and alternative dialysis access established.

16.1.4.5. *Aneurysm*

Nursing staff need to be aware that if weakening of the vein wall occurs either at the anastomosis or in a repeated inappropriately punctured area, then an aneurysm can result. The dilated area must not be cannulated and needs to be carefully monitored by the nursing staff. If the aneurysm increases in size, immediate referral for the surgeon's opinion is vital as the aneurysm may require surgical repair.

16.1.5. *Assessment of the patient and their AVF*

The ongoing documented assessment of an individual patient's AVF or synthetic graft is an invaluable tool whereby the trend in access performance can be monitored and the subtle changes that may indicate a deterioration in AVF function identified.

A holistic approach should be adopted and the patient assessment should aim to encompass the following points:

- the patient's medical diagnosis and other associated health problems such as diabetes, hypertension, hypotension, heart disease, peripheral vascular disease, smoking and obesity. This can help identify those patients that are already predisposed to vascular problems and therefore potential access difficulties.

- identification of anticoagulant or antiplatelet medications as these will affect haemostasis at the puncture sites and assist the prevention of thrombus formation within the fistula or graft.
- the psychological effect that fistula cannulation has on the patient, for example, the degree of apprehension, anxiety or needle phobia exhibited. This can highlight those patients that require the expert skills of an experienced nurse needler to reduce anxiety levels. It also inspires a confident, trusting relationship between the patient and staff.
- a summary of the patient's AVF or graft access history, such as date created; date of first cannulation; previously failed AVF or grafts; and surgical revisions.
- description of the AVF or graft characteristics to help determine the degree of cannulation difficulty. For example, is the bruit strong or faint; is the vessel long, short, wide, narrow, straight, tortuous, shallow, deep, branched, mobile, palpable or non-palpable; is the surrounding skin bruised or swollen?
- whether local anaesthetic is used, and if so, is it injected or topically applied?
- the cannulation approach adopted for the arterial and venous sites: area, rope-ladder or buttonhole puncture; does the patient have single-needle dialysis?; needle gauge. This facilitates continuity of the cannulation approach.
- identification of the post-needle-removal bleeding time and whether haemostatic agents were required, for example, Surgicell. The general trend in a patient's bleeding time can highlight stenosis development.
- a dated diagram of the patient's fistula vessels. This can be lightly traced directly from the patient's arm before needle insertion, thus providing an accurate visual record. This enables the identification of puncture sites, needle direction and areas of cannulation difficulty.
- the results of measurable indicators of AVF blood flow such as compression data, venous pressure readings and percent recirculation. The availability of these results in one document facilitates their comparison and broadens the assessment process. These indicators could be performed every three months, for example, at staggered intervals or more frequently if indicated.

- dates referred for other investigations, such as fistulagram, doppler studies and access clinic.

These criteria need to be phrased so that an objective assessment is performed that is both specific and sensitive. The advantages of this kind of assessment not only allow for the identification of problems at an early stage, but also can be used to identify those patients who are more difficult to cannulate, and as an audit tool. It is imperative that all needling staff are educated regarding cannulation and that their level of competence is regularly assessed. The cannulation skills of staff can then be matched appropriately to the patient's degree of cannulation difficulty.

Including all the parameters would produce a comprehensive document reflecting a thorough assessment that may not be realistic to perform each time a patient attends for his/her dialysis session. However, it could be conducted regularly, for example every three months, and more frequently if clinically indicated. This does not replace the general assessment of the patient conducted by the nurse each time he/she attend for their treatment, nor the identification of problems within the patient's nursing care plan.

16.2. Vascular Access Catheters

Although a functioning AVF is the haemodialysis access of choice, not all patients requiring haemodialysis have one. Therefore, alternative access to their blood circulation is necessary. This is achieved by the insertion of a vascular access catheter by a suitably trained competent individual into a major vein.

The catheters can be temporary or permanent. Temporary catheters are inserted under local anaesthetic and held in position with sutures. A permanent catheter is inserted under local anaesthetic with sedation or general anaesthetic. It is tunnelled subcutaneously with a Dacron cuff which secures the catheter's position once healed, and help prevent infection.

The usual sites for catheter insertion are:

- internal jugular vein — most commonly advocated
- subclavian vein — may cause subclavian vein stenosis or thrombosis, this may well affect future AVF performance

- femoral vein — associated with higher recirculation rates, these catheters require removal after 48 hours to minimise problems of thrombosis and pulmonary emboli (11). However, femoral catheters are easier to insert and do not require a chest X-ray to confirm their position.

Arrythmias are not uncommon during catheter insertions, particularly in acutely ill patients, due to heightened myocardial irritability. Therefore, the patient having a chest catheter inserted should be cardiac monitored.

The site chosen and whether the catheter is temporary or permanent is dependent upon the patient's clinical condition and the urgency of haemodialysis treatment.

Many types of central venous catheters are available including:

- dual-lumen catheter — this has separate identified arterial and venous lumens within one catheter
- single-lumen catheter — one lumen available for arterial supply and venous return. A double pump dialysis machine is necessary and the patient's dialysis time should be increased to compensate for this. The following formula can be used to calculate the average extracorporeal blood flow:

$$\frac{\text{Arterial blood flow} \times \text{Venous blood flow}}{\text{Arterial blood flow} + \text{Venous blood flow}}$$

or this simplified version:

$$\frac{\text{Arterial blood flow} + \text{Venous blood flow}}{4}$$

- twin single-lumen tunnelled catheter — a relatively new system, one catheter is identified for the arterial supply and the other for venous return. Sited independently, these catheters can reduce recirculation whilst maintaining high blood flow rates (12)

16.2.1. *Nursing care post-catheter insertion*

Once the catheter is in position, the nursing care involves monitoring for acute complications associated with the catheter insertion, maintaining catheter patency, observing for signs of infection and educating the patient regarding the care of his/her catheter.

An enlarging haematoma in the neck can seriously compromise the patient's airway without prompt intervention. Therefore, the insertion site must be closely inspected.

If the catheter insertion was difficult and the patient is to have imminent haemodialysis, the nursing staff may consider reducing extracorporeal heparin dose or undertaking a heparin-free dialysis. This necessitates careful monitoring of the extracorporeal circuit for evidence of coagulation.

If chest pain, dyspnoea, angina, arrhythmias, hypotension, clouding of consciousness or sudden deterioration in the patient's condition occurs, initiate prompt action to identify and treat life-threatening complications.

Before dialysis initiation, check that the catheter is secure and observe for signs of exit site infection. Cleanse and redress the site according to unit procedure. The cleansing solutions commonly used are chlorhexidene or sterile saline. There are several types of dressings available that can be used to cover vascular access catheter insertion sites. The choice of dressing should prevent the introduction of infection, be conformable, comfortable for the patient and cost effective. The dressings commonly used are semi-permanent film dressings or non-filamentous gauze island adhesive dressings.

Sutures are removed from the insertion wounds of tunnelled catheters between two to three weeks. It is imperative that sutures are secure for non-tunnelled catheters. Therefore, a suitably trained competent member of the nursing staff will re-suture the catheter as necessary. Other measures include:

- ensuring the catheter limb is clamped prior to removal of, for example, the catheter cap; this prevents the introduction of an air embolus
- extracting and discarding 4 ml of blood from each limb of the catheter prior to use, assessing outflow and inflow resistance; this removes the heparin lock and any clots that may have formed, identifies potential problems of insufficient outflow or elevated inflow pressure
- reheparinising each limb of the catheter, following a 10-ml saline flush; this prevents occlusion of the catheter lumen due to thrombus formation

The quantity of heparin (5000 IU/ml) instilled into the catheter limbs is equal to the respective lumen size, thus preventing systemic heparinisation of the patient. Therefore, it is important to document the patient's catheter lumen volumes for future reference.

The patient needs to be aware of the following:

- avoid getting the catheter dressing wet as this can introduce infection
- ensure the catheter limbs are securely taped to prevent pulling at the insertion site
- should the catheter site become painful or swelling become evident, contact the nephrology department.

16.2.2. *Later complications associated with vascular access catheters*

The main complications associated with the ongoing care of vascular access catheters are:

- infection
- thrombosis
- catheter malposition

Each of these complications will now be considered in more depth.

16.2.2.1. *Infection*

Vascular access catheter infections can be localised to the insertion site, the catheter tunnel or the catheter itself. Alternatively, systemic catheter-related sepsis can develop. As with AVF, nurses must be vigilant for signs of infection, and if present, initiate a prompt medical referral as antibiotic cover will be required. Depending upon the severity of the infection, vascular access catheter removal may be indicated.

The relatively new technique of endoluminal brushing with subsequent culture of the brush may eliminate a suspected case of catheter infection and thus prevent removal of a non-infected catheter (13).

16.2.2.2. *Thrombosis*

A mural thrombosis can occur within the catheterised vein itself, a catheter thrombus can form within the internal lumen of the catheter and a fibrin sleeve can develop on the exterior surface of the catheter. Any one of these thromboses are not apparent until aspiration problems occur or total occlusion of the catheter is encountered.

It is important to thoroughly assess the patient and rule out dehydration as a cause of the outflow insufficiency. Venous phlebography and Doppler ultrasound scanning can be utilised to aid diagnosis.

A catheter thrombus usually responds to thrombolytic therapy. Typically, this is urokinase (5000 IU/ml) instilled only into the volume of the lumen as prescribed by the doctor. Having been left *in situ* for anything from 30 minutes to four hours, depending upon local unit policy, the drug is aspirated from the catheter along with any residual thrombus.

Percutaneous fibrin sleeve stripping could be considered as a prophylactic measure to remove the fibrin sheath from within the catheter lumen prior to occlusion once the extracorporeal blood flow begins to deteriorate. However, further investigation of this new technique is required.

Venous thrombosis and unresolved catheter occlusion require removal and replacement of the catheter. Some renal units advocate oral anticoagulant therapy for their patients with tunnelled vascular access catheters, unless clinically contraindicated, to prevent thrombus formation and maintain catheter patency.

16.2.2.3. *Catheter malposition*

A deterioration in blood flow or aspiration problems can be attributed to catheter malposition, either the catheter is lying against the vessel wall or is kinked. This can temporarily be resolved by altering the patient's position.

An X-ray of the catheter can confirm the problem. If the problem cannot be corrected by repositioning or changing the catheter over a guidewire, then re-insertion of a new catheter is required.

16.3. Peritoneal Dialysis Catheters

As with haemodialysis, successful peritoneal dialysis (PD) is dependent upon problem-free access.

Access to the peritoneal cavity is achieved by the insertion of a permanent silastic catheter under anaesthetic through the patient's abdominal wall into his/her pelvis. The catheter is tunnelled through the patient's abdominal wall with two Dacron cuffs that are positioned subcutaneously and prè-peritoneally. Ideally, the catheter is inserted two to four weeks prior to the initiation of PD therapy to allow adequate healing of the surgical wound and catheter tunnel.

16.3.1. *Nursing care post-peritoneal catheter insertion*

The specific post-operative nursing care of the access involves maintaining a patent PD catheter, facilitating the healing process and educating the patient with regard to the care of his/her catheter:

- flush the catheter according to unit policy until PD treatment begins; this prevents occlusion of the catheter due to fibrin formation

If blood-stained fluid is evident in the catheter post-operatively, then automated flushes can be performed using a PD machine. This allows rapid small volume exchanges to be performed as directed by the renal team. Otherwise, the catheter is flushed at 24 hours and 72 hours post-op. The frequency of flushing can be gradually extended to between two to four weeks until the catheter is used, provided no problems are encountered.

- redress the catheter site and surgical wound according to unit policy, remove sutures between two and three weeks, assess for signs of infection; this facilitates healing and prevents or detects infection

Sterile saline is generally used to cleanse the catheter exit site whilst the wound is healing and when the patient is in hospital. However, once the sutures are removed, many units teach the patient to clean their exit site daily with cooled boiled water.

- ensure the catheter is taped securely to the patient and not stretched across his/her abdomen, this immobilises the catheter, prevents trauma to the catheter tunnel and cuff extrusion

Patient or carer education is vital to facilitate problem-free PD access, as once trained, the patient will perform his/her treatment independently at home. As a result, the frequency of patient to staff interaction is greatly reduced.

The patient needs to be aware of the following in relation to the care of his/her PD catheter:

- how to clean and redress the catheter exit site
- ensure the catheter is immobile and not stretched across the abdomen
- monitor for signs of exit site infection, and if present, report promptly to the nephrology department
- avoid submersion in water without waterproof protection of the catheter and exit site, showering is permitted as the water runs off
- knowledge of how the catheter extension line can become contaminated and the appropriate action to take should this occur
- accurately record your PD fluid drainage in and out in order to identify outflow drainage problems

16.3.2. *Complications associated with peritoneal dialysis access*

The main complications associated with PD access are as follows:

- infection
- leaks
- catheter migration
- drainage problems

Each of these will now be explored in more depth.

16.3.2.1. *Infection*

Peritoneal-dialysis-catheter exit site infections can seriously compromise the patient's dialysis treatment. Therefore, prompt treatment is imperative.

Exit site infections can be localised to just the exit site, involve the catheter cuffs, tunnel or tissues around the tunnel and exit site. Broad spectrum antibiotic therapy will be initiated and therapy changed, if necessary, once sensitivities are known.

Unresolved exit site infections necessitate the removal of the catheter. Should this occur, the patient will require support and preparation for temporary haemodialysis treatment until further PD access is suitably established.

16.3.2.2. *Leakage of dialysate fluid*

Dialysate fluid leaking at the exit site can provide a focal point for infection and is unpleasant for the patient.

Leaks can be treated by resting the catheter. Patients may need to transfer to automated PD to facilitate smaller exchange volumes at an increased frequency. This will reduce intra-abdominal pressure, maintain adequate dialysis delivery and allow resting of the catheter periodically. Persistent leaks require complete rest with temporary haemodialysis, and in severe cases, removal of the catheter.

16.3.2.3. *Catheter migration*

Typically, this presents as an outflow drainage problem. Suspected catheter migration can be confirmed on X-ray. These catheters can be repositioned under fluoroscopic guidance but usually require removal and replacement.

16.3.2.4. *Drainage problems*

Catheter drainage problems can be attributed to constipation and obstruction of the catheter by fibrin or omentum.

The nursing staff will need to perform a thorough patient assessment to establish if constipation or fibrin formation is a problem. Constipation is treated with appropriate laxatives and fibrin with heparinised dialysate flushes. If these measures are not successful, then further referral for investigation is required.

A plain X-ray can determine the catheter position and contrast media films can detect further obstruction or occlusion.

Persistent fibrin occlusion problems can be resolved by instilling urokinase (5000 IU/ml in 50 ml of 0.9% saline) as prescribed by the doctor and left *in situ* for one hour. This is followed by further dialysate flushing of the catheter.

Obstruction of the catheter caused by omentum will require a partial omentectomy, which is sometimes performed when the catheter is inserted, so as to avoid this potential problem from occurring.

By collaboratively working together, the patient, nursing staff, medical and surgical teams can identify, initiate and evaluate the appropriate care and interventions required to aid the effective and efficient survival of the patient's dialysis access. This will improve the patient's quality of life and hopefully reduce his/her number and/or duration of hospitalisation episodes.

Acknowledgement

I would like to acknowledge the assistance of Joanna Dockree for her secretarial support in preparing this chapter.

References

1. Culp, K., Taylor, L. and Hulme, P.A. (1996). Geriatric hemodialysis patients: A comparative study of vascular access. *ANNA J*, **23**, 583–590, 622.
2. Twardowski, Z.J. (1995). Constant site (buttonhole) method of needle insertion for hemodialysis. *Dial Transpl*, **24**, 559–560, 576.
3. Windus, D.W. and Delmez, J.A. (1991). What can be done to preserve vascular access for dialysis? *Seminars in Dialysis*, **4**, 153–154.
4. Schwab, S.J. (1994). Assessing the adequacy of vascular access and its relationship to patient outcome. *Am J Kidney Dis*, **24**, 316–320.

5. Brunier, G. (1996). Care of the hemodialysis patient with a new permanent vascular access: Review of assessment and teaching. *ANNA J*, **23**, 547–556.
6. Krönung, G. (1984). Plastic deformation of cimino fistula by repeated puncture. *Dial Transpl*, **13**, 635–638.
7. Kaufman, J.L. (1991). What can be done to preserve vascular access for dialysis? *Seminars in Dialysis*, **4**, 160–162.
8. Windus, D.W. (1993) Permanent vascular access: A nephrologists view. *Am J Kidney Dis*, **21**, 457–471.
9. Prinse-Van Loon, M.M., Mutsaers, B.M.J.M. and Verwoert-Meertens, A. (1996). Integrated and specialised care of arteriovenous fistulae improves quality of life. *European Dialysis and Transplant Nurses Association — European Renal Care Association Journal*, **22**, 31–33.
10. Aldridge, C. *et al.* (1993). Haemodialysis recirculation detected by the three sample method is an artefact. *European Dialysis and Transplant Nurses Association — European Renal Care Association Journal*, **19**, 2–5.
11. Bander, S.J. and Schwab, S.J. (1992). Central venous angio access for haemodialysis and its complications. *Seminars in Dialysis*, **5**, 121–128.
12. Prabhu, P.N. *et al.* (1997). Long-term performance and complications of the Tesio twin catheter system for haemodialysis access. *Am J Kidney Dis*, **30**, 213–218.
13. Tighe, M.J. *et al.* (1996). An endoluminal brush to detect the infected central venous catheter *in situ*: A pilot study. *Br Med J*, **313**, 1528–1529.

INDEX